# GEOGRAPHY FOR BUSINESS STUDIES

# GEOGRAPHY FOR BUSINESS STUDIES

H. ROBINSON, B.A., M.Ed., Ph.D.
*Head of Department of Geography, Geology and Environmental Studies, The Polytechnic, Huddersfield*

SECOND EDITION

MACDONALD & EVANS LTD
8 John Street, London WC1N 2HY
1972

*First Published June 1965*
*Reprinted November 1965*
*Reprinted (with corrections) June 1966*
*Reprinted (with revised statistics) April 1967*
*Reprinted (with revised statistics) May 1968*
*Reprinted (with revised statistics) June 1969*
*Reprinted (with revised statistics) June 1970*
*Second Edition February 1972*
*Reprinted August 1973*
*Reprinted (with amendments) March 1974*



ISBN: 0 7121 0711 8



*Also by H. Robinson:*

AFRICA, MONSOON ASIA AND AUSTRALASIA
ECONOMIC GEOGRAPHY
HUMAN GEOGRAPHY
LATIN AMERICA
MONSOON ASIA
PHYSICAL GEOGRAPHY

*Printed in Great Britain by*
*Fletcher & Son Ltd, Norwich*

# PREFACE TO FIRST EDITION

I HAVE called this book *Geography for Business Studies* since its primary aim is to cater for students studying geography for the Ordinary National Certificate and the Ordinary National Diploma examinations. Many other institutions, among them the Institute of Bankers, the Institute of Transport, the Town Planning Institute, the Chartered Institute of Secretaries, the Institute of Costs and Works Accountants, the Institute of Grocers, the Co-operative Union Ltd., the Royal Society of Arts, and the Civil Service often require the study of Economic and Commercial Geography, and students preparing for the examinations of such bodies will find this book helpful. It should also meet the needs of Sixth Form students in grammar schools who wish to pursue an introductory course in Economic Geography.

The text is essentially an introductory study to Economic and Commercial Geography. Although the student has probably passed geography at "O" level before embarking upon a professional course, I have thought it wise (as a result of painful experience) not to assume that the student knows all the basic principles, possesses a grounding in geographical ideas, or is even familiar with the fundamental facts of geography. Part One is, therefore, in a sense, an introduction to some of the basic concepts of geography. Parts Two, Three, and Four cover the usual topics of Economic and Commercial Geography.

Colleges of Commerce and Technology have a fairly free hand in devising their syllabuses for O.N.C. and O.N.D. courses and this makes the writing of a text to cover all the contingencies a difficult task, but it is hoped that this book will go a long way towards meeting the requirements of these manifold courses of study. Although efforts are being made to bring these and the geography syllabuses of the different professional examinations into step—a desirable move in many ways, especially from the teacher's point of view when he or she has to teach a class of students having different aims and requirements—it would be unfortunate if all courses were put into a geographical straightjacket. There is much to be said for, and much to be gained from, a flexibility of content, treatment, and approach, even though this entails administrative and classroom difficulties.

There are many approaches to economic and commercial geography. Usually, however, they take the form of either systematic treatment or regional treatment and most books follow either one approach or the other. Here I have adopted the systematic approach but interspersed the text with regional examples dealing, usually, with the more important regional aspects. How successful this is remains to be seen, but it is an attempt to give the student some brief regional studies in addition to Economic Geography of the more systematic kind. The student can elaborate the brief notes given by recourse to other texts and books of reference.

An attempt has been made to make the text as up to date as possible in both its coverage and its factual content. Mention has been made of new developments, for example, the growing use of pipelines and the development of nuclear energy, while certain topics, such as conservation and tourism, have been given more extended treatment than is usual in books at this level. Facts and figures are as recent as it is possible to make them. A variety of graphical techniques have been introduced and explained to familiarise students with the different methods of graphical representation. Questions, largely taken from the papers of various examining bodies, have been given at the end of appropriate chapters and additional exercises are in Appendix I. A guide to the answering of questions is given in Appendix II.

I am grateful to various authors, editors, and publishers for their kind permission to use quotations and maps: to the *Institut National D'Etudes Demographiques* of Paris for permission to reproduce Figure 2; to Miss W. M. Simmons who allowed me to use Figures 14, 16, and 126 which are taken from her book on *The British Isles*; to the Editor of *Progress* for permission to reproduce Figure 154; to Professor Sir Dudley L. Stamp and Messrs. Faber & Faber for permission to use Figures 25, 26, and 27 taken from *Our Developing World*; to the Economic Intelligence Unit and the Clarendon Press for permission to use quotations from the *Oxford Economic Atlas*; to the Danish Embassy for Figure 57; to John Bartholomew for permission to use the Atlantis projection for Figure 148; to various Examining Bodies for their permission to use past examination questions; and to many authorities for the use of visual illustrations, individual acknowledgment being given in the text. Finally, I wish to express my great indebtedness to my colleague R. Riley, B.Sc.Econ., for his kind help and to my life-long friend F. S. Hudson, B.A., F.R.G.S., who read the MS. and suggested innumerable improvements. For Mr. Hudson's constructive and kindly help on this, and many other occasions, I shall always be most grateful.

H. R.

# PREFACE TO SECOND EDITION

It is now six years since this book was first published and, although revised statistics have been included in each of the reprints, it was felt that a thorough revision of the text was desirable. Economic Geography is a dynamic subject and changes take place with astonishing (and irritating) rapidity. Attempting to be up to date is like following a will o' the wisp; however, the attempt has been made. In most cases the most recent figures have been incorporated into the text. In a number of cases there have been texual elaborations and some significant developments, *e.g.* containerisation, have been introduced. All the line illustrations have been re-drawn and metric equivalents have been introduced into the text.

I have been gratified at the success of this book, which has justified my claim that there was a need for a text of this kind. I am most grateful to the publishers for the frequent revisions they have allowed me to make in the past and especially for this opportunity to produce a new and enlarged edition.

H. R.

# CONTENTS


| Chapter | | Page |
|---|---|---|
| | PREFACE TO FIRST EDITION | v |
| | PREFACE TO SECOND EDITION | vii |
| | LIST OF ILLUSTRATIONS | xiii |
| | LIST OF TABLES | xvi |
| | PART ONE | |
| | MAN, THE EARTH, AND HIS LIVELIHOOD | |
| I. | THE WORLD OF BUSINESS | 1 |
| | Modern Geography | 1 |
| | Economic Geography | 3 |
| | One World | 5 |
| | Production, Manufacture, and Distribution | 6 |
| II. | THE EARTH'S PEOPLES | 8 |
| | The Races of Mankind | 8 |
| | Distribution of Population | 10 |
| | The Growth of World Population | 18 |
| | The Population Problem | 20 |
| III. | HOW MANKIND LIVES | 24 |
| | Simple Societies | 25 |
| | Complex Societies | 30 |
| | Non-Primary Occupations | 35 |
| IV. | THE WEALTH OF THE WORLD | 37 |
| | What is Wealth? | 37 |
| | Natural Resources | 38 |
| | Latent Resources | 39 |
| | The Earth's Natural Resources | 41 |
| | Conservation | 47 |
| V. | GEOGRAPHICAL FACTORS | 51 |
| | Natural Factors | 53 |
| | Human Factors | 57 |
| | Man versus Nature | 59 |
| VI. | GEOGRAPHICAL REGIONS | 65 |
| | Equatorial Forest Regions | 65 |
| | Tropical Summer Rain Regions | 69 |
| | Tropical Trade Wind Regions | 70 |
| | Tropical Monsoon Regions | 71 |
| | Hot Desert Regions | 72 |
| | Mediterranean Regions | 74 |
| | Humid Sub-Tropical Regions | 76 |
| | Temperate Continental Regions | 77 |
| | Mid-Latitude Desert Regions | 79 |
| | Cool Temperate Maritime Regions | 80 |
| | Cool Temperate Eastern Marginal Regions | 81 |
| | Cold Continental Regions | 82 |
| | The Tundra Regions | 84 |
| | PART TWO | |
| | FARMING AND FOODSTUFFS | |
| VII. | FARMING AND FOOD SUPPLIES | 87 |
| | Feeding the World's People | 87 |

| Chapter | | Page |
|---|---|---|
| | Land Availability and Use | 90 |
| | Ways of Increasing Food Production | 91 |
| | New Sources of Food | 92 |
| | The Problem of Diet and the Distribution of Food | 93 |
| VIII. | FACTORS AFFECTING FARMING | 95 |
| | Agricultural Influences | 95 |
| | Limits of Crop Production | 100 |
| | Irrigation | 101 |
| | Methods of Irrigation | 103 |
| | Irrigated Areas | 104 |
| | Animal Husbandry | 107 |
| | Factors Affecting the Livestock Industry | 110 |
| | Nomadism | 112 |
| | Transhumance | 113 |
| | Systems of Farming | 115 |
| IX. | TYPES OF FARMING | 118 |
| | Farming Method | 118 |
| | Types of Farming | 119 |
| X. | PLANT FOODSTUFFS: CEREAL CROPS | 131 |
| | Wheat | 131 |
| | Barley | 134 |
| | Oats | 135 |
| | Rye | 136 |
| | Maize | 136 |
| | Millets | 138 |
| | Rice | 139 |
| XI. | PLANT FOODSTUFFS: OTHER VEGETABLE PRODUCTS | 144 |
| | Vegetable Foodstuffs | 144 |
| | Fruits | 148 |
| | Beverages | 152 |
| | Sugar | 160 |
| | Spices and Essences | 166 |
| XII. | ANIMAL FOODSTUFFS | 169 |
| | The Domestication of Animals | 169 |
| | Cattle | 171 |
| | Mutton | 177 |
| | Pigs and Pig Products | 178 |
| | Poultry | 179 |
| | Honey | 179 |
| XIII. | THE FISHING INDUSTRY | 181 |
| | Marine Life | 181 |
| | Methods of Fishing | 183 |
| | The Major Fishing Grounds | 185 |
| | The Major Fishing Countries | 186 |
| | Industries based on Fishing | 189 |
| | Whaling | 190 |
| | Miscellaneous Marine Products | 192 |
| | Fish Conservation | 193 |

PART THREE

INDUSTRIAL MATERIALS AND MANUFACTURE

| | | |
|---|---|---|
| XIV. | INDUSTRIAL CROPS | 195 |
| | Cotton | 196 |

| Chapter | | Page |
|---|---|---|
| | Flax | 199 |
| | Jute | 201 |
| | Hemp | 202 |
| | The Hard Fibres | 203 |
| | Other Vegetable Fibres | 203 |
| | Synthetic Fibres | 204 |
| | Vegetable Oils | 205 |
| | Rubber | 210 |
| | Tobacco | 214 |
| XV. | HAIR, HIDES AND SKINS | 217 |
| | Wool | 217 |
| | Silk | 221 |
| | Hides and Skins | 225 |
| | Horn and Ivory | 227 |
| XVI. | FOREST PRODUCTS | 229 |
| | The World's Forests | 229 |
| | Distribution of the Forests | 231 |
| | Timbers and Their Uses | 233 |
| | Lumbering in the Coniferous and Tropical Forests Compared | 234 |
| | Wood Pulp and Paper | 236 |
| | Other Forest Products | 238 |
| XVII. | METALS AND MINERALS | 241 |
| | The Importance of Minerals | 241 |
| | The Classification, Occurrence and Exploitation of Minerals | 242 |
| | The Exploitation of Mineral Resources | 243 |
| | Building Stones and Kindred Minerals | 245 |
| | Non-metallic Minerals | 247 |
| | Metals | 251 |
| XVIII. | FUEL AND POWER | 267 |
| | The Difference between Fuel and Power | 268 |
| | The Significance of Energy | 269 |
| | Coal | 273 |
| | Oil | 280 |
| | Natural Gas | 289 |
| | Water-Power | 292 |
| | Nuclear Energy | 298 |
| XIX. | MANUFACTURING INDUSTRY | 301 |
| | The Meaning of Manufacture | 301 |
| | Types of Manufacture | 302 |
| | The Industrial Revolution | 303 |
| | The Location of Industry | 303 |
| | Features of Modern Industry | 310 |
| | The Industrialisation of Backward Countries | 311 |
| XX. | THE IRON AND STEEL INDUSTRY | 313 |
| | The Importance of Iron and Steel | 313 |
| | The United Kingdom | 316 |
| | The United States | 322 |
| | The European Coal and Steel Community | 323 |
| | The Soviet Union | 326 |
| | Monsoon Asia | 328 |
| | Australia: Africa | 330 |
| | Latin America | 331 |

| Chapter | | Page |
| --- | --- | --- |
| XXI. | OTHER MAJOR INDUSTRIES | 332 |
| | Engineering | 332 |
| | Ship-building | 338 |
| | Textile Manufacturing | 343 |
| | The Chemical Industry | 354 |
| | The Plastics Industry | 360 |

PART FOUR

## THE MECHANISM OF COMMERCE

| | | |
| --- | --- | --- |
| XXII. | TRADE AND COMMERCE | 363 |
| | The Beginnings of Trade | 363 |
| | The Bases of Trade | 366 |
| | Types of Trade | 369 |
| | The Pattern of Trade | 370 |
| | Great Britain's Place in World Trade | 372 |
| | The Balance of Trade | 374 |
| | Trade Restrictions and Trading Blocs | 376 |
| | The Mechanism of Exchange | 379 |
| XXIII. | THE TOURIST INDUSTRY | 382 |
| | The Development of Tourism | 382 |
| | The Why of Tourism | 384 |
| | The Factors of Tourism | 385 |
| | Types of Resorts | 387 |
| | The Economic Importance of Tourism | 389 |
| | The Tourist Industry in Britain | 390 |
| XXIV. | LANGUAGE AND COMMUNICATION | 394 |
| | Communication | 394 |
| | Communication Devices | 395 |
| | The Importance and Value of Communications | 396 |
| | The Languages of Commerce | 398 |
| XXV. | THE WORLD'S HIGHWAYS | 401 |
| | Modes of Transport | 402 |
| | Speed and Cost | 404 |
| | Containerisation | 405 |
| | Roads | 406 |
| | Railways | 409 |
| | Ropeways and Cableways | 414 |
| | Pipelines | 415 |
| | Inland Waterways | 416 |
| | Ocean Transport | 423 |
| | Ship Canals | 427 |
| | Airways | 431 |
| XXVI. | SETTLEMENTS AND URBANISM | 439 |
| | Factors Influencing Settlement | 439 |
| | Site and Position | 440 |
| | Types of Settlement | 442 |
| | The Growth of Towns | 445 |
| | Classification of Towns | 447 |
| | Million Cities | 453 |
| | APPENDIXES | 457 |
| | I. Additional Questions | 457 |
| | II. Examination Hints to Candidates | 459 |
| | INDEX | 463 |

# LIST OF ILLUSTRATIONS


| *Fig.* | | *Page* |
|---|---|---|
| 1. | Distribution of world population | 11 |
| 2. | World area and population compared | 13 |
| 3. | Distribution of population in North America | 16 |
| 4. | Distribution of population in South America | 17 |
| 5. | Graph of world population growth | 22 |
| 6. | The economies of mankind | 24 |
| 7. | Tillage | 29 |
| 8. | Forestry | 33 |
| 9. | Graph of economic activities | 36 |
| 10. | Soil erosion | 42 |
| 11. | The rabbit pest | 43 |
| 12. | Strip-planted farm | 45 |
| 13. | Natural fertility | 48 |
| 14. | Agricultural productivity | 49 |
| 15. | The uses of coal | 52 |
| 16. | The uses of petroleum | 54 |
| 17. | Graph of water use, United States | 56 |
| 18. | T.V.A. scheme | 60 |
| 19. | Geographical regions | 66 |
| 20. | Amazonian selva | 68 |
| 21. | Desert landscape, Arizona | 73 |
| 22. | Prairie landscape | 78 |
| 23. | Coniferous forest or taiga | 83 |
| 24. | Tundra | 84 |
| 25. | Arid areas | 88 |
| 26. | Mountainous areas | 88 |
| 27. | Total negative areas | 89 |
| 28. | Crop limits in Europe | 96 |
| 29. | Terracing | 97 |
| 30. | Indus valley irrigation | 105 |
| 31. | Artesian basins, Australia | 106 |
| 32. | Irrigation in the United States | 108 |
| 33. | Transhumance | 114 |
| 34. | Intensive agriculture, Japan | 122 |
| 35. | British agriculture | 126 |
| 36. | Intensive horticulture | 128 |
| 37. | World wheat production | 132 |
| 38. | Wheat production on the Canadian prairies | 133 |
| 39. | Types of cereal | 135 |
| 40. | Distribution of wheat and maize | 139 |
| 41. | Rice fields, West Java | 140 |
| 42. | Rice-producing areas of the Monsoon Lands | 141 |
| 43. | World rice production | 142 |
| 44. | Banana plantation | 147 |
| 45. | Wine production in Europe | 150 |
| 46. | Apple farming, Annapolis valley | 151 |
| 47. | Tea-plant | 153 |
| 48. | Tea-growing areas | 154 |
| 49. | Coffee *fazenda* | 156 |

| Fig. | | Page |
|---|---|---|
| 50. | Coffee growing in Brazil | 157 |
| 51. | Cacao in West Africa | 159 |
| 52. | The Cuban sugar bowl | 161 |
| 53. | Sugar-cane plantation | 163 |
| 54. | Cane- and beet-sugar areas | 165 |
| 55. | Dairying and meat production | 171 |
| 56. | Cattle in Argentina and Uruguay | 173 |
| 57. | Dairying in Denmark | 175 |
| 58. | Dairy farm, New Zealand | 176 |
| 59. | World distribution of pigs | 178 |
| 60. | World fishing grounds | 182 |
| 61. | Types of net | 184 |
| 62. | Netting fish | 187 |
| 63. | Fisheries of north-western Europe | 188 |
| 64. | Whaling factory ship | 191 |
| 65. | The United States cotton belt | 197 |
| 66. | Mechanical cotton-picker | 198 |
| 67. | Distribution of jute, hemp, and flax | 201 |
| 68. | Terylene: man-made fibre | 204 |
| 69. | Diagram of a coconut | 207 |
| 70. | Distribution of vegetable oils | 209 |
| 71. | Rubber tapping | 212 |
| 72. | Rubber in Malaya | 213 |
| 73. | World sheep distribution | 218 |
| 74. | Wool exports and imports | 220 |
| 75. | Sheep mustering | 221 |
| 76. | Sheep-rearing in Australia | 222 |
| 77. | Fur-trapping in the northern forests | 226 |
| 78. | The world's main forest areas | 231 |
| 79. | Floating logs | 234 |
| 80. | Elephants moving teak logs | 235 |
| 81. | Paper-mills, Cornwall, Ontario | 237 |
| 82. | Quarrying Purbeck limestone | 245 |
| 83. | China Clay mining | 246 |
| 84. | Potash works in Israel | 250 |
| 85. | Iron-ore mining, Kiruna | 251 |
| 86. | World iron-ore reserves | 252 |
| 87. | Iron ores, Britain | 253 |
| 88. | Movements of iron ore in Europe | 256 |
| 89. | Movements of iron ore in North America | 257 |
| 90. | Ferro-alloy metals | 257 |
| 91. | The copperbelt | 259 |
| 92. | Distribution of non-ferrous metals | 261 |
| 93. | Bauxite mining, Guyana | 262 |
| 94. | The Witwatersrand | 264 |
| 95. | Sources of inanimate energy | 270 |
| 96. | Energy patterns | 272 |
| 97. | World production of primary sources of energy | 272 |
| 98. | Types of coal | 274 |
| 99. | Coalfields of Britain | 275 |
| 100. | Coalfields of the United States | 277 |
| 101. | Coalfields of Europe | 278 |
| 102. | Coalfields of the Soviet Union | 279 |
| 103. | Cross-section of an oilfield | 281 |
| 104. | Fractionating tower | 284 |

*Fig.* *Page*
105. Refineries in Great Britain . . . . . . . . . . . 285
106. Oil in the Middle East . . . . . . . . . . 286
107. Gas and oil pipelines in Europe . . . . . . . . . 290
108. Hwang-ho: power and irrigation . . . . . . . . . 293
109. Nuclear power in Britain . . . . . . . . . . 298
110. Industry in Europe . . . . . . . . . . . 304
111. Industry in Japan . . . . . . . . . . . 307
112. Industrial regions of North America . . . . . . . 309
113. Iron and steel in Britain . . . . . . . . . . 317
114. Iron ore imports . . . . . . . . . . . 318
115. Iron and steel in Europe . . . . . . . . . . 324
116. Industrial regions of the U.S.S.R. . . . . . . . . 327
117. Damodar valley, India . . . . . . . . . . 329
118. Fiat factory, Turin. . . . . . . . . . . 337
119. Lancashire cotton industry . . . . . . . . . 344
120. Textiles in West Germany . . . . . . . . . 348
121. Textiles in Italy . . . . . . . . . . . 349
122. Textiles in the United States . . . . . . . . . 351
123. Chemicals in Britain . . . . . . . . . . 357
124. Chemicals in France . . . . . . . . . . 359
125. Direction of Britain's trade . . . . . . . . . 372
126. The trade of the United Kingdom . . . . . . . . 374
127. E.E.C. and E.F.T.A. . . . . . . . . . . 377
128. Trading blocs of the world . . . . . . . . . 378
129. Bruges . . . . . . . . . . . . 386
130. Unterwasser . . . . . . . . . . . . 388
131. English resorts . . . . . . . . . . . 391
132. World cables . . . . . . . . . . . 395
133. Languages of commerce . . . . . . . . . . 398
134. The roads of Britain . . . . . . . . . . 407
135. Italian autostrada . . . . . . . . . . . 409
136. Diesel-electric train . . . . . . . . . . 410
137. The railways of Europe . . . . . . . . . . 411
138. The railways of Australia . . . . . . . . . 413
139. Barges on the Rhine System . . . . . . . . . 417
140. The canals of central Europe . . . . . . . . . 418
141. The Rhine waterway . . . . . . . . . . 419
142. The Volga system . . . . . . . . . . . 420
143. Iroquois lock . . . . . . . . . . . 422
144. Main sea routes . . . . . . . . . . . 426
145. The Suez Canal . . . . . . . . . . . 428
146. Section along the Panama Canal . . . . . . . . 430
147. The St Lawrence Seaway . . . . . . . . . 432
148. World air routes . . . . . . . . . . . 436
149. New York: its site and situation . . . . . . . . 441
150. Vancouver: aerial view . . . . . . . . . . 443
151. A conurbation . . . . . . . . . . . 444
152. The "new towns" of Britain . . . . . . . . . 446
153. The Port of London . . . . . . . . . . 448
154. Europoort . . . . . . . . . . . . 450
155. Vällingby . . . . . . . . . . . . 452
156. The million cities of the world . . . . . . . . 454

# LIST OF TABLES


| Table | | Page |
|---|---|---|
| I. | Growth of population by continents | 18 |
| II. | Natural resources | 44 |
| III. | Agricultural statistics | 126 |
| IV. | Maize production | 138 |
| V. | Rice (paddy) production | 143 |
| VI. | Tea: chief producers | 155 |
| VII. | Coffee: chief producers | 157 |
| VIII. | Cocoa: chief producers | 160 |
| IX. | Raw sugar production | 164 |
| X. | Sugar-beet production | 166 |
| XI. | Cattle and beef production | 174 |
| XII. | Butter and cheese production | 177 |
| XIII. | Fish catch | 186 |
| XIV. | Cotton production | 199 |
| XV. | Cotton: chief importers | 199 |
| XVI. | Flax fibre production | 201 |
| XVII. | Jute production | 202 |
| XVIII. | Rubber: chief producers | 214 |
| XIX. | Tobacco production | 216 |
| XX. | Sheep population | 219 |
| XXI. | Raw silk production | 224 |
| XXII. | Estimated distribution of world forests | 230 |
| XXIII. | Wood-pulp and newsprint production | 238 |
| XXIV. | Iron-ore production | 254 |
| XXV. | Iron-ore imports | 255 |
| XXVI. | Copper ore production | 258 |
| XXVII. | Lead ore production | 260 |
| XXVIII. | Zinc ore production | 260 |
| XXIX. | Tin ore production | 261 |
| XXX. | Bauxite production | 263 |
| XXXI. | Gold production | 263 |
| XXXII. | Silver production | 265 |
| XXXIII. | Coal production | 276 |
| XXXIV. | Crude petroleum production | 288 |
| XXXV. | Natural gas production | 291 |
| XXXVI. | Hydro-electric power potential | 295 |
| XXXVII. | Steel production | 325 |
| XXXVIII. | Ships: tonnage launched | 339 |
| XXXIX. | Tourism in Britain | 392 |
| XL. | Merchant ship types | 424 |
| XLI. | The world's ship canals | 429 |

PART ONE

# MAN, THE EARTH, AND HIS LIVELIHOOD

## *Chapter I*

## THE WORLD OF BUSINESS

### *MODERN GEOGRAPHY*

GEOGRAPHY may be defined, in a broad way, as "the exact and organised knowledge of the distribution and organisation of phenomena on the surface of the earth." This is a good description of geography; but let us amplify it a little. First, geography is concerned with the surface of the earth on which mankind lives; secondly, it is concerned with the phenomena of that surface, phenomena embracing man-made or cultural things as well as natural or physical things; thirdly, it studies the spatial distribution of these human and physical phenomena and their mutual relationships; and, fourthly, this study is not undertaken in a loose and random way, but scientifically, for it analyses and organises the phenomena and their relationships.

Geography deals with many things: with the shape and size and movements of the earth; with the distribution and position of continental land-masses and bodies of water; with the rocks and relief of the earth's surface; with the temperature, pressure, wind, and rain of the earth's surface and the varying climates to which they give rise; with the earth's garment of vegetation and the distribution of animals; with the races of mankind who inhabit the earth and the distribution of population; with the varying activities of man and the different ways in which he makes a livelihood; and, finally, with the responses, relationships, and interactions between man and his environment.

Clearly, the field of geography is large; this is inevitable, however, because geographical study embraces the entire world. Phenomena are legion, while the relationships between the phenomena are extremely intricate and far-reaching. No one person today can know all there is to know in the study of geography and so that the geographer can find out more and more about the world in which he lives, he, like other scientists, specialises; in other words, he devotes his attention to some particular aspect or problem of the geographical environment. Hence there are many branches of geography: mathematical, physical, biological, historical, political, social, and economic. The specialist geographer chooses one of these branches and devotes his energies to research or systematic scientific enquiry into it or, perhaps, into one small part of it.

We, in this book, are chiefly interested in economic geography; in other words, we are going to be specialists in a small way. But, before we proceed to describe what economic geography is about, let us ask, and satisfy ourselves about, one question: why do we study geography and what is its value to us? First, we should realise that we are living in a world of increasingly rapid transport and new means of communication; this means, in effect, that the world is getting smaller. Distances between place and place are becoming less in the sense that the travelling time is increasingly reduced. Jules Verne's hero, using every device possible, managed, over a hundred years ago, to get round the world in eighty days; today we can get round it in as many hours. Moreover, as a result of this greatly accelerated speed of travel, and of improved communications, peoples living in different parts of the earth are becoming more closely linked together.

Secondly, we should remember that anything that happens anywhere in the world, such as a war in South-east Asia, a revolution in South America, a famine in China, a new oil find in North Africa, or an economic recession in the United States, may have, and often does have, repercussions throughout the world. In other words, few, if any, countries can live these days in complete isolation and shut their eyes to what is going on in the world around them.

Thirdly, many countries, of which Britain is a typical example, live by what they make in factories and export; simply and brutally they must "export or die." They cannot produce their manufactured goods without the raw materials they import; and these, as well as much of their food, have to be paid for by exports. It is important then to know how other peoples live, what are their wants, what kinds of markets they present, what kinds of raw materials they have for sale, what languages they speak, and how mutual exchange can best be undertaken. Thus, from a purely selfish point of view, we should study geography. For those people who intend to become bankers, insurance agents, salesmen, advertising agents, journalists, or hold responsible positions in transport and communication, the study of geography is of considerable value and sometimes even of prime importance.

Finally, since the Second World War several large, multi-nation organisations, such as the European Economic Community, Nato, and the United Nations, have come into being. These are, as the late Professor F. Debenham said, "slowly and painfully at times, seeking a way to the unity and freedom without which civilization cannot hope to survive.

That unity can come only from a full understanding of the three pillars of geography—*the peoples of the world, the places where they live*, and *the work that they do*."* Here, plainly enough, is the answer to our question: not merely for our own sakes but for the sake of the world and its peoples at large we should study geography.

## *ECONOMIC GEOGRAPHY*

Let us now turn to the question: What is economic geography? What is its field? What is its function? Briefly, in a broad way, economic geography is concerned with the ways and problems of making a living. It deals with basic resources and with human activities in relation to these resources. Expressing this in another way, economic geography is concerned with human exploitation of the earth's natural resources, the production of commodities, whether raw materials, foodstuffs, or manufactured goods, and their transportation, distribution, and consumption. The study of these matters may be said to constitute the field of economic geography.

What is its function? Economic geography seeks to provide answers to a number of questions:

(*a*) Where can economic activities be carried on?
(*b*) Where are economic activities carried on?
(*c*) Why are economic activities carried on?
(*d*) When are economic activities carried on?
(*e*) How are economic activities carried on?

It will help us in our study of economic geography if we continually bear these questions in mind and repeatedly ask where? why? when? and how? in considering any aspect of economic geography. For example, in dealing with the production of a commodity, say, the mining of tin, we should ask ourselves: where is it carried on? why is it carried on? when is it carried on? and how is it carried on? In studying a manufacturing industry, the woollen textile industry for instance, we must satisfy ourselves about the where, why, when, and how of the industry. In examining a port, such as Singapore, we want to know where it is, why it is there, when it was developed, and how it functions. And so on. The complete answer to any problem of economic geography involves answering these various questions.

Many people think economic geography is a dull subject, but this is far from being the case. Learning facts may be tedious (but one must always learn a certain number of facts in any subject), but even some of the facts

* *Introduction to Illustrated World Geography*, edited by Frank Debenham and William A. Burns, George Rainbird Ltd., 1960, p. xi.

are extremely interesting. Economic geography can be a very fascinating study; here are some questions which come within the scope of economic geography:

Why did man introduce the horrible killing disease of myxomatosis into rabbits?

Why were giant frogs introduced from the Hawaiian Islands into Queensland?

Why do we study the antics of the herring and the capers of the cod?

Where do sarsaparilla, chewing gum, locust bean, ambergris, quinine, Egyptian tobacco, birds' nest soup come from?

How many acres of forest land does it take to produce an edition of *The Sunday Times*?

What are avocado pears, coriander, sea cucumber, balata, tagua nuts, menhaden, kok sagyz, yerba maté, vanadium?

How does it come about that most of the world's coffee is grown in the Americas when the plant is a native of Africa?

What effect will the building of a Channel Tunnel have on British trade with the continent?

What are fish-farming, hydraulic-mining, milpa-agriculture, geothermal energy?

Why is Pontefract famous for pomfret cakes and liquorice allsorts? Zanzibar for cloves? Burton for beer?

Why are false teeth made in Israel? diamonds cut in Amsterdam? shirts made in Albany, New York? sausage skins produced in China?

What are sago, tapioca, vodka, vanilla, cochineal, *pâté de foie gras*, caviar made from?

Why are some countries very highly developed yet others extremely backward?

Why has India more cattle than any other country in the world? And why does this constitute a great problem to India's economic progress?

How many species of the world's animals have been domesticated? What is the best transport animal? Which is the most useful animal to man?

How significant are the economic resources of the tundra lands, the Sahara, Antarctica to man?

How may inventions influence trade? How may wars affect commerce? How does religion influence economic activities and trade?

When is a hardwood not a hard wood?

Where do Ecuador's 18 million bunches of bananas go to each year? And why are they wrapped in polythene bags and shipped green?

Why is over 30% of all the rhubarb grown in Britain cultivated in the small triangle made by Bradford, Leeds, and Wakefield?

These questions clearly indicate that economic geography is no dry-as-dust subject. Moreover, economic geography is not a dead or static study: conditions in the world are constantly changing, so the subject is very much alive. Indeed, much of the interest in economic geography comes from this constant change; for example, it is interesting to follow the changes taking place in the use of commodities, in industries, in the growth of countries, and in the flow of trade.

## *ONE WORLD*

The day when man lived a self-sufficient, circumscribed life and seldom travelled beyond a few miles of his home has long since gone. Our present-day lives are closely affected by what goes on elsewhere in the world and, day by day, the world grows smaller, due to the shrinkage of time and distance resulting from improvements in communication and transport. Also, the earth is becoming more and more interdependent with every people and every country being affected to a greater or lesser degree. Even peoples and countries who wish to stand aloof are drawn irresistibly into the maelstrom of global development. The primitive isolation of the New Guinea aborigines is rudely shattered by the oilmen seeking petroleum or the anthropologist who is intent on measuring the aborigines' skulls. The happy Buddhist monks turning their prayer-wheels in timeless Tibet are suddenly disturbed by Chinese soldiers who burst in to take over their country. The Eskimoes living on the extreme fringe of the habitable world are suddenly confronted with American scientists and servicemen who wish to set up their radar-detection screens. And so on.

It is surprising how many people are still unaware of the extent to which their lives and the life of the country in which they live are linked with and dependent upon the outside world. A few years ago a French writer amusingly described how the everyday existence of the average Frenchman was intimately bound up with the world environment. He wrote: "On awakening, our mythical Frenchman washes with soap made from material brought from the Congo, and wipes his face on a towel of Louisiana cotton. He dresses in a shirt and collar of Russian linen, a vest of Australian or South African wool, and a tie of Japanese silk. His shoes are of leather from the Argentine, tanned with chemical products from Germany. In his dining room, furnished with a Dutch sideboard of Hungarian wood, he uses utensils compounded of copper from Rio Tinto, pewter from Detroit and silver from Australia, and eats bread from France, Rumania or Canada according to season. His eggs are from Morocco, his frozen meat from the Argentine, preserved peas from California, jam made in England from French fruit and Cuban sugar, and his excellent coffee from Brazil. So on through the day. Finally he goes to sleep under a quilt of Norwegian eiderdown, dreaming that France is a great country, sufficient to itself, which can mock at will the rest of the universe."*

This quotation, which could well apply to many countries, serves to show how even such a country as France, which is well-developed and highly self-supporting in so many ways, owes so much to the rest of the world. This interdependence of countries, one with another, is of course the basis of international trade. The more sophisticated an individual's or a country's life becomes, the more varied and more exacting do their wants become. These, in turn, result in both a greater volume of trade and a greater complexity of trade. Since the peoples and countries of the world are, by and large, developing and their wants are becoming increasingly sophisticated, world trade is not only growing in volume and complexity but these peoples and countries are becoming more and more dependent upon each other. Many thinking people at the present time believe that the world's problems will only be solved when all countries realise that they are mutually dependent upon each other. Hence the efforts of many to bring about "One World."

## *PRODUCTION, MANUFACTURE, AND DISTRIBUTION*

As we have already indicated, economic geography is concerned with the exploitation of the earth's natural resources, the production of various types of commodities, the distribution of these goods throughout the world, and their consumption. Economic geography, therefore, may be said to concern itself with three stages of economic production:

1. The production of raw materials, whether they be foodstuffs, such as wheat, meat, or fruit, or industrial raw materials, such as coal, timber, cotton, or wool; the production of such simple, unprocessed commodities from the rocks, the soil, or the sea is called *primary production* and mining, forestry, farming, fishing, and hunting are known as *primary activities*.

2. The processing, fabrication, and manufacture of primary commodities; this involves the preparation, working, and changing of primary produce into states or forms required by the user or consumer, for example, iron ore is turned into sheet steel, wheat is converted into flour, cotton is spun and woven into cloth, and timber is made into

* Delaisi, M. *Contradictions du Monde Moderne*. Quoted from W. A. Gauld's *Man, Nature, and Time*, 1946, p. 5.

furniture; the making of metal goods, textiles, furniture, cigarettes, etc., is known as *secondary production*.

3. The ways and means by which raw materials reach the manufacturer and manufactured goods reach the consumer; this aspect of economic activity is known as distribution and it involves the use of systems of transport, techniques of communication, and the services of merchants, brokers, bankers, financiers, etc.; these various activities are often called *tertiary activities*.

The more detailed study of these three stages will form the substance of much of this book.

EXERCISES

1. "Few, if any, countries can live these days in complete isolation." Explain what this means.

2. What is the purpose of studying Economic Geography? How might it help the explorer, the banker, the politician?

3. What is meant by the saying "Britain must export or die?"

4. "Economic geography deals with basic resources and with human activities in relation to these resources." Attempt to amplify this statement.

5. Using a dictionary or reference book find out the difference between Economic Geography and Commercial Geography.

6. Explain the meaning of the following terms: primary production, secondary production, tertiary activities. Give examples of each.

# *Chapter II*

# THE EARTH'S PEOPLES

Some account of the peoples of the world—the races to which they belong, their total numbers, where they live, and the standards at which they live—is necessary to help us understand what follows in the ensuing chapters in this book. You may think a study of people and population out of place in a book devoted to business geography; such a study is not as irrelevant as may appear and after you have read this chapter you will be able to see more clearly why people are important, why the distribution of population is important, and why the living standards of people are important from the viewpoint of economic geography.

## *THE RACES OF MANKIND*

The peoples of the earth belong to different races. In a very general way, the peoples of the world may be divided into four groups according to the colour of their skins—white, brown, yellow, and black people. This is not a very good, or scientific, way of grouping the different peoples of the earth, but colour of skin is obvious and provides us with an easy method of separating one group from another. But how unsatisfactory skin colour is as a criterion of race is shown by the fact that both the Negroes of Africa and the Blackfellows of Australia are black, or at least have a very dark skin colour, and yet they belong to two quite separate racial groups. Moreover, most of the brown-skinned peoples of the world belong to the same race as the white-skinned Europeans.

Clearly skin colour is far from being a reliable guide to race. How, then, can we distinguish race? There are various physical characteristics which can be used to differentiate mankind, but before we mention these let us try to answer one question: why are the peoples of the earth different in colour? It would seem that skin colour is a response to sunshine. In tropical regions where there is brilliant sunlight skin colour is usually darker; this may be Nature's way of protecting man from the adverse effects of the sun's rays. In general, skin pigmentation becomes lighter as one goes polewards, and Professor Fleure explained the bleached pink and white complexions of the peoples of northern Europe as a natural adjustment to too little sunlight in these dull, cloudy lands. The bleached skin enabled man to get the maximum benefit from what little sunshine there was. But the whole problem of skin colour presents a difficulty, and we are not quite sure of its significance.

The anthropologist bases his division of mankind upon the careful measurement of man's physical characteristics, such as the shape of his head, his stature and body proportions, the shape of the nose, and the nature of the hair. Some peoples are, on average, tall, others short, and others of medium size. Some have long heads, some broad heads, while other peoples have heads of an intermediate index. Again, hair differs in colour (fair, brown, black), texture (lank, wavy, frizzy), and in section (round, oval, flat). Using these different physical criteria in conjunction with skin colour, the anthropologist has been able to distinguish several racial groups.

## CLASSIFICATION OF RACES

Anthropologists differ among themselves as to exactly how many distinct races there are, but most recognise five types: the Caucasoid (white), the Mongoloid (yellow), the Negroid (black), the Australoid (black), and the Bushmen and Hottentots of the Kalahari Desert area of South-west Africa. The last two types are numerically small and are dying out. The first three groups, on the other hand, are flourishing and expanding rapidly in their numbers. There are more members of the Caucasoid group than the other two, although the Mongoloid peoples come a close second.

The peoples of the Caucasoid group include not only white Europeans and people of European descent found in other parts of the world but the Arab peoples and most of the peoples inhabiting the Indian sub-continent. The latter, though brown-skinned peoples, belong to the "white" group. The Caucasoid group has a very wide extension and, in fact, three sub-divisions, sometimes called sub-races, are recognised: the Nordic peoples, tall, muscular, fair-haired, blue-eyed, fresh complexioned, who are characteristic of northern Europe; the Alpine peoples, stocky, of medium height, brown hair, hazel eyes and rather sallow skin; and the Mediterranean peoples, usually slight in build, olive-skinned, dark-haired, and dark-eyed. In the Indian sub-continent representatives of all these three sub-types are present since throughout the region's history these racial sub-groups have invaded the sub-continent in turn.

The Arab peoples of North Africa and the Middle East belong to the Mediterranean sub-group. So do most Jews, although it is difficult to convince an Arab he is of the same race as a Jew. The antagonism between the two arises out of cultural, rather than racial, differences. These relate mainly, though not entirely, to religion. This leads us to make another point: the relationship between language and race. There is no necessary relationship between the two. Language is a cultural feature; it is something that is acquired and which can be changed. Race, which implies inherited physical characteristics, is unchangeable; we inherit our race at birth, the physical characteristics being passed on from father to son.

The Mongoloid group is found mainly in Central and East Asia and is typically represented by the Chinese, although even they vary among themselves. The native American Indians (Amerinds) are an early offshoot of the Mongol group. The Polynesian peoples are probably a subgroup, having undergone racial intermixture. The Negroid peoples are chiefly confined to Africa south of the Sahara, although negro groups are found in the Americas as a result of the early iniquitous slave trade.

### RACIAL INTERMIXTURE

Although we have noted these main racial groups, it should be emphasised that there has been a great deal of racial intermingling and it is very doubtful whether there is such a thing as a pure race. The people of Britain, for instance, are of very mixed ancestry and the historian H. A. L. Fisher aptly called the British a group of "energetic mongrels."

One of the most interesting human experiments going on in the world today is in Latin America. Here racial intermixture has been taking place for several centuries and interbreeding between the three racial groups (Whites, Negroes, and Amerinds) has given rise to several hybrid types. These hybrids, in turn, are intermixing with the result that the entire region is a great melting pot of races. In due course a new racial "type" is likely to emerge from this large-scale mongrelisation.

One final point should be made with respect to race. While we can recognise different physical differences between racial groups, it seems very unlikely that races differ mentally. No race can claim mental superiority. Formerly, it used to be said that certain races were mentally inferior to others, and especially to the white race. There are no grounds for this belief. If peoples in the past were backward, it was because of cultural isolation and stagnation and not through having any inferior intelligence or mental equipment.

## *DISTRIBUTION OF POPULATION*

If we examine a map of the world showing the distribution of population we are immediately conscious of the irregularity of this distribution: some areas are very densely peopled, others very thinly peopled. For example, it will be seen that the islands of Java and Sumatra, though next to each other, show greatly differing densities: the one is extremely densely peopled—in fact it has one of the highest average population densities in the world—while the other has relatively few inhabitants. Again, if we study a population map of the United States, we find that the eastern half of the country, east of the 100 degree W. meridian, has a relatively high density with very high densities in the north-east, whereas the western half of the country, except for local high densities along the Pacific seaboard, is relatively sparsely peopled. Even within the British Isles there are very wide differences in the density of population between part and part; compare northern Scotland with the central valley

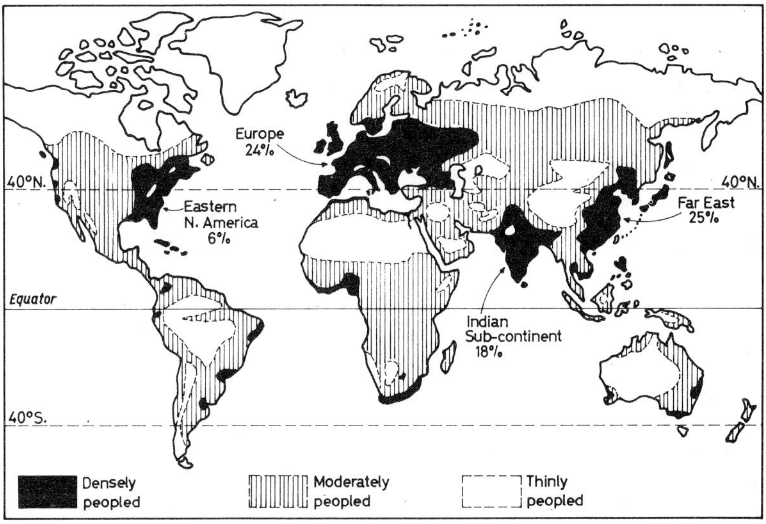


FIG. 1.—DISTRIBUTION OF WORLD POPULATION

Note the four areas of dense population, all of which are in the northern hemisphere and lie mostly in temperate latitudes. Note, also, the location of the thinly peopled areas: relate these to ice-cap, tundra, forest, desert and high altitude areas.

of Scotland or the Welsh Uplands with the English Midlands. Such inequalities in the population distribution are features of great interest and one is tempted to ask: why are people so irregularly distributed? Why, for instance, is Java overcrowded when Sumatra next door is under-populated? In a moment we shall try to offer some explanation for this patchwork of population, but first let us look at the general pattern of world population distribution.

The population map (Fig. 1) shows quite clearly that mankind is massed in four main regions of the earth:

1. In Western and Central Europe, especially in Britain, France, the Low Countries, Western Germany, and Italy.

2. In the east central part of North America, *i.e.* eastern United States and south-eastern Canada.

3. In the Indian sub-continent comprising Pakistan, India, Bangladesh, and Ceylon.

4. In the far east of Asia, especially in eastern China, Korea, Japan, and Taiwan.

Outside these four main areas of population concentration, mankind is much more thinly spread, although here and there are clusters of fairly dense or very dense population, *e.g.* in Java, in Egypt, in Nigeria, around Buenos Aires in Argentina, in south-eastern Brazil, in California, and around Sydney in New South Wales. There are many areas, often of great size, which are extremely thinly peopled, where, in fact, there is a density of under two, and sometimes under one, per square mile.

While it is true that there are over-populated areas carrying too many people in some parts of the world and under-populated areas capable of supporting more people in other parts of the world, it may be said that, in general, the world distribution of population reflects the suitability of the environmental conditions for settlement. But, in addition to the physical geographical factors, certain human factors, *e.g.* of an historical and political kind, have sometimes influenced settlement of areas and population densities within areas. Let us now come back to the question we posed a moment ago and examine the reasons which explain the inequality of the distribution of population.

## FACTORS GOVERNING POPULATION DISTRIBUTION

Mankind usually congregates where natural and other conditions most easily offer a supply of food or provide a means of earning a livelihood. Where conditions are difficult, *i.e.* where the possibilities of securing a food supply are limited and where the chances of earning a living are closely restricted, man is not tempted to settle down and reproduce his kind.

The chief factors influencing the distribution of population are: (*a*) accessibility, (*b*) relief, (*c*) climate, (*d*) vegetation, (*e*) water supplies, (*f*) soil, and (*g*) mineral deposits.

The accessibility, or position in relation to means of transportation, of a land area helps to determine whether or not it is capable of being occupied. Until man devised some means of water transport, islands proved to be inaccessible; so far as we know New Zealand, for instance, was unoccupied until the Maoris reached it by canoe. Other areas, such as the vast interior plateau of Brazil, remain scantily peopled partly because they are largely inaccessible.

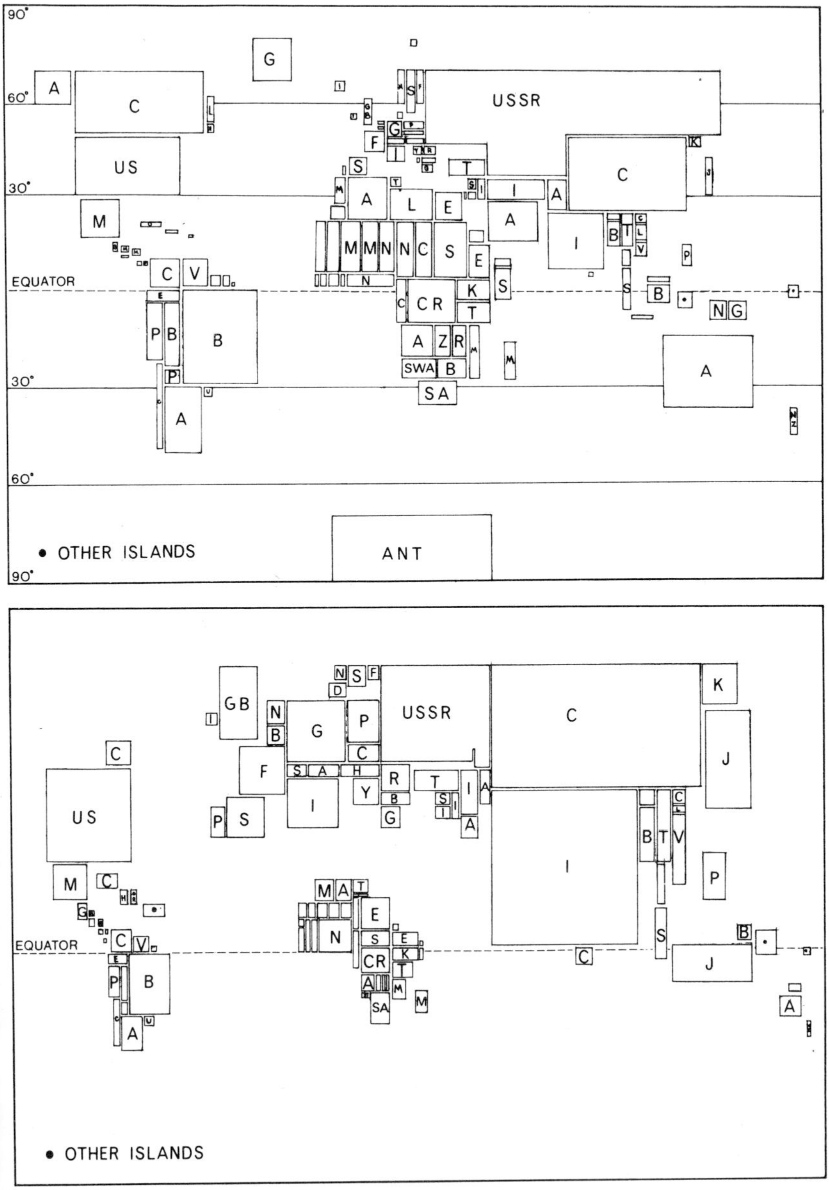


FIG. 2.—WORLD AREA AND POPULATION COMPARED

The unevenness of the world's population distribution. The upper "map" shows land areas and the lower one the relative sizes of population on them. Note the wide discrepancies.

High mountains and rugged terrain may, as they certainly do in many parts of the world, limit settlement. The Himalayas, the southern Andes, and parts of the Rockies, as well as many other mountains have very few inhabitants; this is due to their ruggedness, their usually unfavourable climate, their handicaps to transport, and their poor, thin, stony soils which place a limit upon agriculture. Many high, inaccessible areas, such as Tibet, have relatively few inhabitants; on the other hand, we should note that the cool plateaus of the central Andes are much more densely peopled than the adjacent Amazonian plains which are hot, wet, and thickly forested.

Climatic effects upon population distribution are twofold: direct effects, such as cold and aridity, which repel human settlement, and indirect effects, such as those affecting crop production and disease. The lands of perpetual snow and ice are shunned by man except for small specialised groups such as Eskimoes. At the other extreme, the hot deserts are thinly peopled unless water is available in quantity for irrigation, as in Egypt and coastal Peru, or unless mineral wealth provides an attraction in spite of the shortage of food and water as in the case of the gold deposits of the Western Australian desert or the nitrate and copper deposits of Northern Chile.

Vegetation, which is largely a response to climate, may retard human occupation of an area; for example, there are many who believe that the rain-forest of the Amazon Basin is the primary obstacle to the occupation of that region: certainly the forest may be said to be *the* dominating feature of the region in so far as man must come to terms with it before he can do anything; the vastness and thickness of the selva is such that, so far, man has been curbed and confined by it.

The presence or absence of water is a basic factor affecting population distribution. Man, and animals and plants, too, cannot live long without water, hence the climatic and vegetational deserts of the world are also the human deserts. If water is available in desert areas providing drink and moisture for the cultivation of crops, then man will inhabit the hottest and most desolate deserts—think of the crowded oases to be found in the heart of the Sahara and the "little Egypts" of the arid Peruvian and Chilean desert.

Soils, by their fertility or infertility, may encourage or discourage dense settlement. Fertile alluvial soils or loess-type soils will, if climatic conditions are favourable, give rise to dense agricultural populations as in the middle and lower Ganges valley, the Great Plain of North China, and the Börde zone of central Germany. Conversely, the poor, leached lateritic soils of hot, wet tropical regions often compel the people to resort to shifting cultivation. This is one of the reasons why Sumatra has such a low density of population. Java, on the other hand, has large areas of fertile volcanic soils.

Finally, the presence or absence of mineral wealth is a notable factor influencing population distribution at the present day. The coalfields, yielding power for industrial developments, have acted as magnets and become seats of some of the highest population densities in the world. Mineral wealth has attracted man to areas which, on other counts, are highly undesirable for human settlement: to Kalgoorlie in the hot, waterless Australian desert, for gold; to Chuquicamata, 10,000 feet (300 m) up in the Andes, for copper; to Uranium City in the Barren Lands of northern Canada, for uranium; and to the equatorial forests of New Guinea for oil.

## NON-PHYSICAL FACTORS

It will be clear that physical factors may encourage or discourage settlement. But, important as are the physical influences affecting the distribution of population, we must note that man himself has helped greatly to change the pattern of population distribution and density throughout the world. Wars have affected populations; a good example of this is provided by Paraguay in South America which, during the War of the Triple Alliance, 1865–70, lost more than half of its population. It has been estimated that out of a total population of just over half a million only 221,000 were left alive after the war, and of these a mere 10% were males; as a result, not until recent times has the normal ratio between the sexes once again been reached. This disastrous war had the dual effect of drastically restraining population growth and holding back the country's progress.

Religious persecution throughout history has resulted in numerous population movements and readjustments. Many early Christians were slain, French Huguenots fled France, the Pilgrim Fathers went to America, and the Mormons moved to Utah. More recently several million Jews in Germany and Poland were exterminated, and as a corollary of this the number of Jews living in Israel has jumped from 84,000 in 1922 to over 2 million in 1971.

Political factors have sometimes played a part. In the inter-war period, Mussolini, the Italian dictator, for political reasons, did his utmost to stimulate Italy's population growth, even though the country was suffering from severe pressure of population. The Soviet Union, in order to open up, develop, and populate its Asiatic territories has drafted millions of Russians into Turkestan and Siberia.

Social factors, such as ancestor worship in China, polygamy among Moslem peoples, early marriage among the Hindus, etc., have also greatly

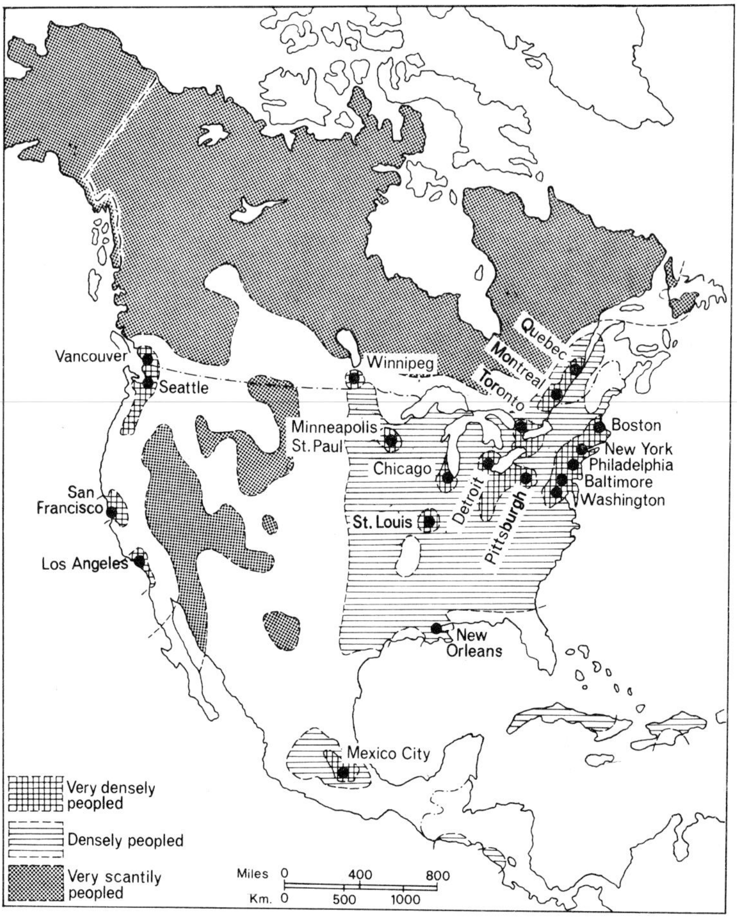


FIG. 3.—POPULATION OF NORTH AMERICA

Population in N. America is very unevenly distributed. Nearly all Canadians live along the narrow southern fringe of Canada, where the climate and soils are kindest. The densely peopled part of the United States lies in the east where the moist climate and deep, rich soils favour cultivation, where rich deposits of coal, oil and iron ore have formed the basis for industry, and where the indented Atlantic coastline, facing and nearest to Europe, has been utilised for port construction. Three other areas of dense population occur: around Mexico City on the Central Plateau of Mexico; in California, concentrated around Los Angeles and San Francisco; and in the Seattle–Vancouver area.

All the northern continents show a broadly similar pattern: a major zone of concentration which fades into a zone of relatively sparse population with pockets of denser population (*see* Fig. 1).

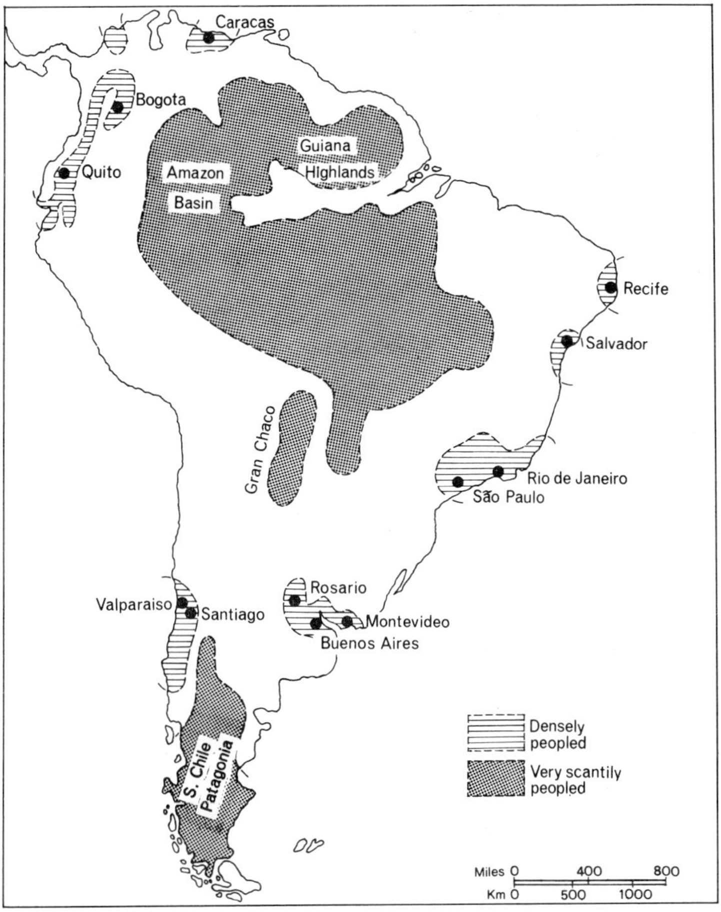


FIG. 4.—POPULATION OF SOUTH AMERICA

The map brings out several notable features in the pattern of settlement: (i) a definite tendency for the population to congregate in clusters; (ii) a distinct tendency for the more densely peopled parts to lie on or near to the coast; and (iii) vast inland areas of very sparse population.

This pattern is the outcome of historical, physical, and economic factors. Favoured areas may be seen from the map, while the hot, wet, forested plains of Amazonia, the savanna lands, and semi-arid Patagonia are very sparsely peopled: settlement, where it has occurred, is chiefly along the river valleys since these offered the only easy means of penetration.

This population distribution pattern is common to all the southern continents. It is repeated in Australia and in the case of central and southern Africa but it is less clearly evident in the latter.

influenced both the distribution and the density of population in particular areas.

Thus it may be concluded that a full and proper understanding of the distribution of the world's population lies as much "in the history, traditions and aptitudes of the people"* as in the determinants of the physical environment.

## *THE GROWTH OF WORLD POPULATION*

If the figures in the following table are studied it will be seen that: (*a*) the world's population has increased enormously during the past three hundred years, and (*b*) the rate of growth has varied widely between continent and continent.

TABLE I

*Growth of Population by Continents (in millions)*

| | *1650* | *1750* | *1850* | *1960* | *1970* |
|---|---|---|---|---|---|
| Asia | 330 | 479 | 749 | 1556 | 2227 |
| Europe | 100 | 140 | 266 | 588 | 619 |
| Africa | 100 | 95 | 95 | 225 | 283 |
| Central and South America | 12 | 11·1 | 33 | 200 | 260 |
| North America | 1 | 1·3 | 26 | 190·5 | 225 |
| Australasia | 2 | 2 | 2 | 15·4 | 18 |
| Total | 545 | 728·4 | 1171 | 2774·9 | 3632 |

Perhaps the single outstanding fact about world population is the rapidity of its growth. Three hundred years ago there were approximately 500 million people in the world; today there are more than 3600 million. Population, today, is growing more rapidly than at any time in the past. In fact, it is estimated that at the present rate of expansion the world's population is just about doubling itself every fifty years. Every day there are 150,000 more babies born into the world. Or, in other words, a population equivalent to that of Huddersfield, Luton, or Stockport is added to the world total *every day*.

According to a United Nations report the world's population is expected to grow faster during the 1970s than ever before in spite of the growing use of birth control measures. The most recent estimates indicate that the total world population will rise from about 3632 million in 1970 to 4457 million in 1980. Most of this increase, as in the next decade, will

* POUNDS, N. J. G., *An Introduction to Economic Geography*, John Murray, 4th edition, 1970, p. 117.

take place in the less developed parts of the world which show annual birth rates of 2·4% and 2·5% as against rates of 1·1% and 1·2% in advanced countries. By 1980, it is calculated that there will be some 3247 million in developing regions and only 1210 million in advanced countries.

If the present trend of population increase continues, and there is little to suggest that it will be interrupted or changed, by the year 2000 there will be approximately twice as many people in the world as there are now. Such large numbers are, quite clearly, going to create numerous problems.

## CAUSES OF GROWTH

What factors affect population growth? There are, in fact, many, and they are often complex. Periodically throughout human history major cultural developments have permitted substantial growth in world population. The discovery of the art of crop growing, which ushered in the Neolithic Revolution, stimulated population, growth for agriculture gave rise to an assured food supply. But the growth of population was relatively slow until about two hundred years ago. Then, the "revolutions" in agriculture, science, and industry made possible a vastly increased measure of economic productivity which was accompanied by a rapid and continually accelerating increase in the numbers of people.

In the past, plagues, such as the Black Death, and diseases, such as tuberculosis, have checked population growth. Progress in medicine, in hygiene, and in sanitation have greatly increased the chances of survival and led to growth in numbers. One example will illustrate the importance of the struggle against disease. In Ceylon in the past, malaria, carried by mosquitoes, led to the deaths of many people and this kept the population down. In recent years, malaria has been all but eliminated by using insecticides to kill the mosquitoes. As a result, the expectation of life has almost doubled within the short space of 25 years and the population of Ceylon is now increasing rapidly.

Other so-called natural checks, such as flooding, famine, and war, have curbed population growth in the past. For instance, it is estimated that the flooding of the rivers and famine have been responsible for the deaths of about 100 million people in China since 1800 A.D. Flood-control measures and improved communications enabling food to be quickly brought into stricken areas have greatly lessened the deaths resulting from such catastrophes.

Social and political factors may predispose towards population growth. We have already referred to such social customs as ancestor worship among the Chinese, polygamy among the Moslem peoples, and early marriage among the Hindus which encourage population growth.

Sometimes national policy aims deliberately at stimulating the increase in the population; for example, in the inter-war period, both Fascist Italy and Nazi Germany encouraged the procreation of children by offering state bounties and even medals to prolific mothers!

## *THE POPULATION PROBLEM*

Some areas, as we have indicated, are already over-populated but, apart from local or even regional over-population, there is a very real threat of global over-population, especially if the current high birth rate and recent trends in overall growth continue. There are two main problems with respect to the population problem: (*a*) the problem of easing local overcrowding, and (*b*) the problem of controlling overall growth.

There are a number of possible remedies which may be adopted to ease, if not to solve, the problem of local overcrowding:

1. *Pressure may be relieved by emigration.* In the past large numbers of people did migrate from over-populated countries; for example, between 1901 and 1913 Italy exported more than half a million of its inhabitants annually. But the days of large-scale emigration appear to have gone for good. Few countries nowadays open their doors widely to immigrants. Where immigration is allowed, it is usually reserved for scientists, technicians, and people with capital; impoverished, illiterate peasants, or labourers are not welcomed.

2. *Development of natural resources.* A more positive approach to the problem is to develop the natural resources of a country so that it is capable of supporting more people. Improved agricultural production, irrigation schemes, exploitation of mineral wealth, fishery development, etc., are some of the ways by which increased numbers might be supported. Developments of this kind, however, usually require much capital for their realisation, and all too often the most overcrowded countries are also the most poverty-stricken. Such countries are largely dependent upon the goodwill and beneficence of the world's rich uncles, particularly Uncle Sam.

3. *Industrialisation.* Industrialisation and the production of manufactured goods for export offers another possible alternative. Industry is capable of absorbing surplus rural labour up to a point and the export of manufacturers provides a source of income. But industrialisation is at best merely a palliative and not a cure for over-population; nor is industrialisation always economically feasible or socially desirable. Countries lacking in industrial raw materials, power resources, and technological know-how are likely to find industrialisation difficult and of limited applicability.

From the point of view of controlling overall population growth, there is only one possible practical solution: birth control. Birth control, either by moral restraint or the use of contraceptives, is not an easy matter. It has to meet the challenge of religious, social, and even economic and political factors. For example, the use of contraceptives is contrary to the teaching of the Roman Catholic Church; the Hindu fathers many sons to ensure a survivor to perform the last rites for him; the Oriental peasant sees his many children as potential workers and therefore as an economic asset; and many states view a large population as a yardstick of power. Clearly it is not likely to be easy to educate people to practise birth control; and, in any event, it must be looked upon as a long-term method.

## ECONOMIC ASPECTS OF THE POPULATION PROBLEM

The size of a country's population is of economic importance for two main reasons:

1. The factor of production depends very much upon the amount, availability, and quality of human labour, and the supply of this labour is itself closely related to the number, age composition, and education of the population.
2. People are consumers as well as producers, and the standard of living bears a close relationship to the numbers sharing the national income, that is, the total of goods and services produced by the economically active population.

Also of great importance is the size of a country's population in relation to other factors of production, such as the natural resources, physical handicaps, availability of capital, etc. The total economic output of a country will fall short of what it might be if there is insufficient manpower to make full and effective use of the non-human factors of production. On the other hand, if the population is too large in proportion to the non-human factors of production, then living standards, accordingly, will be much lower than they might be had a better balance between the two groups of factors obtained. Theoretically, there is an optimum population for every country: this optimum population is reached when the labour force is just sufficient to make the best possible use of the available resources. But this fine balance is not necessarily constant: any increase in resources—an improvement in soil fertility, new mineral finds, realisation of power potential, stock of capital—will probably permit or require increased supplies of labour; it follows, therefore, that the level of the optimum population will be raised.

Over-population does not depend merely upon the total number living in a country, nor upon the density of the population. For example, a population density of 60 persons to the square mile may mean over-population in one area but in another under-population. Much indeed depends upon the available resources, the capacity of the land to support population, and the degree of cultural development. Moreover, a country is not necessarily over-populated simply because it is incapable of providing sufficient food to support its people; Venezuela provides an illustration for, although its population is small, about 10 million, and there is no shortage of cultivable land, it does not produce enough food to support

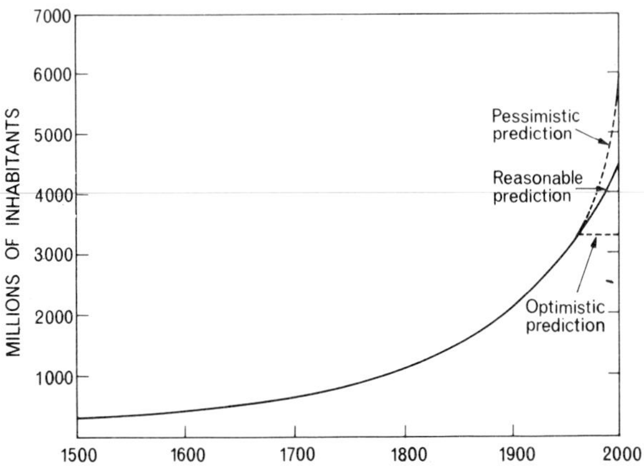


FIG. 5.—GRAPH OF WORLD POPULATION GROWTH

This is an example of a *simple line graph*. Line graphs are commonly used to show continuous changes. If times or dates are involved, they are usually shown on the horizontal scale; time is described as the independent variable. Quantities are indicated on the vertical scale; these are the dependent variables. Line graphs may be used to show absolute or percentage values of items, such as industrial production, trade statistics, population, over a period of time. This graph shows the curve of world population growth over a period of nearly five centuries. Note how the population remained fairly stable during the sixteenth century, how it began to increase during the period 1700 to 1850 (the time of the Agricultural and Industrial Revolutions), and how, since about 1850, population growth has accelerated.

its people and must import considerable quantities. Again, a country may be able, as in the case of Britain, to employ its labour force more effectively in manufacturing industry, exporting its surplus manufactured goods in exchange for foodstuffs.

Conditions vary widely throughout the world. Some countries or areas are over-populated, others under-populated: in the first category fall the Netherlands and Italy in Europe, Japan and India in Asia, and the republics of Haiti and El Salvador in America; in the second such countries as New Zealand, Canada, Paraguay and Nicaragua in America and many parts of Africa.

There are several other aspects of the population problem. In addition to the problem of numbers in relation to food supplies or the other factors of production, there are such questions, often of major national importance, as the rate of population growth or decline, the distribution between the various age groups, the balance between rural and urban population, the proportion of males to females, and the degree to which the population is literate and educated. All these things in their different ways affect the standard of living of a people.

## EXERCISES

1. Most of the world's population is concentrated in four main clusters. How do you account for such an uneven distribution and how is the growth of the world's population affecting the density pattern?

2. Describe the distribution of population in Australia and explain the geographical reasons for its very small population in relation to its size. (*Chartered Institute of Secretaries.*)

3. Write an essay on the population problems of *either* India *or* China.

4. Examine the reasons for the rapid increase of population in parts of Asia and Latin America. What problems does this increase create?

5. Name *two* distinctive areas within the tropics, one with a low and one with a high population density. Describe precisely the location of each and account for its population density.

6. Give an account of the distribution of population in *either* Scotland *or* Ireland. State the geographical reasons which have produced this population pattern. (*East Midlands Educational Union Examinations.*)

7. Discuss the relationship between annual rainfall and density of population, using Australia and South America as your examples.

8. Describe the distribution of population in Brazil and examine the problems arising from this distribution in relation to: (*a*) the ability of the land to supply food for the people; (*b*) natural resources; and (*c*) space for development.

9. "The combined population of the three largest cities in Australia is well over one-third of the total population of the country." Comment on this statement, analyse the factors responsible, and assess future trends.

10. To what extent can the major features of the world population distribution be understood in terms of the physical environment?

11. What geographical factors influence the present and may determine the future distribution of population in Australia?

12. What recent changes have there been in the geographical distribution of population in the United Kingdom? How can you account for these changes?

# Chapter III

# HOW MANKIND LIVES

In order that man may live certain basic needs must be satisfied: these are water, food, clothes, and shelter. Most of his life is spent, directly or indirectly, in supplying these needs. The ways and means he adopts to satisfy these essential wants are many, and vary from place to place and between people and people. In some environments the physical conditions make life extremely difficult; nature is niggardly and limits the available resources by which man can live. On the other hand, nature

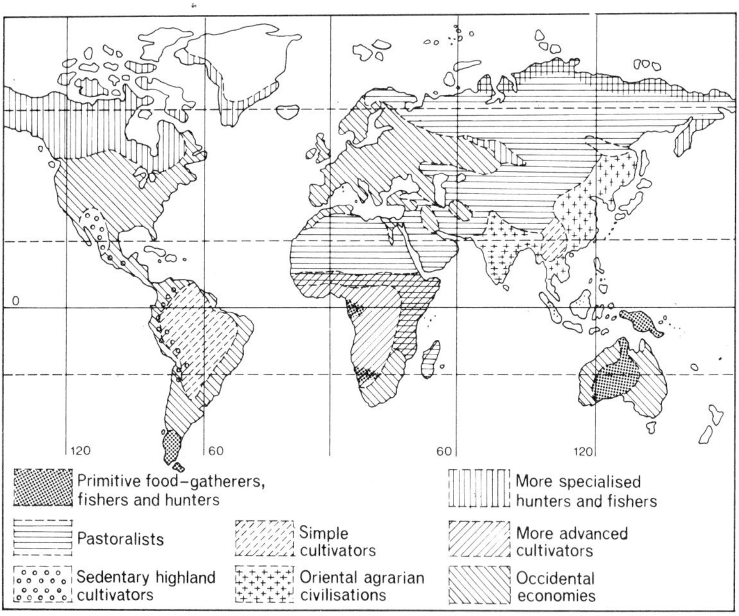


FIG. 6.—THE ECONOMIES OF MANKIND

This map, which shows the principal economies of mankind is, it should be remembered, generalised. Much of Asia is indicated as being under pastoralism but obviously there are localities where cultivation takes place, as in the oasis settlements of Central Asia. Moreover, this map takes no account of mining or industrial activities which clearly are widespread in Europe and the United States and occur elsewhere in the world.

offers much less difficult conditions in some areas and man finds it much easier to live; nature is more bountiful and offers a wider range of resources.

While some peoples have learned little about using the resources which nature provides, others have learned much. The more man learns concerning the ways of using what his environment offers, the easier will the conditions of his life become. If man merely lives off the land, that is merely collecting the fruits of nature, his existence will be hazardous. But once he has learned how to make implements and tools with which he can cultivate the ground, his food supply is more assured for he can produce and store food. Man's ability to produce food relieves him of the necessity of moving around constantly from place to place; it enables him to live a sedentary life: he can live and work in one spot.

Once man has adopted a sedentary mode of life, other things follow: he can have a permanent dwelling and acquire possessions. Furthermore, under settled conditions, larger groups can live together. This, in turn, allows specialisation of labour to take place; in other words, one man can do one job, another a different job. Hence in a community, one man will grow crops, another will make tools, another will weave cloth, another will build houses, and so on. Under these conditions new skills can be developed and learned and, together with the increased leisure which the division or specialisation of work brings, developments in art and literature and thought can take place. In this way, the early and great civilisations of the past developed.

Let us now look at the different ways in which man, both in the past and at the present day, has lived and earned his livelihood.

## *SIMPLE SOCIETIES*

### COLLECTORS AND HUNTERS

Man is a social animal, that is, he prefers to live in groups and from the very earliest times men have congregated together. In early prehistoric times, when man lived by collecting and hunting, he was compelled to live in small groups—perhaps limited to the family group—simply because the available food supplies were incapable of supporting more than a few people. The size of the groups, or in other words the density of the population, was limited by the productivity per square mile of land. During the Ice Age, which lasted until about 10,000 years ago, man frequently dwelt in caves which provided shelter. Subsequently, as the ice melted, the ice-sheets retreated, and the climate became warmer, man began to move into the lowlands where there were patches free from forest, to the lake shores, and to the sea coasts. Remains of pile-dwellings around lake shores and shell middens along the coasts provide evidence of these early settlements.

In the beginning man was a food collector and petty hunter; he gathered wild fruits, nuts, seeds, roots, eggs, grubs, wild honey, etc., and undertook the snaring of small fry, the hunting of animals, and fished for his food supply. Some groups of people still follow this way of life; usually they are to be found in the more remote and inaccessible parts of the earth's surface. Primitive, backward groups, true, are relatively few and scattered nowadays and, year by year, are either being squeezed out of existence or being brought to a higher level of economic life. These small human groups existing by collecting, hunting, and fishing are represented by some of the Eskimo and Indian tribes of North America, the Yaghans of Tierra del Fuego, the Bushmen of the Kalahari Desert, the pygmies of the Congo forest, the Semang of Malay, the forest dwellers of New Guinea, and the Australian Blackfellows.

While it is customary to describe these groups as "primitive" and "backward," it does not follow that they are necessarily savage or unintelligent. The Bushman, for example, is a clever and intelligent fellow in spite of his primitive mode of life: he is so intelligent that he has been able to adapt himself to a hostile natural environment in which he lives quite successfully. The Australian Blackfellow, too, has a high natural intelligence, a rich language, and a complex social organisation. On the other hand, some of these primitive groups, such as the Yaghans, live a miserable and wretched existence and appear to be stunted physically and mentally.

Sooner or later such primitive groups come into contact with modern civilisation and such contact has often brought unfortunate and disastrous results. White man's diseases have frequently created havoc and resulted in many deaths. Guns and strong drink have also had unfortunate repercussions. But there have been other effects of a different kind. For instance, the contact of the Eskimoes of St Lawrence Island in the Bering Sea with Western civilisation has led to very considerable changes in their way of life and there has been a steady shift from reliance upon the local environment to reliance upon imports from the nearby Alaskan mainland. Now, instead of hunting seals for oil which provided them with a fuel for heating, lighting, and cooking, they import mineral oil; as one of the Eskimoes said: "You know, blubber used to be the most valuable thing we had—we used it for heat and light. Now we throw it into the ocean."

Like these Eskimoes, there are other primitive groups whose lives have been radically changed by contact with the white man. There is the case of the Jivaro Indians of the tropical forests of Ecuador who have a reputation as head-hunters and who have lived outside government control. When, however, an oil company began to drill for oil in the Ecuadorean jungle, some of the Jivaros came to be employed by the company and were thereby introduced, if in a casual manner, to Western civilisation. Within a very short time they had discarded their loin-cloths for American jeans and were smoking American cigarettes.

One of the most interesting facts of our present day and age is that, in spite of the marvels of science and technology and the highly organised economic and social life of modern civilisation, there are people still living exactly, or almost exactly, as they did 10,000 years ago.

## PASTORAL PEOPLES

Somewhere around 5000 B.C., perhaps earlier in some areas and certainly later in others, the first great cultural revolution took place. This revolution, ushered in by the domestication of animals, the cultivation of crops, and the beginnings of town life, introduced the Neolithic, or New Stone Age, period.

The domestication of animals meant that man, instead of hunting animals for his livelihood, began to herd them. Some wild animals, such as the horse, cattle, sheep, and goats, are more amenable to domestication than others; these were the creatures he tamed and herded. Such animals were especially useful to man, not only because they were relatively docile creatures and so fairly easily tamed, but because they were also food animals, yielding milk as well as meat. Such animals were herbivorous, *i.e.* grass-eating, and they wandered over the grasslands in search of pasture. In the beginning man wandered with them, going whither they would; later he organised or systematised their movements so that definite seasonal migrations from area to area took place. Such wandering movement is called nomadism.

Formerly most pastoral people were nomadic; a few groups remain so still, though the numbers engaged in nomadic herding are rapidly declining. The Bedouin of Arabia continue to wander with their camels and horses from oasis to oasis. They breed camels and horses which they sell to the settled peoples living around the desert margins. Until quite recent times the Kirghiz, who live on the grassy plains of Central Asia, were nomads rearing horses and sheep, but their migratory mode of life has been changed during the past generation or so by the Soviet Government which has "collectivised" animal herding.

A people similar to the Kirghiz are the Kalmuk of the high plateaus of the Altai region in east Central Asia. The Kalmuk, however, are semi-nomadic. In the Altai region good pastures are available throughout the year, hence widespread seasonal migrations are not necessary and most of the Kalmuk dwell in semi-permanent encampments. Until recently the Kalmuk and other steppe-dwellers of Asia were purely pastoralists and did not cultivate the land; but now many grow crops in addition to rearing animals.

Pastoral nomadism is not confined to the semi-desert and steppe lands for on the tropical grasslands of the Sudan and the East African Highlands Negro tribes, such as the Fulani and Masai, rear herds of cattle or, in the drier areas, flocks of goats. Pastoralism is also the way of life of peoples such as the Lapps, Samoyeds, and Tungus who live on the northern margins of Eurasia where the tundra meets the taiga. In this zone reindeer are the animals herded.

A point of interest worthy of note is the absence of pastoral nomadism in the Americas and in Australia. In both these continental areas the early inhabitants failed to domesticate the native animals; this may well have been related to the fact that the bison and the caribou were not domesticable and also because the horse, cattle, sheep, and goat were absent. Paradoxically enough both the Americas and Australia now possess vast numbers of cattle and sheep, all descended from stocks brought in by European colonists.

### PRIMITIVE CULTIVATORS

Many peoples are simple cultivators, growing crops upon which they subsist. Often the primitive cultivator is a food-collector and hunter also. Even herders will sometimes grow crops if conditions are suitable and they sojourn a while in a given spot. This shows the inadequacy of trying to classify human activities too rigidly; and this is a matter constantly to be borne in mind, for people, particularly at the primitive level, are seldom exclusively food-gatherers, fishers, hunters, herders, or cultivators.

Simple cultivation is either of the shifting kind or the sedentary type. Shifting cultivation, which is widely practised in the inter-tropical zone, in both the forest and grassland areas, involves the clearing of the land of its vegetation growth, usually by the "slash and burn" technique, the planting of seeds and cuttings either by crude hoeing or simply by making holes in the ground with a digging-stick, and subsequently harvesting the crop. Such primitive tillage knows little or nothing of the plough, the rotation of crops, or the use of fertilisers. After two or three years of cropping, the soil becomes exhausted and the cultivator clears a fresh patch of land.

Sedentary or settled cultivation is usually of a more developed kind. Simple cultivation of this kind varies widely, however, ranging from the backward methods pursued by many tropical cultivators to the skilled highly-intensive tillage practised by Oriental subsistence farmers. The careful and productive methods followed by peoples such as the Chinese and Japanese compares very favourably with commercial horticultural practices in the West.

The Boro, who live in the far west of the Amazon Basin, are cultivators of the simplest type. They combine tillage with collecting, hunting, and

[*Courtesy: Bolivian Embassy.*

FIG. 7.—TILLAGE

This photograph shows Bolivian peasants tilling their fields. It is an example of primitive peasant agriculture. Simple, even crude, implements of this kind are still usual in backward Andean countries in South America—not that modern tractors would be of much use on the steep mountain slopes! Note the terracing in the background and the stony soil in the foreground. Only by the arduous and painstaking construction of terraces, known as *andenes*, was it possible for the Andean Indians to cultivate the steep slopes. Similar terracing has been carried out in the hilly lands of South-east Asia. *See* Fig. 29 and Fig. 41.

fishing. They live in a difficult and restricting forest environment. Patches of land are cleared of trees, a job undertaken by the menfolk, the ground is then broken by means of wooden stakes, a procedure which may be likened to a primitive form of ploughing, a job also carried out by the men, but for the rest cultivation is the task of the womenfolk. A simple digging-stick is used to score holes in the ground, and in these seeds, tubers, and cuttings are planted. The principal crop is manioc; from the roots or tubers of the plant cassava is made. Other crops, grown in smaller quantities, include sweet potatoes, yams, pumpkins, and peppers. The cleared patches quickly become exhausted and unproductive, and after two or three seasons new plots must be cleared. Thus the Boro are primitive shifting agriculturalists living a largely self-sufficing existence.

The Hausa of northern Nigeria, who live in a tropical grassland environment, provide a good example of a people living by cultivation but practising a more highly developed form of tillage under settled conditions. Agriculture is still of a primitive type, but the people often resort to irrigation. They grow a wide variety of crops including maize and millet, which are primarily food crops, and cotton, tobacco, groundnuts, and shea nuts, which often form export crops. Farming, too, is often more diversified for many of the Hausa have flocks and herds.

The cultivators of the Far East, notably the Chinese and Japanese, have developed agriculture of the simple kind to its highest level of efficiency. Though sometimes growing crops for export, basically farming is for subsistence purposes. Cultivation here is distinguished by the use of irrigation, the rotation of crops, fertilisation of the soil, and careful hand tillage. Although enormous labour is expended upon the land, the return is high. Tillage in this region is simple only in the sense that it is small-scale and unmechanised and for subsistence purposes.

## *COMPLEX SOCIETIES*

### COMMERCIAL FISHING

This, the production of fish and other marine products for sale, differs from primitive fishing, which is almost entirely for subsistence, in its scale and organisation. It is true that some commercial fishing is small-scale and the production units are small but, by and large, fishing on a commercial basis is a highly organised activity employing specialised craft and techniques; moreover, it is becoming increasingly specialised.

Most commercial fishing is confined to certain rivers, notably the "salmon streams," a few inland waters, such as the Great Lakes, Victoria Nyanza, and the Caspian, and the continental margins of higher latitudes, more particularly where the shallow continental shelves are well-developed. Most of the world's major fisheries are to be found in the northern hemisphere, although commercial fishing is an expanding industry in many southern hemisphere localities. Exact figures of those directly employed in commercial fishing are unknown but relatively small: if the fishers of China and Japan be included, perhaps some 5–6 million are so employed. But the significance of fishing as an economic activity varies considerably from country to country: in some, as in Iceland and Norway, the proportion of the population engaged in fishing is relatively high. Fishing, too, may be a seasonal or part-time occupation.

Under commercial fishing are also included the collecting of various marine products such as sponges, coral, bêche-de-mer, mother of pearl, pearls, etc. Whaling and sealing are commonly looked upon as branches of the fishing industry. We might add, also, that the fishing industry gives rise, directly and indirectly, to numerous other industries, *e.g.* canning, pickling, fish oil extraction, the making of fertilisers on the one hand, and the manufacture of fishing nets, barrel-making, and boat-building on the other.

## COMMERCIAL GRAZING

Large-scale commercial grazing is the predominating form of land use in the drier parts of the earth's surface where both crop growing and forestry are impractical. The grazing areas are usually in the interiors of the more recently settled continents where rainfall is light and seasonal in its distribution; where, in fact, moisture is insufficient to support tree growth and usually inadequate to support crops. In certain areas, however, grazing is carried on in humid regions but where this happens it is usually because tillage has not yet been started. On the other hand, in many areas, *e.g.* the humid pampas of Argentina and much of the Canadian prairies, erstwhile grazing lands have now become important areas of crop cultivation.

Commercial grazing is virtually restricted to cattle and sheep rearing, and to the production of meat, dairy produce, wool, and hides for export. It is an economic activity which has developed largely within the past hundred years and has been closely related to the industrialisation of Western Europe, the growth in population, and the increase in the standard of living of peoples in the West. But it is also closely related to a series of scientific and technical developments, such as the invention of the tin can and refrigeration.

Ranching is an extensive activity requiring large areas of territory. The density of livestock is usually low due to the limited amount of grass and forage in areas of low and largely seasonal rainfall. The livestock range freely over wide areas, though their movements may be directed from area to area. Periodically roundups are carried out. Commercial grazing is economic in the use of man-power: a few men on horseback, sometimes with the help of dogs, can control large herds. Where commercial grazing takes the form of animal rearing for dairy products, the farm units are much smaller and a much greater labour force is needed. This aspect of grazing is really more closely related to agriculture and is better thought of as such.

Commercial grazing is the characteristic form of economic production in the drier temperate grassland areas: the Great Plains of North America, the pampas of Argentina, the veld of South Africa, and the interior lowlands of Australia are the major areas of commercial grazing. The tropical grasslands are potentially important as future cattle areas, but apart from certain areas of the Brazilian, Venezuelan, and northern Australian grasslands, they are still very much under-developed.

## COMMERCIAL AGRICULTURE

Commercial agriculture embraces a very wide variety of different types of farming activity; these we shall deal with in Chapter IX. But in all cases a large proportion of the crops grown or other farm products produced go to market, and much may be exported. Production for sale, as distinct from production for family consumption, is one of the principal features differentiating commercial agriculture from primitive subsistence agriculture.

Commercial farming may concern itself with the production of a single crop, *e.g.* wheat, coffee, cotton, or rubber, in which case agriculture is described as *monoculture*, *i.e.* one-crop or single-commodity cultivation. Farming of this type is usually large-scale and often employs a considerable paid labour force: this is especially true of plantation agriculture. At the other extreme is the type known as *mixed farming*, which is typically small-scale family farming in which crop-growing is associated with livestock keeping and in which the two aspects are well integrated.

Commercial farming differs from primitive subsistence farming in a number of respects: (*a*) some or all of the produce is for sale; (*b*) scientific practices, such as crop rotation, are consciously applied; (*c*) fertilisers, either natural or artificial, are usually much used; and (*d*) farming practice is mechanised to a greater or lesser degree.

Many people are involved in commercial agriculture, but, as a human activity, it looms largest in the "newer" lands of the world since such countries are still, to a very large extent, producers of primary products. It is in such countries that the proportion of people engaged in commercial agriculture is greatest.

As a human activity commercial farming is becoming increasingly specialised and certain kinds of intensive commercial agriculture, *e.g.* market-gardening, fruit-growing, are frequently very highly specialised.

The application of science and mechanisation to agriculture is revolutionising the industry and reducing drastically, in many instances, the need for manpower.

## FORESTRY

Some people—in total their numbers are relatively small—earn a livelihood through forestry. Exploitation of the forests extends from the Equator to the Arctic Circle and includes the tropical evergreen, temperate deciduous, and coniferous types of forest. In some cases forestry is associated with agriculture; indeed, in some instances, it may be an integral part of the farming system: this is so in parts of Britain. In some

[*Courtesy: Forestry Commission.*

FIG. 8.—FORESTRY

This photograph shows young spruce trees being lifted before being planted out in a Perthshire forest. The practice of afforestation is virtually a development of the present century. Previously man cut down the forests with little or no thought for the future. Now it is realised that forests are a valuable resource and re-planting is taking place. In Britain over one million acres (400,000 ha) of new forests have been planted since 1919, the year in which the Forestry Commission was set up to deal with the country's forest problems. Today Britain produces about 10% of its timber needs but sometimes as much as £400 million are spent a year on timber imports.

countries, *e.g.* the Scandinavian countries, lumbering may be a seasonal occupation, the farmers cultivating their land in summer and working in the forests in winter.

Formerly much more extensive areas of the earth's surface were forested, but the need for cultivable land and of timber for constructional purposes, fuel, etc., led to great inroads upon the forests. World demand for timber continues to grow, moreover, and there is a very real danger of a timber famine in the future.

The lumbering industry is of great importance, particularly in the softwood (coniferous) forest areas of the world. So far there has been little exploitation of the tropical forest areas except in one or two places where the valuable teak and mahogany are found. In the coniferous forest lands lumbering formerly was primarily a seasonal occupation, but with the growth of afforestation and forest regeneration techniques the work is becoming increasingly less strictly seasonal in its character.

Included in forestry is the collection of forest products, *e.g.* wild rubber, nuts, medicinal plants. This aspect of forestry, which is essentially a gathering economy, comes close to primitive collecting and may, in fact, overlap with it.

## MINING

The satisfaction of human needs with respect to food and clothing are met by man's use of plants and animals, but for most of his other basic needs—tools, implements, weapons, dwellings—he relies upon the materials of the earth's crust to a very large extent. Man's cultural development has been so closely bound up with these crustal materials that the historian has divided human history into the Stone Age, the Bronze Age, the Iron Age, and the Steel Age. The search for, and winning of, mineral wealth have carried men to the ends of the earth. The Phoenicians went in search of tin, the Conquistadores in search of gold, and twentieth-century man has gone in search of oil.

Thus, mining, the exploitation of the earth's mineral wealth, is a human activity going back many thousand years; but it becomes more and more important with every passing decade.

Mineral wealth is widely but irregularly distributed over the face of the earth. It is a form of economic production which is often mixed in with cropping, grazing, and forestry, since mineral occurrences are highly localised. Because of this localisation, we often get concentrations of population in areas which otherwise are unpopulated or very scantily populated.

Mineral wealth is the basis of much of modern industry, hence large numbers of people are engaged in mining either for fuels, such as coal and oil, for metals such as iron and copper, for non-metallic minerals such as salt and sulphur, or for building materials such as sandstone and marble. Mineral wealth is exhaustible, hence mining is a "robber economy." A given deposit sooner or later becomes exhausted; and once the material has been removed, it has gone for good.

## *NON-PRIMARY OCCUPATIONS*

### MANUFACTURE

In contrast to all the foregoing types of activities and forms of production, manufacturing is a secondary activity. Manufacture uses the primary raw materials from the sea, land, field and forest to process them and convert them into commodities more useful to man.

In the widest sense manufacturing has a world-wide occurrence. Even the most primitive communities engage in some form of manufacture, *e.g.* the making of tools, pots, cloth, etc. But such "domestic" and "cottage" industries stand contrasted with the highly complex manufacturing activities of modern communities. Modern industry, organised on a factory basis, using mechanised plant, and powered by some source of energy, is concentrated in relatively few localities, *e.g.* in Western Europe, parts of the U.S.S.R., in Japan, south-east Australia, parts of South Africa, south-east Brazil, and the north-east U.S.A. But, besides these major concentrations, manufacturing industry is widely scattered throughout the world in small units usually located in towns.

More people are employed in manufacture than in any other activity apart from agriculture, and in the more advanced countries manufacturing is the whole-time occupation of the greater proportion of the population; this is true of countries such as Denmark and Holland, which we tend to think of as being more particularly farming countries.

### COMMERCE

While the majority of mankind is engaged in primary production of one kind or another or in the secondary activity of manufacturing, a large, and increasing, number of people, especially in the advanced countries, are concerned with other jobs which do not add directly to the total volume of goods produced. Such people as, for instance, those engaged in buying and selling, in banking, financing, and insurance, in transportation and communication, are equally as necessary as "producers" in modern society. Such people may be said to be employed in the organisation of production or in commercial activities. They organise, facilitate, and transact exchange.

Commercial activities may be said to fall into two broad groups:

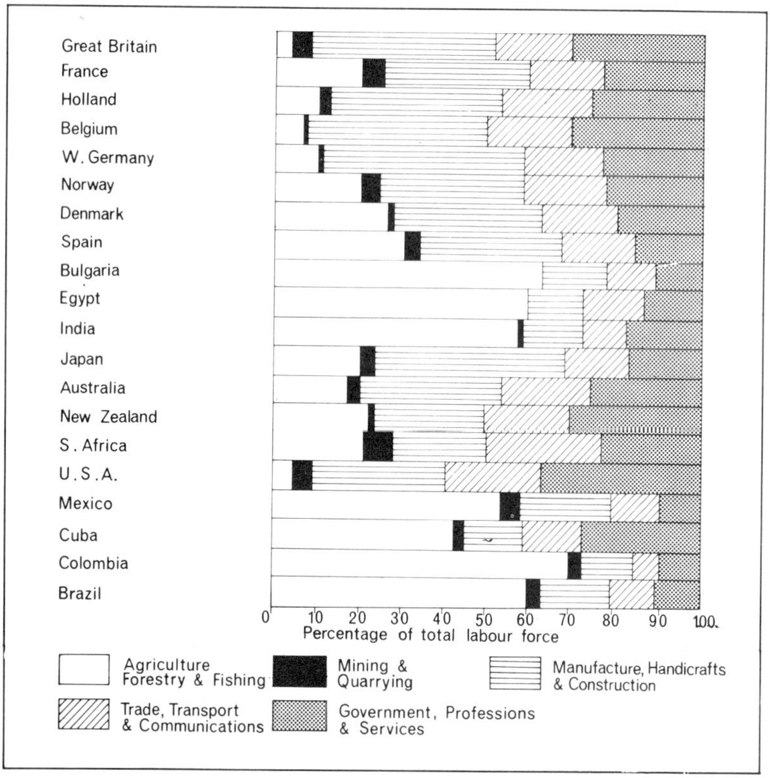


FIG. 9.—GRAPH OF ECONOMIC ACTIVITIES

Fig. 37, showing the chief wheat-producing countries, is a simple bar graph. This particular graph is a variety of bar graph but it is known as a *percentage bar graph* because it shows a series of bars of uniform length which are divided up on a percentage basis.

The graph shows a number of selected countries; in each case the proportions of the active working population engaged in the five main categories of employment are indicated. Compare, for example, the occupational differences between Great Britain and India or between Belgium and Holland.

(*a*) trading, and (*b*) transporting. Those who are engaged in the former act as middlemen between the producer and the consumer; they include merchants, brokers, insurance agents, shippers, bankers, and financiers. Those engaged in transport engineer the carriage and movement of goods from one place to another. Included in transportation are the communications services. Activities of this kind are usually called tertiary activities. To this group of human activities may also be said to belong the professions and public services—the services performed by scientists, teachers, doctors, ministers of religion, administrators, etc.

# Chapter IV

# THE WEALTH OF THE WORLD

## WHAT IS WEALTH?

To most people wealth means money. If someone is lucky and scoops the "pools" he wins a large sum of money and overnight becomes rich or, as we commonly say, wealthy. Although the terms "money" and "wealth" are loosely used to mean the same thing, to the economist they have very different meanings. Money, in the economic sense, is merely a medium for buying things. It is the "things" that constitute wealth, "things" in this case meaning land, goods, and services.

Let us explain this a little more fully. Supposing a millionare with a wallet full of £5 notes became stranded in the middle of the Sahara; before very long he would become thirsty and hungry and, when the sun went down, cold and miserable, too. In such a predicament his money would be of no value: he could neither eat nor drink his notes, nor could he buy food or drink or shelter with them. The time would shortly come when he would be quite willing to give every penny he possessed for a drink of water, for such a drink would probably save him from death. In such a situation, therefore, money is of no use and has no value. Money has use and value only when it can buy, that is can be used as a medium of exchange, for things that are wanted.

Let us take another example. Supposing this time a Japanese came to Britain and went into a shop to buy a suit. If he offered to pay for it with Japanese yen, he would be told politely that his money was not acceptable. Thus, although the yen is the accepted medium of exchange in Japan, it would not normally be accepted in Britain. The shopkeeper would demand that the Japanese pay for his suit with English pounds since yen were valueless to him. If, however, the Japanese had a beautifully made jade necklace with him and offered to exchange it for a suit, the shopkeeper might be tempted to strike a bargain. In such a case, he would be exchanging the value of the suit, *i.e.* the cloth and the tailoring, for the value of the trinket, *i.e.* the jade and the workmanship.

We can now see a little more clearly that money, a medium of exchange, is not the same as wealth and that it is only goods or services that possess real value. Thus it is not money, but the things that money can buy, that matters. "Things," therefore, constitute real wealth.

## *NATURAL RESOURCES*

Let us proceed a stage further and elucidate what we mean by "things." We said above that "things" meant goods and services. Food, drink, clothes, houses, tools, equipment, machines, fuel, etc., are all "things." The services and skills of other people who make all these goods and who minister to our wants are also classed as "things." Almost all the goods we desire and possess are, first, made or manufactured through the services of people and, secondly, they are extracted, processed, or fabricated from the earth's natural resources. Bread, for example, is made from the wheat grown in the soil; tables and chairs from the timber grown in the forests; suits from the wool grown on sheep; and tools from iron ore mined from the ground. These, and all other things, depend for their origin upon the natural resources of the earth: the animals, the plants, the soils, the minerals, and the waters of the earth. Basically, therefore, it is these resources, which can be turned into useful products, that constitute the real wealth of the world. Since nature provides these essential resources, we call them natural resources.

The earth possesses many kinds of natural resources or "wealth." These resources, which are available for man's use, fall into two main groups: (*a*) organic or living resources, and (*b*) inorganic or non-living resources.

In the former group are those resources derived from natural plant and animal life; these include forests, natural pastures, wild-life, fish, and other marine life. The forests have provided man with a variety of products including fruits, nuts, leaves, bark, fibres, gums, and valuable juices such as rubber. They are of most value at the present time, however, for the timber that they yield. Where natural grasslands are to be found, man has often used them for grazing animals, especially cattle and sheep. Living in the forests and on the grasslands are many wild creatures that man has used for food or from which he has secured hides, skins, hair, furs, and horn. Some of the wild creatures he tamed and domesticated long ago so that, in addition to herding them for food or raw materials, he could use them as draught animals or as beasts of burden. The lakes, the streams, and the seas possess a variety of life in the form of fish and other marine creatures which man has used to great advantage.

Soil, which is composed of both organic and inorganic matter, falls between the two groups but it is a resource of fundamental importance. Plant life cannot live without soil, for it is at once a medium for seed germination, an anchorage for growing plants, and a source of moisture and plant nutrients. Natural vegetation is dependent upon it, but more important these days is the fact that it is used by man to grow a great variety of crops.

The non-living, or inanimate, resources embrace air, water, mineral fuels, metals, non-metallic minerals, and building stones. Man has come to use these because he can make or manufacture useful products from them and because some, such as coal, oil, and natural gas, provide him with a source of power. Air and water, which we take so much for granted, are man's most important resources, for without either, man and all other organic matter could not live; but, in addition, air provides windpower and nitrogen for the manufacture of chemicals while water provides moisture for crops and energy in the form of hydro-electric power.

## *LATENT RESOURCES*

We have been talking about natural resources and their importance to man. One thing we have clearly established: it is these earth resources, and not money, which constitute *real wealth*. Before we look a little more closely at the various resources which the earth provides, one point needs emphasising. Although nature has provided a very wide range of resources for human use, such resources do not become of significance and value to man until he has reached a particular stage of cultural development. The resources, true, are there, but they remain hidden or unused until man requires them, *i.e.* until he has a use for them. Resources of this kind may be described as latent resources. Only when man finds a use for them do they come to have any real meaning or value to him and only then are they realised. This idea can best be illustrated by examples.

Early primitive man used stones and flints with which to fashion his weapons. These sufficed, but clearly a stone axe is not as effective as a cutting tool as a metal axe. Why, therefore, did early man not make his axes, or for that matter his other tools, of metal? Why, when metal was available in the earth's crust, was it not used? In the first place, early man, because of his limited experience, could not always recognise metals or metallic ores (the ones that were distinctive, shiny, and attractive by their colour, such as gold, were the first to be used); in the second place, he had no technology by which he could extract metal from its ore or fashion metal into tools. Hence, until he had learned to distinguish ores from ordinary rock, and until he had discovered techniques of smelting, he had no use for metal.

Another example is provided by rubber. Rubber trees have always existed in the hot, wet tropical forests and, although the native Indians of the Amazon basin were aware of the rubber tree and the peculiar properties of its sap and even used it to some extent, rubber to Europeans at any rate had little utility. Man discovered it could be used as an eraser to rub out pencil marks, but this was its sole use for a long time. The importance and value of rubber only became fully appreciated during comparatively recent times. Its use was intimately linked with, once again, technology: until the method of vulcanisation (*i.e.* the mixing and heating of rubber with sulphur) was discovered, rubber had little utility, and until the invention of the pneumatic tyre and the advent of the motor car there was little demand for rubber. These developments, however, resulted in a rapidly increasing demand for rubber. And once it became common other uses were found for it.

A third example is provided by petroleum. Oil has been known in various parts of the world from very early times. It was familiar in the Near and Middle East, for example, in Biblical times, when it was called bitumen (natural seepages of oil occurred on the land surface). But, apart from its occasional use as a waterproofing material—remember Moses' ark was coated with bitumen—its potentialities as a fuel were not realised until about a century ago. Oil was present in the ground (though in unknown quantities) but its use and exploitation could not take place until, first, a use had been found for it as a fuel; secondly, the invention of the internal combustion engine which could use it as a fuel; thirdly, until technology had discovered the process of distillation; and fourthly, until the science of geology had become sufficiently advanced to be able to locate petroleum deposits. In this case a chain of events and circumstances made possible the use of petroleum; this would have been impossible two centuries ago.

And so it is with many resources. The Russians, unable to procure natural rubber, turned to explore their own resources and found a good substitute in the *kok sagyz* plant. The use of beet for the extraction of sugar was not possible until a German scientist found a way of separating out the sugar content. Uranium had no value until the atom could be split and nuclear power became a practical possibility. The metal titanium, which is fairly abundant, has many advantages over aluminium and steel: its strength–weight ratio is higher, it is very ductile, and it is resistant to corrosion, but its use is hindered by extremely high production costs; but if this can be solved along with certain technical problems, it is very likely that it will become of great importance in the future. Running water, again, though used as a source of power to turn mill wheels from at least medieval times, was not used for the generation of hydro-electric power until electricity was discovered; hence its energy potential remained largely latent until about one hundred years ago.

A useful exercise would be to find out when the following were discovered and how and why they came to be used, for all were originally latent resources which have now come to be realised: aluminium, chromium, nickel, asbestos, guano, helium, nitrogen, natural gas, heavy water, copra, kapok, carnauba wax, esparto grass.

## *THE EARTH'S NATURAL RESOURCES*

Basically, there are six: the soil, vegetation, animal life, minerals, air, and water. Let us look at these in turn.

### SOIL

Soil is of prime importance to man and almost all living organisms because the bulk of the world's food is grown in the earth's superficial covering of soil or comes from animals which feed on the vegetable matter growing in the soil. The small remainder comes from the waters of the earth's surface, especially the oceans which yield their quota of fish and other marine products. The soil is also the source of the vegetable raw materials, *e.g.* cotton, tobacco, vegetable oils, etc., which are used in industry. The forests, too, grow in the soil. Clearly, therefore, soil is a fundamental natural resource.

Soil is a substance of much greater complexity than most people realise. It is not something dead, lifeless, immovable, inexhaustible; rather should we look upon soil as being dynamic and destructible. Soil is slow to form but easily lost. Under certain conditions the soil can be lost by being washed away by torrential rain or blown away by strong wind, such soil loss being termed soil erosion (*see* Fig. 10). It is one of the tragedies of human history, especially recent history, that much soil erosion has taken place. This has meant a loss of land fit for cultivation or grazing or forest growth, a loss we can ill afford with the rapid growth of the world's population and the ever-expanding acreage of urban areas (which frequently swallow up first-class agricultural land). More soil is, in fact, now being lost each year than nature makes; thus the soil is a shrinking asset. Obviously this cannot be allowed to go on indefinitely; only drastic action on the part of man can be really effective in reversing this trend, however.

### VEGETATION

Cultivated crops apart, natural pastures and forests form important natural resources. Many animals, especially the world's beef cattle and wool sheep, depend upon the natural pastures of the prairies, pampas, steppes, and savannas. Apart from the domesticated animals, many wild grass-eating animals (herbivores) also depend upon nature's pasture lands. Many areas of these natural grasslands have been over-grazed by both domesticated and wild animals leading to a reduction in their carrying capacity. Here, too, there is need of some control to preserve the earth's pasture lands.

[*Courtesy: U.S. Information Service.*

FIG. 10.—SOIL EROSION

A scene in the Western Plains of the United States. Bare, ploughed fields, even when protected by a wind-break of trees, are periodically attacked by wind erosion after long, dry spells. Wind is one factor causing soil loss; torrential rain is another. Can you name other contributing factors?

At one time forests covered a much larger area of the earth's surface than they do today. The need of land for cultivation and the demand for timber for constructional purposes, shipbuilding, charcoal, fuelwood, etc., and more recently for pulp (for paper and rayon) have greatly depleted the forest cover; in fact, there are now extensive areas, *e.g.* in Europe, the U.S.A., and China, which were formerly under thick forest but which are now practically treeless.

The world's forests supply man with two groups of commodities:

1. Timbers, both hardwoods and softwoods of various kinds used for constructional work, furniture, pulp-making, and, sometimes, fuel.

2. Forest products such as vegetable oils and waxes, rubber, nuts, cork, medicines, gums, resin, etc., used for food or industrial raw materials.

Notwithstanding the growth of synthetic products and substitute materials, the demand for timber is greater than ever before and continues to increase. It will be apparent that the forests constitute a great source of wealth which we can ill afford to lose and yet, in the world as a whole, more timber is felled than grows, and the area under forest is rapidly

[*Courtesy: Australian News & Information Bureau.*

FIG. 11.—THE RABBIT PEST

The rabbit (which breeds very quickly), when uncontrolled, can be a serious pest. This remarkable picture illustrates vividly the damage rabbits can do. Through the middle of the photograph runs a rabbit-proof fence. To the left, every scrap of pasture has been nibbled away by rabbits; but the protected land to the right offers plenty of fodder for stock. In 1952 it was estimated that the rabbit was robbing Australia of 25 million sheep. The cost of keeping the rabbits in check amounted to some £60,000 annually. Myxomatosis practically eliminated the rabbit from Australia and now enables the rabbit pest to be kept under control. The extinction of the rabbit has also enabled the natural pastures to recuperate.

shrinking. Once again, unless decisive action is taken (happily there are signs of this at least in some areas) the world will suffer from timber hunger in the future.

**ANIMALS**

Man makes use of the animal world for a variety of reasons: he uses animals to provide food (milk, meat, eggs); wool, hair, fur, feathers,

## Table II

# NATURAL RESOURCES

- POWER
  - DESTRUCTIBLE
    - Wood
    - Peat
    - Coal
    - Petroleum
    - Natural Gas
    - Alcohol
  - INDESTRUCTIBLE
    - Wind
    - Running Water
    - Tides
    - Geothermal Heat
    - Solar Heat
- RAW MATERIALS
  - MINERAL
    - Iron
    - Other Metals: *Copper Tin, Lead, Zinc, etc.*
    - Ferro-Alloy Metals
    - Precious Metals
    - Uranium
    - Fertilisers *Nitrate Potash Phosphate*
    - Sulphur
    - Salt
    - Limestone
    - Sand
    - Clay
    - Building Stones
  - VEGETABLE
    - NATURAL
      - Timber
      - Fibres ***Kapok Raffia, etc.***
      - Gums, Resins
      - Rubber
      - Oil Seeds and Nuts
      - Vegetable Wax
      - Yerba maté
      - Quebracho and Barks
      - Camphor
      - Cork
      - Medicinal Plants
      - Seaweed
    - CULTIVATED
      - Fibres: *Cotton Flax, Hemp, etc.*
      - Rubber
      - Oil Seeds and Nuts
      - Tobacco
      - Flowers (*for perfume*) *etc.*
  - ANIMAL
    - NATURAL
      - Wool and Hair
      - Hides and Skins
      - Furs
      - Horn and Ivory
      - Oil
      - Tallow
      - Manures (Guano)
      - Pearls
      - Sponges
    - DOMESTICATED
      - Wool and Hair
      - Hides and Skins
      - Furs
      - Silk
- FOODSTUFFS
  - MINERAL
    - Salt
  - VEGETABLE
    - NATURAL
      - Wild Fruits and Nuts
      - Roots
      - Leaves and Shoots e.g. *Bamboo*
      - Mushrooms
    - CULTIVATED
      - Cereals: *Wheat Rice Maize, etc.*
      - Sugar
      - Beverages *Tea Coffee Cacao*
      - Vegetables
      - Fruits
      - Vegetable Oils
      - Spices
  - ANIMAL
    - Meat: ***Beef Mutton Pork***
    - Dairy Produce
    - Eggs
    - Honey
    - **Fish and Marine Products**

hides, skins, horn, etc., which he uses as industrial raw materials; and for transportation purposes. Today, as a result of breeding, man can very largely control the numbers of his domesticated animals. Wild-life is in a much less happy state. In the past, hundreds of species have become extinct and it is reckoned that the world is losing one or two animal types every year. In some cases this is a matter of relatively little importance

[*Courtesy: U.S. Information Service.*

FIG. 12.—STRIP-PLANTED FARM

Strip-cropping (*i.e.* cultivating alternate strips of varied crops and fallow), contour-ploughing (*i.e.* following the contours when ploughing instead of making furrows at right-angles to the slopes), and tree-planting (to act as wind-breaks and to help bind the soil together), assisted by crop rotations are among the methods now being used in the United States (and elsewhere in the world) to prevent soil erosion and maintain crop yields.

and often of zoological interest only; but in others, especially where the animals have some direct importance to man, the gradual extinction of a species is important. Many of the game of Africa are being lost, due partly to the shrinking of the area of their natural habitat and partly to illegal poaching, *e.g.* in the case of the elephant which is killed by the natives for the sake of its ivory tusks. Many of the animals of Africa, if properly husbanded, could be made to yield valuable food supplies.

A number of cases of wanton and stupid killing of animals readily spring to mind—bison, otter-seal, walrus, whale. The bison, which roamed the North American prairies in their tens of millions, were reduced to the verge of extinction within less than half a century. The whales, hunted for their oil, have had their numbers drastically reduced. Fortunately, man is now beginning to appreciate the importance of wild-life and notable steps have already been taken to help preserve some of the declining animal populations.

## MINERALS

The crust of the earth contains a variety of mineral resources of great use to man. These minerals are indispensable in modern civilisation. For present purposes we can distinguish four main groups of minerals: mineral fuels, such as coal and petroleum; metals, such as iron, tin, and gold; non-metallic minerals, such as salt, sulphur, and nitrate; and constructional materials, such as granite, sandstone, and clay.

Some minerals occur in abundance; others are very scarce. Building stones, gravel, and clay are widely spread over the earth's surface and are abundant except locally. Metals, which occur chiefly in the older rocks, are widely but unevenly distributed; some, such as iron, are relatively plentiful while others, such as cobalt, are of limited occurrence. Some minerals have been used for a long time and consumed in appreciable quantities whereas others (*e.g.* nickel, vanadium) are of recent discovery, of limited occurrence, and fairly scarce. Certain minerals, though of great value for particular purposes, are not consumed in large quantities. As time has gone on man has discovered more and more minerals and found ever-increasing uses for those he knows.

Minerals differ from most of the other natural resources in that once mined they are lost to the earth's crust forever. In other words, minerals are exhaustible natural resources and the earth's stock of mineral wealth is gradually being used up. Since minerals are irreplacable, it is important that they be used sensibly, especially those that are in limited supply.

## AIR AND WATER

Air and water are the most precious of all the natural resources. At first sight this may seem strange, but a little careful thought will soon show that this is so. Without air neither man nor animals nor plants could breathe and live. The other planets in the Solar System appear to be lifeless and one important contributing factor explaining this is that, unlike the earth, most of them have no atmosphere. As you know, nothing can live in a vacuum. This is the principle behind the canning and bottling of fruit for example: air is excluded from the container and, so long as it remains airtight, the fruit will keep, for it is shielded from bacteriological action—bacteria cannot live without air. Life, of whatever kind, needs air.

Water, again, is indispensable for life. Without it no plants or trees could grow and if there were no plants there would be no crops, no animals (for they live directly upon plant life or indirectly upon it), and therefore no man. Without water the earth, like the moon, would be a lifeless planet. But, quite apart from its crucial importance in this respect, water has a multitude of other uses: it provides power, transport, is used in greater or lesser quantities in all industries, and has an amenity value for man.

## *CONSERVATION*

We have already referred to the fact that the earth's natural resources can be classified into inexhaustible and exhaustible resources. Air and water are the only two truly inexhaustible resources. Short of something catastrophic happening, the earth will always have its envelope of air around it. Although man is now extracting nitrogen from the air to make nitrates for fertilisers there is no fear of the nitrogen ever becoming used up, especially since nitrogen is constantly being replenished in the atmosphere. Again, although carbon monoxide and other noxious gases are constantly being liberated into the air by man (as a result of his industrial processes) there is little chance of the atmosphere becoming poisoned (except very locally) to such an extent that life would be extinguished. The volume of the atmosphere is so great that it is able easily to absorb all these gaseous constituents.*

Water, like air, is also inexhaustible. Though there may be local shortages, there is no prospect of overall water supplies ever running short. Unlike some of the earth's natural resources, water can be used, and re-used but never used up. This does not mean, however, that water should be wasted. In many areas where water is in relatively short supply, it is necessary to use water sparingly and to conserve it. In countries such as Britain, where normally there is an abundance of water, one is apt to take it for granted, to use it lavishly, and to give no thought to its conservation.

The other resources—soil, vegetation, animal life, mineral wealth—are not inexhaustible. Soil, natural vegetation, and animal life can be so managed that they could become inexhaustible, but until very recent times little or no attention has been paid to this matter. Soil, as we have already noted, can be easily lost: it needs very careful use and management for its maintenance. If farmers neglect the soil it gradually becomes exhausted, useless, and may even be completely lost. The point is that soil must be carefully tilled, continually replenished, and periodically

* Radioactive fall-out could be an extremely dangerous menace, however.

rested (not necessarily by allowing it to lie fallow), if it is to remain in good tilth, fertile, and productive. And the good farmer pays due attention to the requirements of his land. Unfortunately, there are too many farmers the world over who are content to "mine" the soil, to rob it of its fertility, and do not, or cannot, care for their soil properly. The offenders in this respect are not always primitive

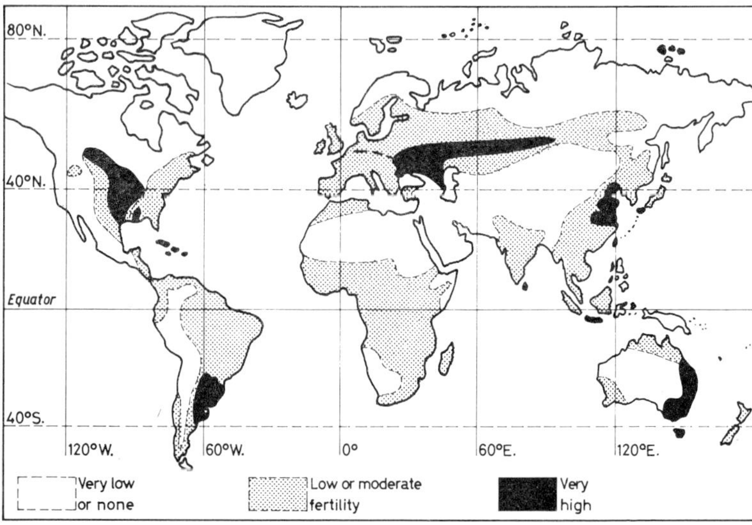


FIG. 13.—NATURAL FERTILITY

The map, based upon the Food and Agriculture Organisation's appraisal of natural conditions, indicates the natural fertility rating of the different areas of the earth's surface. Note how the interior temperate grassland areas with their black earths and dark brown soils form the regions of highest natural fertility. The mountain areas, with their thin, stony, immature soils, and the desert areas (hot or cold) are regions of low fertility.

backward peoples; many so-called educated European and American farmers have contributed to soil deterioration and soil loss (cf. the American dust-bowl).

Materials of biological origin—foodstuffs, fibres, forest products, etc.—can be constantly replaced without difficulty, provided soil fertility is maintained and new planting keeps pace with cropping. Similarly, animal life can be maintained or even further developed providing man adopts a rational approach towards animal husbandry. Hence, both vegetable and animal life could be turned into never-ending resources with due care and proper management.

Mineral matter which is mined from the earth's crust is, on the other hand, definitely exhaustible. Once used, the mineral wealth is gone. Moreover, the earth's store of mineral wealth, which is relatively limited, cannot be expected to meet indefinitely the great and constantly rising demand. Already many deposits of certain minerals have been exhausted and there is a certain anxiety about some mineral wealth, more especially

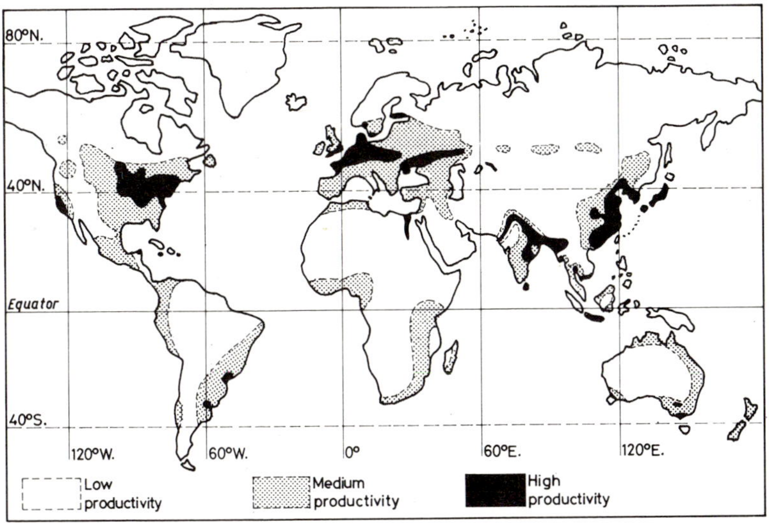


FIG. 14.—AGRICULTURAL PRODUCTIVITY

Compare this map with the map of natural fertility. Can you see any correlations? Can you explain why agricultural productivity is so high in Western and Central Europe?

the fossil fuels. Increasingly it is becoming apparent that man will have to adopt a conservationist policy to ensure that the valuable mineral wealth will last as long as possible. The conservation of minerals can be undertaken in various ways: (*a*) care must be taken not to use minerals wastefully; (*b*) improved methods of extraction must be adopted; (*c*) efficient recovery of minerals must be practised; and (*d*) substitute material should be used wherever possible.

But conservation, the careful, rational, intelligent use of resources, should be applied to all the earth's natural resources, not merely to the mineral wealth but to the soil, the forests, animal life, and water supplies. The achievement of this will entail the removal of human ignorance and carelessness and a measure of control and co-operation among the states of the world.

## EXERCISES

1. Money and wealth are usually thought of as being synonymous; show that this is not so.

2. What is understood by the phrase "the natural resources of the earth"?

3. Explain the meaning of the terms: latent resource, inorganic resource, conservation practices.

4. Consider the general question of soil erosion. Explain the circumstances under which it may occur and describe methods of preventing its occurrence.

5. What is meant by the term "conservation"? Discuss the problems of conservation as they affect Britain and one other area.

6. Attempt to justify the assertion that the soil is the earth's most valuable resource.

7. Why is the conservation of mineral resources of particular importance? How may the conservation of minerals be undertaken?

8. "Man can largely control the numbers of his domesticated animals but wildlife is in a much less happy state." Elaborate this statement.

## *Chapter V*

# GEOGRAPHICAL FACTORS

In Chapter III we described the different ways by which man lives and in Chapter IV we noted the various natural resources which have enabled man to follow varying occupations. It may be said, in general, that the capacity of any area or country to support its people is related basically to two sets of conditions: (*a*) natural, (*b*) human. Expressed in a different way, this means that Nature supplies the resources and materials, but that man supplies the thought, organisation, and labour to utilise them for human use and consumption.

A country may provide its means of livelihood in a number of different ways:

1. By applying its labour force to develop its own natural resources so that it can produce directly what it requires, *e.g.* Britain and her coal, the United States and her beef, and China and her rice.
2. By applying its labour to its own resources and exporting its surplus production in return for commodities which it cannot itself produce, *e.g.* Britain and her iron and steel goods, Sweden and her timber, Cuba and her sugar.
3. By providing raw materials for some other country to work up in return for foodstuffs and manufactured goods, *e.g.* Malaya and her rubber, Bolivia and her tin, Australia and her wool.
4. By providing labour to process and fabricate raw materials not available at home and imported from other countries, *e.g.* Britain and her cotton manufacturing, Switzerland and her engineering products.
5. By providing labour and transport for the carriage of goods, whether foodstuffs, raw materials, or manufactured goods, from producing to consuming centres, *e.g.* Britain's air lines, Norway's mercantile marine, Syria's oil pipelines.
6. By providing services of various kinds for which a financial return is given, *e.g.* banking, insurance facilities, tourist trade; with the money thus earned countries can purchase foodstuffs, raw materials, and manufactures.

Whichever way of livelihood is adopted that livelihood is very closely related to natural and human conditions. Nature offers the essential material resources for human use, but the degree to which these resources

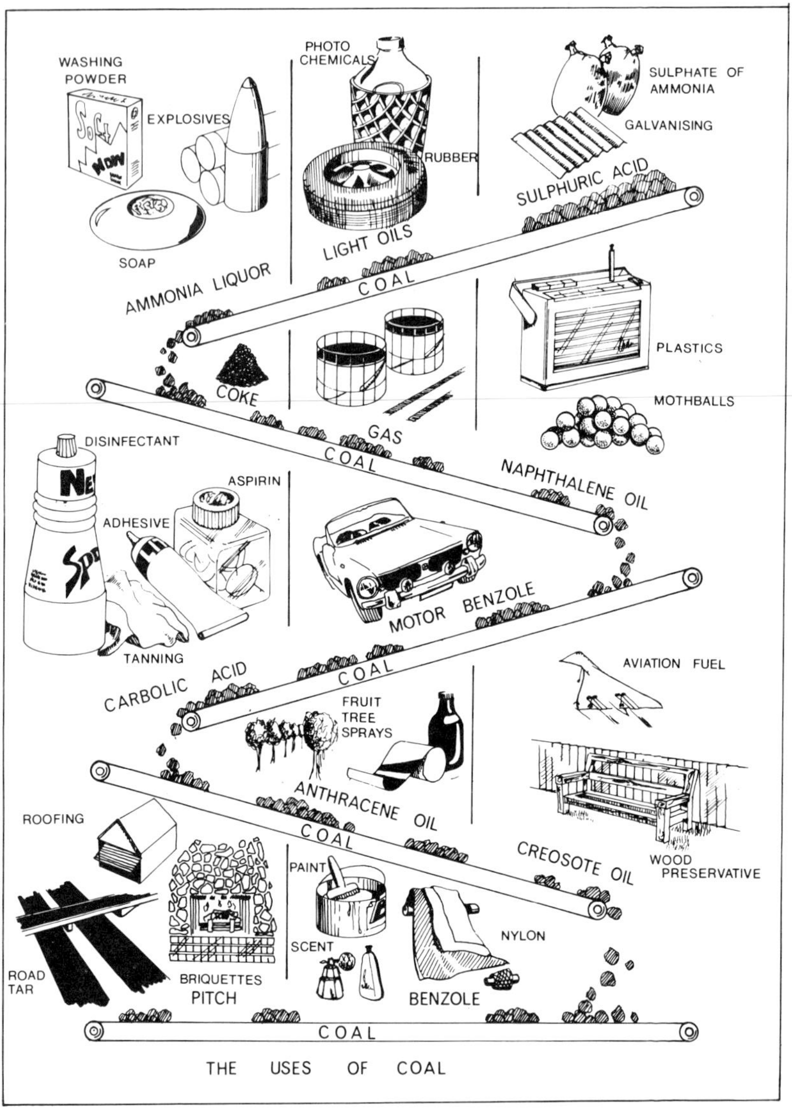


FIG. 15.—THE USES OF COAL

Coal was first used (and still is primarily used) as a fuel but the chemist has got to work on this raw material and has found a multitude of uses for it and its by-products. The diagram shows some of the products derived from coal. Coal-gas and chemicals are the most important derivatives but natural gas is now beginning to challenge the gas manufactured from coal. Some petrol is made from coal by hydrogenisation process.

are utilised for man's benefit depends upon the human response to the environmental conditions and this response depends, in turn, upon man's cultural development.

From the foregoing it will be clear that economic development in any area or among any people is the outcome of the resource endowment and the human culture. Both natural conditions and human conditions (both are geographical factors) influence and control economic development. Let us, therefore, now look a little more closely at the influences and controls exerted by the environment and by man.

## *NATURAL FACTORS*

The natural conditions which largely determine a country's capacity for supporting its people fall into four groups:

1. Food supply or possibilities for procuring food.
2. Supplies of raw materials.
3. Sources of power.
4. Facilities for transport.

### FOOD SUPPLIES

Except for a few small groups of people, such as the Australian Blackfellows, who subsist by collecting and hunting, no large groupings of people can live purely and simply off the bounty of Nature. Some are largely dependent upon a single basic food resource such as, for example, the Icelanders who eat large quantities of fish caught in home waters, or some Arabs who subsist mainly upon milk and milk products from their animals. In such cases, a particular diet is dictated by the inability of the lands they dwell in to produce much in the way of food crops. But where Nature has provided favourable soil and climatic conditions man is able to grow crops and rear animals for food. The kinds of crops he grows and the kinds of animals he rears are, however, fairly closely determined by the environmental conditions. Thus Nature sets certain limits or controls upon particular kinds of production. And, normally, man ignores such natural limitations only at considerable risk. Many mid-western farmers in the United States flew into the very teeth of Nature (by planting wheat on marginal land) only to bring disaster upon themselves.

### RAW MATERIALS

All raw materials may be classified, as we have previously noted, into three types: animal, vegetable, and mineral. Animal and vegetable resources depend upon climate and soil. The geographical distribution

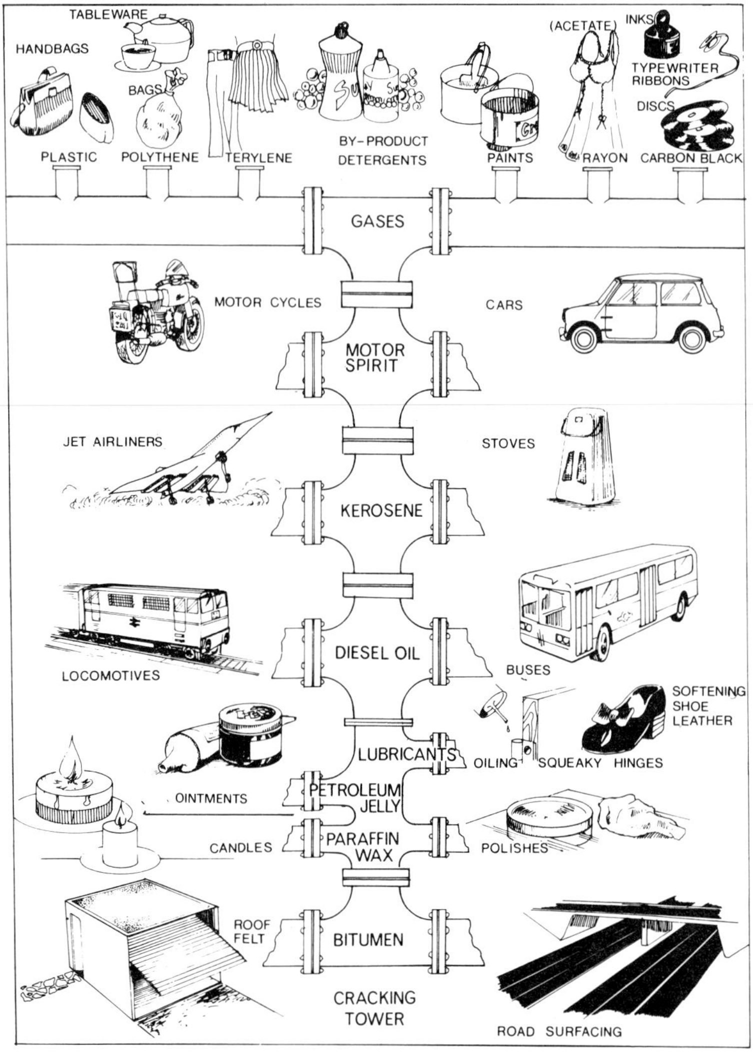


FIG. 16.—THE USES OF PETROLEUM

Oil, like coal, has varied uses. The "cracking" of crude petroleum yields a variety of products—bitumen, wax, heavy oil, paraffin, petrol. But petroleum can also be used for making plastics, synthetic fibres, detergents, etc. Petroleum, like coal, is a natural resource of great value and versatility. The petro-chemical industry, which has grown by leaps and bounds during the post-war years, has given us many new products.

of animal and vegetable species is often comparatively restricted; in fact, we shall find that where they are widely spread it is often the result of human interference. Mineral resources depend upon geological structure. The occurrence of specific minerals is frequently very restricted and most are very unevenly distributed. Some, of great importance, such as coal and petroleum, are restricted to sedimentary rocks; others are limited to igneous or metamorphic rocks. Water, which we may regard as a mineral, is of prime importance and its quantity and quality are factors of major significance in influencing settlement, agriculture, grazing, manufacturing industry, and power supplies. Animals, vegetation, and mineral resources vary considerably in their value and utility to man in his business of getting a living but they supply his basic needs of food, clothing, and shelter, and provide the fundamental requisites for his industrial and commercial activities.

## POWER RESOURCES

Originally man was dependent upon his own muscle power for fetching and carrying; it was the sole motive force employed. Today, except in certain areas and instances, it is the power economised most. In some under-developed countries human muscle power is still used for pulling the plough, manhandling goods, and transport, *e.g.* in parts of Asia, Africa, and Latin America. Animal power is still much used in most countries, though decreasingly so. In the more developed countries human and animal power have been replaced by inanimate sources of power such as wind, running water, steam, oil, gas, and electricity. Wind can be used in flat areas, *e.g.* Holland, Argentine pampas, and in exposed situations. Running water has been used to turn wheels from early times but water to be of service as a modern source of power, *i.e.* to generate electricity, must have a "head" or fall of water which is provided naturally by waterfalls, or artificially by impounding water by dams. Steam power requires water and heat, the latter being provided by burning a fuel. Gas is produced from coal but increasingly natural gas, associated with petroleum, is being used. Electricity requires some other power to produce it; consequently it can be generated most easily where other sources of power are available. Coal remains the most important basic source of power and the great industrial areas were, and to a great extent still are, tied to the world's coalfields.

## TRANSPORT FACILITIES

Facilities for communication and transport form a fourth determinant. The conditions, in addition to the existence of power resources, which lead to easy, and therefore cheap, transport are: level terrain or easy

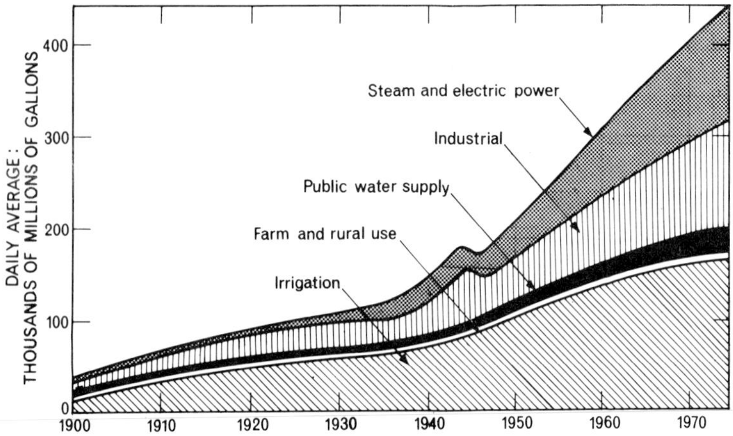


FIG. 17.—GRAPH OF WATER USE, UNITED STATES

People in Britain tend to take water very much for granted; it is only during occasional long, dry summers, as in 1947 and 1959, that the problem of water supply becomes serious and attention is focused upon the water problem. But in the arid and semi-arid areas of the world the supply of water is an ever-present problem and every drop is precious.

Water is used in larger quantities and for more things than the average person realises. For example, the production of 1 lb (0·4 kg) of beef requires 30 tons of water; 1 barrel of beer needs 1000 gallons (4545 litres) of water in its making; while a motor car which has virtually no water in it, requires some 200,000 gallons (909,000 litres) in its production!

Water has many uses but from the human point of view water resources may be said to meet eight main needs: (i) domestic requirements; (ii) agricultural needs; (iii) industrial demands; (iv) power resources; (v) transport; (vi) drainage; (vii) fisheries; (viii) amenities. The requirements of water for these differing purposes often conflict, especially so if the available water supplies are restricted: then man has to choose which has priority. The conflict of interests varies, of course, between place and place and is dependent upon a variety of factors such as the amount and seasonal occurrence of the rainfall, the nature of the economy, the development of communications, the standard of living, etc.

The above graph shows the use of water in the United States. Note the large quantities required for the purposes of irrigation, industry, and the generation of power. Note, too, the rapidly growing demand for water particularly for industrial and power requirements: during the decade 1950–60 the demand rose by 50%, and during the past decade 1960–70 the demand almost doubled.

This type of graph is known as a *compound curve graph*. It shows a series of multiple line graphs. The quantity of water used for irrigation has been graphed, then the quantity for farm and rural use, and so on. The final, or topmost, curve shows the total amount of water used at a given date, *e.g.* 200,000 million gallons (909,000 million litres) per day in 1950. In such a compound curve graph it is customary to distinguish different items by distinctive shading.

natural routes through hill and mountain country; land unhampered by natural obstacles and hindrances, such as forests, swamps, deserts, and watercourses; long direct rivers which are slow-flowing, of sufficient depth, and unbroken by falls, rapids, etc.; good natural harbours offering deep, sheltered water unhampered by bars, silting, or strong currents; long coastline in proportion to area with no place very far removed from the sea; freedom from ice during winter, except in so far as it can be used for sledges; freedom from fog which restricts the movement of air, land, and sea vehicles; and easy accessibility to other countries and main world traffic routes. Clearly natural conditions can assist or hinder movement and commercial circulation these days is of great significance to economic development. We shall see how the most highly developed countries are those with the most highly developed and closely integrated communications networks, while the most backward ones are those lacking efficient distribution systems.

## *HUMAN FACTORS*

The capacity of a country to support its people depends also upon the human factor. Social conditions such as the following may, and do, affect the situation:

1. Physical health.
2. Standard of education.
3. Standard of living.
4. Culture.

### PHYSICAL HEALTH

Some peoples are more robust and energetic than others, although these features are due rather to general health, food, and climate than to any innate racial differences. The Chinese, though often slightly built and inadequately fed, toil laboriously for long hours, do strenuous work, and may often be seen carrying quite fantastic loads. The Hindus, Malays. and the South American Indians are often said to be lazy but this is probably due to the climatic and other conditions under which they live rather than to any inferiority in their physical make-up. Cool, temperate climates having distinct seasonal, even daily, changes in the weather conditions would appear to offer the best conditions for a maximum output of energy. Such climatic change also probably acts as a mental stimulus: certainly the most notable advances in modern science, technology, and the arts have come from peoples living in such environments. In contrast, tropical environments tend to be enervating; diseases are more rife, also, in such lands and these weaken people constitutionally. But it should be emphasised that the charge made in earlier times that some races are naturally physically and mentally inferior simply will not stand up to impartial investigation. On the other hand, it can scarcely be doubted that some peoples are better adapted to specific environmental conditions than others; the Negro, for example, is better adjusted to the heat and humidity of tropical regions than is the white man, and accordingly lives more comfortably and is less affected by the climatic conditions.

## EDUCATION STANDARDS

There is no evidence to suggest that any races or racial types are inferior in natural intelligence to any others. Given equal opportunities and facilities for training and education all peoples appear to be of similar intellectual capacity. The leading nations of the present day are those which are best educated; the most backward ones are, in general, those with the highest illiteracy rates. Education is a means of developing and directing natural intelligence. Intelligent enquiry and thought lies behind scientific discovery, mechanical invention, the application of power resources, and the organisation of transport. The better educated peoples are those who have made the greatest advances in science, technology, etc.; hence such peoples are better able to support themselves for they can apply their science and technology to the development of the natural resources and save man the drudgery of manual toil. Skill, foresight, and organisation are essential qualities in economic development and these qualities are all manifestations of trained intelligence. Britain's great development in the nineteenth century was based upon these three qualities. Countries lacking adequate educational facilities are incapable of realising and organising their resources and are apt to squander and waste such resources as they have. Thus the productivity, and therefore the wealth, of a country is closely related to the way in which it marshals, utilises, and organises its resources.

## LIVING STANDARDS

It is a truism that countries which are the most economical in the use of their resources are able to support the most people. Such economical use can be interpreted in two ways: it may involve everybody having less than he requires, which is a very undesirable form of it; or it may involve getting the best results with a minimum of waste, which is a good form of it. Generally speaking, the countries which possess and demand the highest standard of living must be prepared to work hardest; the only alternative is for them to be content to support fewer people than those countries willing to tolerate a lower standard. Usually they are willing to work hardest. Usually, too, their people are well educated, possessing skill, foresight, and organising ability, leading, in turn, to large-scale, highly-mechanised, well-co-ordinated production. The more sophisticated a society becomes the greater and the wider are its wants; these can only be secured and maintained by increased productivity. We are constantly being reminded that increased production is necessary if our living standards are to be maintained and people must be prepared to work diligently and economically to ensure their maintenance. Better living standards, or the promise of them, are a great spur to increased economic effort.

**CULTURE TRAITS**

Cultural features influence man's activities and these may affect economic production and the capacity of a country to support its population. Religion, for example, is a cultural feature of great social significance and sometimes exercises very powerful influences on such things as animal husbandry, food habits, and human attitudes with respect to work, progress, and the like. Buddhism, for instance, forbids the killing of animals and therefore predisposes towards a vegetarian diet and poorly developed animal husbandry. Moreover, so many of the males are demanded for the priesthood that labour may be in short supply. The Hindu caste system has in many ways militated against efficient economic development while the Hindu reverence for cattle still remains as a great drain on the food supply. Islam, too, acts as a great restrictive force, for such economic activities as mining, trading, and money-lending, are forbidden to members of the faith. The Moslem attitude to the pig—an "unclean" animal—results in the virtual absence of the pig from the Islamic world. But there are other factors, too. Different peoples have varying concepts of wealth. For example, to some peoples such as many African Negroes, wealth is counted in terms of the numbers of cattle they possess. Numbers of animals provide direct evidence of social rank; their quality, *i.e.* their economic value, is a very secondary consideration. And, finally, let us note that among some peoples "to work" is looked upon as being completely degrading.

## *MAN VERSUS NATURE*

It has been the purpose of this chapter to show how geographical factors, natural and human, affect economic development. Natural factors, such as geographical position, structure, relief, climate, vegetation, and animal life, influence and sometimes limit and control human activities. On the other hand, man can develop and turn many of the

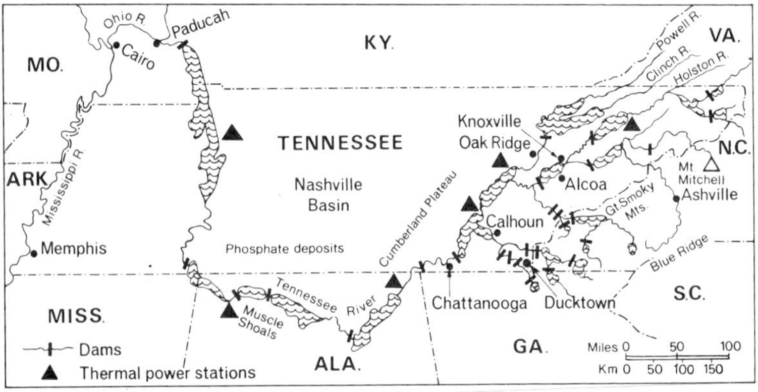


FIG. 18.—T.V.A. SCHEME

The Tennessee Valley Authority scheme in the U.S.A. was the prototype for modern conservation and economic development projects. "In 1933, during a period of world-wide economic depression," writes F. S. Hudson,* "the Tennessee Valley Authority was inaugurated. The measures it was authorised to undertake by Act of Congress were both imaginative and revolutionary, and the area over which it was given a large measure of control was, most unusually, a geographical and not an administrative unit. Indeed, it transgressed the boundaries of seven different Appalachian states, and involved the whole of the Tennessee Basin, an area of 40,000 square miles (100,000 km$^2$), equal to the combined dimensions of Denmark, The Netherlands, and Belgium. Its work has been so successful that it has become a model for other areas, *e.g.* the Damodar Valley in north-east India."

In 1933 the area had a population of over 2 million, mostly "poor whites" and Negroes. It was an area of depressed agriculture, of badly eroded land, of little industrial development, and of human poverty. Moreover, the Tennessee River was choked with silt, practically unnavigable and subject to periodical flooding. The T.V.A. was given power to plan the entire economic and social rehabilitation of the region and its inhabitants, and to introduce flood control measures, generate hydro-electric power, and improve navigation.

Today the region has been transformed. Some thirty large dams have been built on the Tennessee and its tributaries which have controlled the flood-waters, enabled power plants to be constructed, and improved navigation on the river. Soil erosion has been checked by re-foresting the hill-sides and by using contour-ploughing and strip-farming techniques. Agriculture has been greatly helped and farming has become more diversified. The development of hydro-electric power enabled industrial enterprises to be started and now there are numerous factories manufacturing chemicals, plastics, fertilisers, paper, furniture, textiles, etc. An aluminium refinery was built at Alcoa using bauxite deposits worked in the Chattanooga district. The many lakes that were created have led to the growth of a tourist industry. As a result of all these measures, the standard of living of the peoples in the region has risen.

* *North America*, Macdonald and Evans, 2nd ed. 1968, p. 212.

natural conditions to his own advantage. Though unable to alter materially many of the physical conditions of his natural environment, he has at least been able to mitigate and modify some of their effects. By building tunnels and bridges, by drainage and reclamation, by terracing the land, by irrigating and fertilising, by the use of glasshouses and central heating, by air-conditioning and refrigeration, by plant and animal breeding, etc., man has done much to lessen natural controls and influences. As man's scientific knowledge and technical developments increase, the bonds of Nature will be increasingly lessened. Just how far he will be able to burst the bonds only the future can tell. Let us conclude this chapter with two brief accounts, one to show how man has mastered to some extent an environmental control, the other to illustrate how Nature may yet call the tune.

**MARQUIS WHEAT**

Climatically the whole of the Prairie Provinces of Canada lies within the 0° C (32° F) isotherm for three months of winter. Winters are long and hard, summers short, warm, and sunny. When Europeans began to colonise Canada they commenced to raise the crops they were used to growing back home. The immigrants managed to grow wheat successfully in the St Lawrence lowlands, but when they moved into the interior they found that frost killed off the wheat. All the early attempts to grow wheat on the virgin soil of the prairies were unsuccessful due to the killing frosts. In 1860 a Scottish immigrant farmer, called David Fife, wrote home bemoaning the Canadian winter, the shortness of the growing season, and his inability to grow grain. One of Fife's friends, unloading a consignment of Polish wheat which had been shipped from the Baltic port of Danzig (present-day Gdansk) to Glasgow, decided to send Fife a sample of this wheat. He gathered a hatful and parcelled it to Canada. David Fife sowed the wheat, watched it germinate and tended it carefully; he saw it grow and flourish and then, to his dismay, the plants withered and died. All that is except a few—the tradition says all but one plant. Fife saved the seed, a few grains in all, and planted them. They grew and matured. Year after year he re-planted until eventually he built up a good stock of seed. For the first time a wheat, christened Red Fife (after David Fife and the Polish hard red spring wheat), had been produced that would grow and mature on the prairies when every other type failed. Soon the prairies were under a sea of golden grain, every plant descended from David Fife's handful of seeds.

For half a century Red Fife reigned supreme. However, Red Fife required from 120 to 135 days to attain maturity, and due to the short growing period on the northern fringes of the prairies, Red Fife was often spoiled by frost shortly before it could ripen properly. An agricultural scientist, Dr Charles Saunders, decided he would try to produce a wheat which would have the good qualities of Red Fife but which would ripen a few days earlier. After ten years of research and experiment, and sixty years after David Fife's experiment, Saunders produced a variety of wheat which came to be called Marquis. Marquis was produced by crossing the Red Fife strain with Hard Red Calcutta, a variety obtained from India. The new cross-bred variety was a quicker ripening wheat, maturing six to ten days earlier than Red Fife, and also gave a higher crop yield. Because it ripened a few days earlier, Marquis could be sown farther north than Red Fife; the wheat frontier was pushed poleward. By 1928 four-fifths of the wheat grown in Canada's prairies was Marquis, and by growing it the Prairie Provinces were able to increase their wheat production by several million bushels a year. Saunders subsequently became head of the Cereals Division of the Canadian Experimental Farms, and for his great services to agriculture he won a knighthood.

Marquis wheat became a parent for subsequent hybrids. Numerous varieties, such as Red Bobs, Redman, Thatcher, and one named Saunders were produced which reduced still further the number of days required for wheat to mature. Saunders, for example, required an average of 106·9 days to mature and gave a yield of 24 bushels to the acre. More rapidly maturing varieties suited to the northern margins of the prairie wheat belt are Garnet, Reward, Ruby, and Prelude, the last of which ripens in a period of from 85 to 100 days. The poleward extension of wheat cultivation depends upon the success of such varieties. Research by the Canadian Experimental Farms Division (which has some three dozen stations, eight laboratories, and various agricultural colleges) has not been directed solely towards the development of early-ripening varieties of wheat, but also to the production of disease-resistant, especially rust-free, types. But undoubtedly the biggest success of the Experimental Farms Division was Saunders' development of the famous Marquis variety.

This account illustrates how man by care, persistence, scientific enquiry, and experiment, but not forgetting an element of luck, too, can overcome the limiting conditions of the natural environment. This example, by no means an isolated one, shows how man may master what at first appears to be an insurmountable control.

## GROUNDNUTS

One of the most exciting projects of the post-war years was the British Groundnuts Scheme in East Africa. In the immediate post-war years there was a world shortage of fats and the plan, based upon the *Wakefield Report* (1947), was aimed at relieving this shortage. But the plan was also intended to initiate the major economic development of British East African territories. Briefly, the plan was to turn 3 million acres (750,000 ha) of East African bush into farmlands which would produce groundnuts and from which large quantities of margarine and cooking fats would be derived. The Groundnuts Scheme, as the project came to be called, was hailed as a great Imperial dream, and even General Smuts, the Prime Minister of South Africa said: "This is the biggest thing that has happened to Africa since the Boer War ended." And yet, by 1950, some £50 million had been sunk in the scheme, a mere 50,000 or so acres (20,000 ha) had been cleared and planted, and only a few thousand tons of groundnuts had been harvested. The plan had envisaged the clearing by mechanised methods of 2½ million acres (1 million ha) within the first five years and an estimated production of 600,000 tons of groundnuts by 1950–51 at a total capital expenditure of £24 million. What had started as a bold and imaginative scheme ended up as the greatest failure in Colonial history. Why?

The reasons for the dismal failure of the Groundnuts Scheme were many, but among them were: (*a*) exaggerated optimism with respect to the acreage that could be cleared and cultivated in a year; (*b*) lack of appreciation, by those who should have known better, of farming conditions, rainfall, etc., in the Kongwa area of Tanganyika; (*c*) the use of unsuitable tractors and bulldozers for which no provision of spare-parts had been made; and (*d*) the debasing of what was an agricultural operation into a political stunt—a "groundnuts or bust" directive from Whitehall. But whatever political, administrative, and technical difficulties harassed the Scheme, probably the fundamental cause of its collapse was the failure to appreciate the geographical conditions prevailing in East Africa. The thick African bush, the tough roots, the hard red earth, which was like concrete in the dry season, played havoc with the mechanised equipment. No study had been made of the natural vegetation or the soil; incredibly, almost the whole of the vast area to be cultivated was never surveyed, except from the air, before operations began. At one time blame was put on the climate: the rains had failed to materialise and drought had ruined the crop; but vicissitudes in the rainfall are a well-known feature of savanna lands. And there were problems of labour and transport, too. Most of the errors and blunders could have been avoided if only geographers, land-use experts, soil scientists, and botanists had been called in to advise.

The Groundnuts Scheme proved beyond any shadow of doubt that man cannot ride roughshod over Nature. Nature yields herself only to those who approach her in a proper spirit of humility and with patience enough to learn her ways and understand her moods. In East Africa man learned by bitter experience. Failure demonstrated that if man wished successfully to farm the land he must first find out the correct way to manage and cultivate the soil; secondly, take due cognisance of the vagaries of the climate; thirdly, work experimentally and on a small scale until he has gained experience of local conditions; and, fourthly, not think that one has merely to marshal agricultural machinery and let it loose on the land to create a sound system of farming.

### EXERCISES

1. Study Figs. 15 and 16 and write an essay on the uses of coal and petroleum.
2. Show how natural conditions may (*a*) assist, (*b*) hinder, transport and communication.
3. Discuss the following statement: "Countries which are the most economical in the use of their resources are able to support the most people."
4. Show how culture traits may influence human activities and economic production.
5. Select three or four illustrations to show how man has overcome the limiting conditions of the natural environment.
6. Write an essay on: "Water, its importance and use to man."

# Chapter VI

# GEOGRAPHICAL REGIONS

THE geographer commonly divides the earth's surface up into regions, that is into areas possessing fairly distinctive characteristics which allow them to be distinguished from neighbouring areas. The geographer recognises many different kinds of regions, *e.g.* structural, topographic, climatic, vegetational, economic, and cultural regions. Most familiar, however, are the so-called natural regions which are based upon the combination of relief, climate, and vegetation. These regions, strictly speaking, take into consideration only the physical conditions of the environment. But man is a geographical factor or agent and comes into the picture in practically every area of the earth's surface. He occupies the land and, through his activities, sensibly modifies the land surface. Hence, here, we are going to describe the geographical regions of the earth (Fig. 19). These are very similar in their global pattern to the natural regions with which you are probably already familiar.

A study of a map showing the geographical or natural regions of the world shows that similar types may recur in different continents in similar latitudes and in similar land-mass positions (western margin, interior, eastern margin). Two points should be noted, however:

1. Though areas may be grouped into a single type of region, *e.g.* the equatorial forest region, conditions within the various areas belonging to this region, though bearing many close resemblances, do not necessarily agree in detail.
2. Regions seldom have clear-cut boundaries and the lines drawn on a map are largely arbitrary; in fact, as is sometimes said, nature knows no boundaries; rather does one region grade, almost imperceptibly, into another.

Let us bear these two points in mind during the following accounts of the earth's geographical regions.

## *EQUATORIAL FOREST REGIONS*

These occur in equatorial latitudes within a few degrees north and south of the equator. Mostly lowland areas, they are always hot, humid, and wet. The average monthly temperatures are always about 27° C

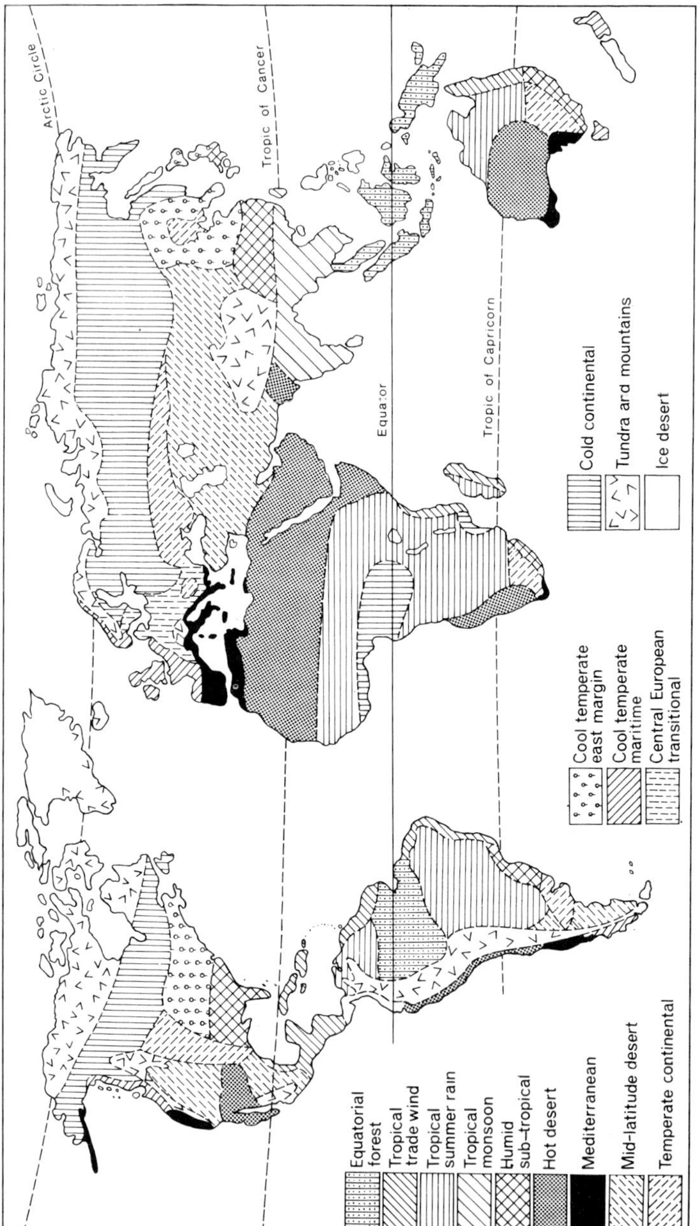


FIG. 19.—GEOGRAPHICAL REGIONS

(80° F). The diurnal, or daily, range of temperature is greater than the annual range. Convection rains fall most afternoons. The average annual rainfall is usually about 80 in. (2032 mm). Two periods of exceptionally heavy rainfall commonly occur a little after the equinoxes. There is no seasonal change or rhythm; the same sequence of weather repeats itself day after day. Such monotonous regularity is not stimulating and the climatic conditions are enervating. This is especially true of interior locations where the air may be still; conditions are pleasanter in coastal localities where sea breezes bring air movement.

The constant heat and plentiful moisture all the year round promote a rapid and luxuriant growth of vegetation which expresses itself in forest. Extensive forests, rich in species, cover vast areas in the basins of the Amazon and Congo and in parts of the East Indies. These equatorial rain-forests, commonly called *selvas*, contain isolated specimens of a great number of different broad-leaved evergreen trees. Often they show a dense undergrowth, sometimes the trees grow in "tiers," while climbing creepers (lianas) and plants using trees as anchorages (epiphytes) or living off trees (parasites) are common. Ground animals are relatively scarce but there are many tree-dwellers, a large number of reptiles, and a wealth of bird and insect life.

Except in a few areas, *e.g.* parts of West Africa and the East Indies, notably in Java, the selvas are thinly peopled and economically little developed. Because vegetative growth never ceases and because soils are frequently infertile, cultivation is extremely difficult. Many of the forest-folk, especially the native Indian peoples of Amazonia, the Pygmies of the Congo Basin, and the aborigines of the East Indian region, are mainly collectors and hunters occasionally growing a few simple crops, such as yams, bananas, and beans, in forest clearings. Some peoples are more particularly shifting cultivators. This primitive system of farming has led to the destruction of large areas of valuable forest. The most notable agricultural development in equatorial regions has been the establishment of plantation farming, best developed in the East Indian region where rubber, oil-palm, tea, coffee, cocoa, and sugar-cane are produced. In Africa, plantation agriculture is less well developed, although in West Africa much cacao is grown on small native-owned plantations. Amazonia is the least developed of all the equatorial forest regions.

In a few places the forest trees, especially the valuable hardwoods, are cut down for export, *e.g.* greenheart and iron-wood from Brazil, and mahogany from West Africa. Many difficulties face lumbering in these regions (*see* p. 235). Among other products collected from the forests are wild rubber, brazil nuts, kapok, kola nuts, medicinal plants, alligator skins, and birds' feathers.

[*Courtesy: Booth Line.*

FIG. 20.—THE AMAZONIAN SELVA

Luxuriant evergreen forest (*selva*) is natural to hot, ever-wet equatorial regions. In the Amazon basin, rivers are almost the only means of communication, and man lives at the sufferance of Nature.

The equatorial forests hold out considerable possibilities for future occupation and development, although large areas are, at present, handicapped by flooding, lack of communications, insect pests, and tropical diseases.

## *TROPICAL SUMMER RAIN REGIONS*

Commonly called the Savanna Lands, these regions, extending through some 10 degrees of latitude, border the equatorial regions on their equatorward side and the hot deserts on their poleward side. Climatically, they are transitional regions coming under the influence of the equatorial belt of convectional rains in summer (which follow the poleward migration of the sun) and under the dry offshore Trade Winds in winter (when the pressure and wind belts move south with the sun). Thus the climate is one of marked seasonal changes. Temperatures are high all the year round, seldom falling below 21° C (70° F) and usually registering between 24° and 32° C (75° and 90° F). The annual range of temperature is greater on the poleward than on the equatorward margins. Rainfall varies widely and there is a gradation from over 60 in. (1524 mm) on the forest edge to 15 in. (381 mm) on the desert margin. Though the climate is usually described as being one of hot, wet summers and hot, dry winters, it is possible to recognise three seasons: a comparatively cool dry season, a hot dry season, and a rainy season.

Not only is there a long dry season but the rate of evaporation is great; these conditions are not suited to tree growth and the characteristic and predominant natural vegetation is grassland. Trees line watercourses giving rise to "fingers" of forest—known as *galeria* forest. Elsewhere umbrella-shaped trees and some such as baobabs, with swollen trunks, in which they store water, together with shrubs, dot the landscape. Grasses grow quickly in the wet season and may reach heights of six feet (1·8 m) and more; the tall grass is commonly called "elephant grass." Grasses are thick and hard-stemmed and when the grass dies down it forms a coarse, dense mat in many places. On the forest margins where there is a scattered tree growth, a "parkland" landscape is characteristic.

The tropical grasslands, especially in Africa, are the home of a rich fauna. The animals are of two chief types: the herbivores, grass-eating and browsing creatures which are fleet of foot, *e.g.* antelopes, giraffe, zebra; and carnivores, beasts of prey such as the lion and leopard which hunt the first group. Ants, locusts, and other insect pests are numerous. The tsetse fly, carrier of sleeping sickness and animal diseases, infests large parts of Africa and prevents cattle-rearing. In the savanna lands of South America, the cattle-tick is a pest.

Man in the savanna lands is traditionally a hunter but in many areas he has become a pastoralist herding cattle, and in some areas a cultivator. Some of the native peoples of the African savannas herd cattle, *e.g.* the Fulani of West Africa and the Masai of East Africa; they count their wealth in terms of the numbers of cattle they possess. Others are primitive cultivators, either of the shifting or sedentary type, growing crops such as yams, millet, groundnuts, beans and peppers. Native cultivation in Africa is almost always done by womenfolk. In some parts of Africa, notably in East Africa, large single-crop plantations established by Europeans, but worked by native labour, produce coffee, cotton, sisal and tobacco. Commercial cattle-ranching, though present in parts of East Africa, is more especially a development of the Australian and South American savanna lands. None of the tropical grasslands is, as yet, highly developed, but they hold out considerable promise for the future, especially as cattle lands and as crop lands, particularly where irrigation is possible, for the cultivation of maize, cotton, and tobacco.

## *TROPICAL TRADE WIND REGIONS*

Some regions in tropical latitudes with east coast locations have temperature and rainfall conditions bearing some resemblance to equatorial regimes on the one hand and monsoon regimes on the other. The climate is described as Tropical Trade Wind type or Tropical East Coast Marine. Winter temperatures are around 21° C (70° F); summer temperatures are typically over 27° C (80° F). Temperatures in summer are tempered by the breezes which blow persistently onshore. Trade winds blowing over continental areas are notoriously dry winds, but in those areas where they blow over great stretches of ocean they pick up great quantities of water vapour. When these winds reach the continental margins, and especially where they meet upstanding relief, orographic rainfall occurs. Since the onshore Trades blow all the year round, the rainfall is fairly evenly distributed throughout the year, although there is a pronounced summer maximum. In this wise, the climate differs from the tropical monsoon type which has clearly marked wet and dry seasons. The precipitation, which is well-distributed and which in amount is 40 in. (1016 mm) and over, is very like that of equatorial regions.

The chief areas of this type are: the Caribbean region, the tropical east coast of Brazil, Madagascar and Mozambique, and the coast of Queensland. Because of the plentiful rainfall, the natural vegetation of these regions is luxuriant forest. Where the forest has been cleared, cultivation is practised; the crops usually grown are rice, manioc, sugar-cane, coffee, bananas, and pineapples. In general, the high temperatures, abundant rainfall, and high humidity are not well suited to white peoples, and though some Europeans live in these tropical east coast regions most of the labour is supplied by coloured peoples, *e.g.* by Negroes in the West Indies, by Hindus in Guyana. In two regions only have Europeans settled in large numbers and undertaken the necessary labour; these are coastal Brazil, which has been colonised by Portuguese, Spaniards, and Italians who work on the plantations, and coastal Queensland where, because coloured immigrants have been excluded, almost all agricultural labour is undertaken by whites.

The degree of progress of these various regions differs considerably and is largely related to the availability of a labour supply. Basically they remain areas producing foodstuffs and raw materials for export; as yet there is very little industrial development and manufactured goods must be imported.

## *TROPICAL MONSOON REGIONS*

Many areas experience monsoon winds in summer but the only area where the climate is completely dominated by them is South-eastern Asia. Northern Australia is strongly influenced by them as is Ethiopia, while the coastal margins of East Africa feel the effects of the Indian Ocean monsoon system. In those lands experiencing the true monsoon conditions the winds are emphatically seasonal, blowing from land to sea in winter and from sea to land in summer. India provides the classic example of an area having a tropical monsoon type of climate. Winters are cool or warm and dry, early summers are hot and dry, and late summers hot and wet. The year is divisible into three seasons instead of the usual four. Temperatures during the summer months are high, commonly between 27° and 32° C (80° and 90° F). The highest temperatures come in the months (April and May) immediately preceding the "burst" of the monsoon which usually occurs in June. The onset of the summer rains, brought by the south-west monsoon, is typically sudden, and violent storms with torrential rains introduce the "wet season." The total annual rainfall varies considerably between place and place, but monsoon regions claim some of the highest rainfall totals in the world. Rainfall is greatest where mountain ranges lie athwart the air streams. For example, at the station of Cherrapunji in Assam over 400 in. (10,160 mm) of rain is recorded each year. The moonsoon rainfall, however, is notoriously fickle: one year there may be more than enough rain and an acute shortage the next. It is this chanciness of the rainfall which has made farming so difficult and life so precarious in India.

Because of the great variability in the amount of rainfall received from area to area (as well as at any specific place) there is a wide variation in the natural vegetation. This ranges from rain forest, through deciduous forest (teak, sal, etc.), to jungle, occupied by elephants and tigers, and thorny scrub vegetation. Much of the original forest and woodland, particularly in the Indian sub-continent and China, has been cleared for cultivation. In Burma, Thailand, and the countries formerly comprising Indo-China, much monsoon forest remains.

Ways of life in the monsoon region are mostly bound up with the cultivation of the soil and the agricultural village is the characteristic form of settlement. In the moister areas the chief crops are rice (the staple foodstuff), sugar-cane, tea, oilseeds; in the drier areas millet, groundnuts, wheat, cotton, and tobacco are grown. The latter crops are also grown during the winter season in the wetter areas.

While in many parts of the monsoon region forest gives way to a wooded savanna type of vegetation, nutritious fodder grasses are few, hence cattle and sheep are relatively unimportant in most areas. Moreover, in many areas, due to the shortage of cultivable land and the pressure of population upon the land, livestock are almost squeezed out of existence. India provides the exception. Cattle are sacred to the Hindus, and, largely on this account India has a large cattle population (about 175 million). There is no large-scale meat industry, however, owing to the religious proscription of the killing of animals, although there is some export trade in hides. In the wetter parts of the monsoon region the water buffalo is important because it is the chief work animal. Pigs are especially important in China (about 210 million).

By and large the tropical monsoon region is one of seething populations, abject poverty, and the agricultural way of life. The economy is basically of the subsistence type and although urbanism and industrialism are growing they are not as yet, except locally, developed to any extent.

## *HOT DESERT REGIONS*

The hot deserts of the world lie in interior and west-coast regions between approximately 15 and 35 degrees North and South of the Equator. Desert landscape is often dramatic. The visual image usually conjured up in one's mind by the mention of desert is of extensive stretches of sand, some of which is piled high in a series of dunes. But we should remember that there are other types of desert, some consisting of bare rock pavements, others of sheets of pebbly material, and yet others of areas of baked clay. To these different types of desert special names are given and in the Sahara the sandy desert is called *erg*, the rocky desert *hammada*, and the stony desert *reg*.

To the geographer, the term desert is linked with rainfall; they are areas of deficient moisture. The hot deserts lie in tropical latitudes—astride the tropics—and come under the influence of the high-pressure systems of those latitudes. Hence, the air is subsiding and flowing outwards. In the Sahara, for example, the air streams flow outwards and prevent any moisture-laden winds coming in from the ocean. Since the air streams are blowing from cooler to warm latitudes and are subsiding,

[*Courtesy: U.S. Information Service.*

FIG. 21.—DESERT LANDSCAPE, ARIZONA

Here, under the hot, dry conditions of the Arizona Desert, is a varied array of succulents (water-storing plants), including the giant organ cactus (*left*) and the prickly pear or agave (*foreground*). Of what use are deserts to man?

they absorb rather than yield moisture. On the Atlantic coast the aridity is further emphasised by the cold Canaries current. In summer, when the sun is overhead, temperatures are very high, averaging about 32° C (90° F). In winter it is usually considerably cooler. Desert areas also have great daily ranges of temperature and even in summer the nights may be bitterly cold. These extreme temperature conditions are due to the clear air and cloudless skies of desert regions. Immediately the sun goes down, the ground, which has stored up heat during the day, quickly radiates it back into the atmosphere, where it is rapidly dispersed due to the absence of the blanketing effect of cloud.

The hot deserts are lacking in moisture to support much plant life. Plants cannot thrive without moisture (though some manage to exist where there is very little) and, in general, desert is marked by the 10-in. (245-mm) isohyet. There is, therefore, too little and too variable rainfall to support more than a few plants, and these are usually specialised types which can withstand drought, *e.g.* cacti, acacias. When, as occasionally happens, a sudden rainstorm occurs, a quick growth of grass and herbaceous plants sprouts up. Because of the lack of vegetation, there is a corresponding lack of animal life.

The chief hot desert regions are the Sahara and its continuations the Arabian and Thar Deserts, the Kalahari Desert of South-west Africa, the great Australian Desert in the western part of that continent, the Atacama Desert of northern Chile which is continued into coastal Peru, and the desert of the Mexican–United States border.

The hot desert regions are of relatively limited value and use. Accordingly they are, in general, very sparsely peopled. Only where there is a major source of water supply, as in Egypt, is population dense. The people of the desert regions fall into three groups:

1. The nomadic camel, sheep and goat herders who wander in search of pasture and frequent the water-holes, people who, in the past, were also often carriers of merchandise and equally as often marauders.
2. The settled people of the oases who grow cereals, vegetables, fruit (notably the datepalm), and sometimes irrigated cotton and sugar-cane and who also often rear cattle, sheep, goats, camels, asses, etc.
3. The dwellers of the mining settlements who tolerate the climatic conditions for the sake of the mineral wealth, *e.g.* the goldfields of Western Australia, the diamonds of the Kalahari, the nitrate deposits of northern Chile, and the petroleum of Algeria, Libya, Saudi Arabia, Kuwait, South-west Iran and Iraq.

## *MEDITERRANEAN REGIONS*

These regions, though of very limited extent, are of great human importance. They occur in mid-latitudes on the western margins of the continents and the climate is sometimes called the western marginal warm temperate type. The Mediterranean regions are areas of marked seasonal changes, but the changes relate to precipitation rather than temperature. Mediterranean regions are commonly identified as regions having warm, wet winters and hot, dry summers; this, as a broad, general statement, is acceptable, although a careful study of Mediterranean regimes shows that it is subject to modification. Winters are typically mild, with no month averaging below 6° C (43° F). Summers are hot, with temperatures over 21° C (70° F), and in interior districts away from the sea they are frequently over 27° C (80° F). A distinguishing feature of Mediterranean regions is their high amount of sunshine; even in winter there is little cloud. Rainfall varies between approximately 10 and 30 in. (254–762 mm), though places in exposed situations sometimes receive appreciably more than the latter figure.

The distinct and rather emphatic seasonal changes—springs and autumns are somewhat brief—are due to the position of the Mediterranean regions in latitudes (30–45 degrees) which come under the influence of cyclonic storms in winter, which bring heavy but infrequent rain, and tropical high-pressure systems in summer. Thus they may be said to have cool temperate conditions in winter and hot desert conditions in summer. They are "transitional" in much the same way as the savanna lands are "transitional." The areas are the coastlands surrounding the Mediterranean Sea, California, central Chile, the south-western corner of Cape Province, south-western West Australia, and the district around Adelaide in Southern Australia.

The natural vegetation is classed as broad-leaved evergreen forest and scrub. During the normal growing season (*i.e.* summer) plant growth is retarded and most of it must take place during winter. The vegetation, therefore, is of a unique type which shows adaptations to the hot, dry summer. Adaptations such as pinnate leaves, small, tough, leathery leaves covered with varnish, compact, shrubby growth, thick bark, and long, widely-spreading roots are all concerned with preventing excessive transpiration on the one hand or tapping sub-surface supplies of water. Much of the original scrub-forest has been cleared and in many areas replaced by a tangled bushy scrub known as *maquis*. In the drier limestone areas a thinner, degenerate growth called *garigue* is found. Bulbous and tuberous plants, which store moisture in their root systems, are common. Because of the summer drought there is little pasture except in marshy areas and on the higher mountain slopes where melting snows support summer pastures.

Small-scale fruit and grain farming, the production of olive oil and wine together with transhumant animal grazing are the traditional occupations of the lands surrounding the Mediterranean Sea. More recently, commercialised intensive fruit-farming and market-gardening, usually by irrigation, have become important. Tourism is another important development of modern growth. In the areas outside Europe a sequence of economic development can often be traced: first man was attracted by mineral wealth, followed by the rearing of animals for their hides and the growing of grain, then by the cultivation of fruit under irrigation, and finally by the development of agricultural industries such as fruit drying and canning, wine-making, etc. None of the Mediterranean lands is a centre of major industrial development.

## *HUMID SUB-TROPICAL REGIONS*

These regions, which have a warm temperate eastern margin location, have a climate variously described as humid sub-tropical, temperate monsoon, or China type (because Northern and Central China provides the type example). The areas falling within this category include northern and central China, most of Japan, the coastlands of New South Wales, Natal in South Africa, southern Brazil, and the south-eastern part of the United States.

Climatic conditions vary appreciably between these different areas. In the northern hemisphere distinct monsoonal effects are apparent since moisture-laden air streams are drawn inland to the continental low pressure areas in summer, while dry cold winds blow seawards from the continental high pressure areas in winter. This seasonal change is apparent in the south-east of the United States, but is much more clearly marked in China where seasonal temperature and rainfall changes are quite emphatic. These northern hemisphere regions have a considerable range of average temperature, often over 4° C (40° F). Summers are usually hot (around 27° C; 80° F), but the winters are cool or even cold (average temperatures for the coldest month are common under 4° C; 40° F). Some rainfall is usually experienced all the year round, but there is a distinct summer maximum. Precipitation totals between 30 and 60 in. (762–524 mm). Northern and Central China really form a part of the "Monsoon Region" of Asia, but the area is differentiated from the remainder of that region by the coldness of the winters. In general, winters in the south-east of the United States are warmer, although occasional invasions of cold continental air may cause periods of low temperature.

In the southern hemisphere the three corresponding areas have climatic conditions which are more nearly alike. The continental land areas here are not extensive enough to develop monsoonal tendencies so that the marked seasonal changes distinctive of the northern hemisphere areas are not apparent. Summers are very warm and moist and winters mild or warm with rain. Although precipitation is well distributed throughout the year, most of the rain comes in summer from the onshore Trades.

Forest is the natural vegetation of all these areas. The forests are of the sub-tropical evergreen or deciduous types, and are luxuriant with a wealth of species. Many of the trees and shrubs have commercial importance: some—oaks, maple, walnut, and pines—for the timber they yield, others for beverages (tea, yerba maté), gums, resins, camphor, etc. In China and the United States much of the forest has been cleared.

Agriculture is general in all the areas. Food crops include wheat, maize, soya beans, rice, tea, and sugar-cane. Industrial crops are important, especially cotton and tobacco in both the United States and China. In the southern hemisphere, because of the mild, moist conditions, cattle-rearing is an important occupation. However, the types of crops and animals raised in all these different areas show variations due to the variable factors of climate, historic and cultural development, and the presence or absence of a coloured population providing a labour force.

Apart from Japan, none of these regions is industrially well developed, although manufacturing is growing rapidly in the south-east of the United States and in New South Wales. In Natal and southern Brazil industry is still in its infancy, but many of the towns now have varied manufacturing activities.

## *TEMPERATE CONTINENTAL REGIONS*

These are the mid-latitude temperate grassland regions. They lie, typically, in mid-continental areas between approximately 30 and 50 degrees. They have strong seasonal temperature contrasts with very cold winters (several months are usually below freezing) in the northern hemisphere and hot summers. Winters are much less severe in the southern hemisphere and the range of temperature is greatly reduced; these more tolerable conditions are due to the narrower width of the southern continents in temperate latitudes and to the fact that maritime influences are nowhere completely absent.

The temperate grasslands have a low total precipitation, about 15 to 20 in. (381–508 mm). Rain, mainly of the convectional type, comes in spring and early summer; there is occasional light snow in the northern hemisphere. Topographically, these grasslands are gently rolling plains. They are grass-covered because of the long period of winter frost and the high rate of evaporation in summer which are hostile to tree growth. The grass tends to be stiff, coarse, and wiry, quite unlike meadow grass. In the drier areas it is sparse and grows in clumps. Generally speaking, these grasslands are lands of vast unbroken horizons, uniform and monotonous in aspect, particularly in late summer when the ground is baked hard, the grass is withered and tawny in colour, and the air is dusty. In early spring they are more attractive, for at that time the grass is green and there are innumerable gaily coloured flowers.

The principal temperate grassland areas in the northern hemisphere are the Great Plains region of North America, the steppe belt of Eurasia, and the Plains of Manchuria; in the southern hemisphere there are also three areas, the pampas of Argentina, the veld of South Africa, and the interior plains of the Murray-Darling Basin in Australia.

[*Courtesy: George Hunter, Toronto.*

FIG. 22.—PRAIRIE LANDSCAPE

An aerial view of the Canadian prairies near Edmonton, Alberta. Note the flat terrain, the endless horizon, the large rectangular fields, and the straight roads, everywhere characteristic of central Canada. On these rich prairie soils large-scale cereal cultivation is carried on. The oil well, now a common feature in many parts of southern Alberta, shows that riches lie below the soil as well as on top of it.

The temperate grasslands are not only of great economic importance but of great interest since they present a sequence of evolutionary development. Five phases are distinguishable:

1. Originally these grasslands were regions of nomadism, either nomadic hunters, such as the Plains Indians of North America who followed and killed the buffalo, or nomadic pastoralists, such as the Kirghiz of the Asiatic steppe who reared animals.

2. The second phase led to the opening up of these grasslands by European peoples who developed large-scale cattle-ranching and sheep-

farming, in the earlier days to provide hides and wool but subsequently to produce meat as well.

3. In the moister and more fertile areas, where rich black prairie soils or chernozems were found, animal-raising was displaced by large-scale grain-farming, animal farming being relegated to the drier areas of the grasslands.

4. In due course, partly to maintain soil fertility and partly because of economic considerations, the extensive cultivation of cereals was replaced in some areas by farming of a more mixed type in which a wider variety of crops came to be integrated with animal husbandry.

5. Industrialisation is beginning to invade these areas which hitherto were essentially concerned with primary activities.

Not all the temperate grassland regions have passed through, nor will necessarily pass through, these stages of development. The southern part of Alberta may be said to have run the gamut of them all, whereas in central Asia the first stage was still in being until a generation or so ago. But, in the main, these grasslands remain fundamentally great stock-rearing and grain-growing areas.

## *MID-LATITUDE DESERT REGIONS*

The temperate desert covers a large area in the heart of the extensive land-mass of Eurasia, considerable areas in the western interior of North America, but is represented in the southern hemisphere only by the Patagonian Desert of South America. These regions, which are extensive plains or plateaus, are deserts simply because of their interior locations remote from the ocean or because they lie, like Patagonia, in a rain-shadow area.

The mid-latitude temperate deserts are areas of wide temperature ranges and low precipitation. Summers are hot with burning sunshine and at this time of the year resemble closely the hot desert areas. But winters are bitterly cold with strong searing winds, for the regions lie in high-pressure areas with out-blowing winds. There is a very low, scanty rainfall which occurs mainly in summer and a powdering of snow in winter. There is little vegetation in these temperate deserts other than a scattered growth of sparse, tufted grass and drought-resisting herbaceous growths.

Sometimes a distinction is drawn between the higher plateau areas of temperate desert such as are characteristic of Iran and the lower plains areas typified by Turkestan.

Life in these areas is clearly limited by the climatic conditions. As in the hot deserts, population is scanty and comprises either nomadic or sedentary groups. In the past, and still to some extent in the present in Asia, tribal groups have eked out an existence herding animals and to some extent carrying trade. Theirs was a precarious existence and from time to time, when probably the rainfall was less than usual and the scanty pastures failed, these nomads moved into and plundered the richer marginal lands—think of the constant Mongol invasions of China. The sedentary groups were those who congregated around sources of water supply—streams coming down from the adjacent mountains—and created oasis settlements such as the famous central Asian cities of Balkh, Bukhara, Ferghana, Samarkand, and Tashkent.

Until relatively recent times Turkestan (Soviet Central Asia) was little developed, being largely a pastoral region with a little crop-growing activity. However, great irrigation schemes and the exploitation of the Karaganda coal deposits have begun to transform the area. Large quantities of cotton are now grown by irrigation and a number of ancient trade route centres have been turned into important industrial cities. In North America irrigation farming and mining are carried on in a number of places, though the desert areas are of little importance. In Patagonia, settlement is thin and sheep rearing with a little cultivation in the sheltered valleys are the main economic activities.

## *COOL TEMPERATE MARITIME REGIONS*

On the western sides of continents in the higher latitudes of the temperate belt (approximately between 45 and 60 degrees North) are regions lying in the track of westerly winds and depressions which bring maritime influences and rain at all seasons. Warm ocean drifts move towards these regions, bathing the coasts in warm water and ameliorating winter cold as well as maintaining ice-free ports. These conditions result in these regions having unusually warm winters for their latitude and equable conditions all the year round.

The chief characteristics of the climate are: the warm winters (around 4° C; 40° F) and cool summers (around 16° C; 60° F), with only slight or moderate ranges of temperature (about 9–11° C; 15–20° F); the plentiful but usually gentle rainfall, anything from about 20 in. in the drier rain-shadow areas to over 100 in. (2540 mm) in exposed situations, which is well distributed seasonally; the day-to-day variability of the weather which is due very largely to the never-ending succession of depressions; and the strong winds, plentiful cloud, and frequent fog and mist which are especially typical of the winter half of the year. In brief, the climate is cool, moist, lacking in sunshine, and changeable. The chief areas having these conditions are North-western Europe (including the British Isles), the north-western coastlands of North America, southern Chile, Tasmania, and South Island New Zealand.

The natural vegetation is deciduous forest, except where altitude or soil conditions promote either coniferous forest or heath and moorland. In most of North-west Europe the forest cover has long since been largely cleared to be replaced by meadow and cultivated land. Because of the cool, moist conditions, this region is especially well suited to dairying, and it has become the world's premier dairying region. Farming generally, however, is very largely of the mixed type. The presence of plentiful coal deposits in North-west Europe, together with the fact that the Industrial Revolution had its origin here, has led to the large-scale development of mining, manufacture, and commerce. This, in turn, led to a rapid growth in population and urbanism with the result that it is a region of high population density and numerous industrial towns with the greatest concentrations of people and cities upon the coalfields.

None of the other regions have become developed to anything like the extent of North-west Europe. In British Columbia and Oregon and southern Chile there are vast areas of virgin forest, very little so far having been cleared. In these lands development is still in the early stages. In British Columbia lumbering, fruit and dairy farming, fishing and mining are all well developed. Southern Chile remains a very much under-developed land but it is rich in timber and water-power resources and the potentialities for the lumbering, dairying, and fishing industries are considerable. New Zealand's South Island and Tasmania are more reminiscent of British Columbia than southern Chile, though they are both still primarily agricultural.

## *COOL TEMPERATE EASTERN MARGINAL REGIONS*

Occupying comparable positions but on the eastern margins of the continents are regions having a climate designated cool temperate eastern margin climate, or a Laurentian type after the St Lawrence region of Canada, which offers the type example. There are only two areas in the world having this type of climate and falling in this regional grouping: the eastern part of Canada and the extreme north-east of the United States and the Amur basin, parts of Manchuria, and northern Japan. In the southern hemisphere the continents do not extend sufficiently far south or are not broad enough for this type of climate to develop. Some would class Patagonia as belonging to this group, but we have preferred to call it temperate desert.

Climatically, these eastern margin lands are regions of extreme conditions. They are much colder in winter and much hotter in summer than corresponding latitudes on west coats. In winter the weather is brisk, calm, and clear with temperatures well below freezing point; snowfall, sometimes of a heavy character, occurs. Summers are usually very warm and rather humid. Precipitation is moderate in amount, about 20–40 in. (508–1016 mm), and fairly evenly distributed, although there is a summer maximum. The eastern Asian area is drier in winter than its American counterpart.

In the north-eastern United States, the Maritime Provinces, Newfoundland, Quebec, and Ontario, the well-distributed, moderate rainfall has fostered dairying, alongside fruit-farming, market-gardening, and general mixed farming. There are still considerable tracts of forest, chiefly coniferous, but with some deciduous forest in the more southerly parts, which make lumbering an important occupation. In various areas of this region industrial development has taken place, aided by coal, water-power, and mineral resources. Off the coasts fishing is of especial importance, and the Grand Banks near Newfoundland is one of the world's great cod fisheries. Cold coastal currents sweep southwards and many harbours in eastern Canada (as in eastern Asia also) are frozen in winter.

Eastern Asia is much less highly developed than eastern Canada. There are many areas as yet only thinly peopled or even unpopulated. The timber and mineral resources have begun to be exploited and there is some rather patchy agricultural development. It would seem that the region offers considerable possibilities though these are only just beginning to be realised by the Chinese, Russians, and Japanese who control the area.

## *COLD CONTINENTAL REGIONS*

The boreal or coniferous forests (*taiga*) of sub-arctic latitudes have a cold continental type of climate. Summers are short with long days: winters the reverse. Only four to five months have temperatures higher than 6° C (43° F), the temperature below which plant life cannot grow. In summer, days may be warm, occasionally hot, especially in interior locations, but winters are long and cold with temperatures below freezing. Precipitation is generally scanty to moderate, usually between 12 and 30 in. (304–762 mm), and occurs chiefly in summer. In winter the ground is snow-covered. The weather in winter is cold, crisp, and calm with occasional blizzards; in summer conditions are usually warm and sunny and pleasant with convectional showers.

Millions upon millions of tightly-packed trees cover vast areas of Eurasia and North America and these boreal forests form one of the outstanding natural features of the earth's surface. The boreal forests are mainly evergreen coniferous forests but on their southern margins, where it is milder and wetter, they may be mixed with deciduous forest. The trees, mainly species of pine, fir, spruce, tamarack, hemlock, etc., show adaptations to cold, snow, and wind through their compact form, conical shape, needle leaves, downward-bending branches and cones instead of fruit. Though species may be mixed, the *taiga* commonly consists of immense colonies of the same species.

[*Courtesy: Finnish Embassy.*

FIG. 23.—CONIFEROUS FOREST OR TAIGA

This is a characteristic view in the Lake Plateau region of Finland. Finland, like Norway and Sweden and a wide belt in northern Russia, lies in the coniferous forest zone. These forests yield most of the world's supply of soft wood much in demand for constructional timber, wood pulp, and paper. The coniferous evergreen forests are vast and impressive but somewhat monotonous and dreary.

The forests provide shelter in winter and hence they form the home of deer, fur-bearing animals, bears and wolves. Along the southern margins of the forest human settlement has long taken place and there is a continuing tendency for man to colonise these southern fringes. The podsol soils, however, are acid and of limited agricultural value, but crops such as rye, potatoes, and flax can be grown and the planting of fodder-grasses has led to the rise of dairying. In the early days hunting and trapping were almost the only occupations of the forest lands, but in recent times lumbering (for softwood timber) and pulp manufacture have become important in districts which possess water-power and have access to ice-free seas.

In North America the boreal forests extend in a broad belt from Alaska to Newfoundland with a "tongue" projecting southwards along the Western Cordillera. In Eurasia they stretch in a vast belt from Scandinavia to Sakhalin and cover enormous areas in Siberia. The cold forests lie almost exclusively in the northern hemisphere. There is no land in the corresponding latitudes in the southern hemisphere and only in the extreme south of Chile is the coniferous forest really present.

[*Courtesy: Swedish Tourist Traffic Association.*

FIG. 24.—TUNDRA

These arctic and sub-arctic lands, snow-covered in winter and ablaze with many low-flowering plants in the brief summer, are of limited value to man. Often ice has swept the soil away to expose bare rock. Where soils exists the sub-soil is permanently frozen giving permafrost conditions. Reindeer can live off the mosses and lichens, hence some people are nomadic reindeer herders.

The coniferous forest lands, apart from their timber wealth and sporadic mineral wealth, offer little to man, and it seems likely that large areas will continue to remain in their natural state for a long time to come.

## *THE TUNDRA REGIONS*

The tundra lands are found on the continental fringes of polar margins. The arctic coastlands of Eurasia, Iceland, the so-called "barren lands" of

Canada, and much of Alaska fall within the tundra region. In the southern hemisphere, it is scarcely represented, though Tierra del Fuego, some of the islands in the Southern Ocean such as South Georgia, and even the Falkland Islands may be said to have tundra conditions.

The tundra lands might be described as having two seasons—day and night. During the long winter night the tundra lands are frost-bound for seven months. Climatic conditions during the winter period are fairly stable, though there may be some blizzards. In the short summer, temperatures rise 11° C (20° F) or more above freezing, and for several weeks it may be warm, but such warmth is largely due to the continuous daylight. Precipitation, coming in the form of snow and rain, is generally light or moderate, often under 10 in. (254 mm). Most of it comes in summer when the weather is unsettled and often windy; strong winds are a characteristic feature of the arctic coastlands.

The soil and sub-soil are permanently frozen in winter, but in summer the surface thaws out though the sub-soil remains frozen. This condition of permanent frost is known as permafrost. Where, as in northern Asia and Canada, great rivers flow polewards, extensive flooding occurs during the spring thaw, for the upper courses unfreeze before the outlets of the rivers. As a result vast swamps and morasses are formed which provide ideal breeding grounds for mosquitoes.

Clearly the climatic conditions are not conducive to vegetation growth. From the point of view of plant life these regions are sometimes called lands of physiological drought in contrast to physical drought. In other words, plants are starved of moisture because water is in the form of ice which is not assimilable. Plants which do grow are of low habit. Mosses and lichens are typical, the former growing in the damper, the latter in the drier, areas. Ground plants such as bilberry, cranberry, and yellow cloudberry, together with grasses, grow in the summer period, and there is also a sudden, wonderful but ephemeral growth of myriads of small flowering plants. Few trees grow, but here and there stunted dwarf willows, alders and birches are found which grow along the ground rather than upwards.

The berry-bearing bushes provide food for large flocks of migrant birds in summer. Large herds of caribou and musk oxen in North America move from the shelter of the forests to the south to these arctic pastures in summer. Apart from the nomadic reindeer herders, the tundra lands of Eurasia have few people, but certain stretches of coast and some river banks have permanent settlements where fishermen and hunters and trappers live. In a few localities mineral exploitation has led to the setting up of "outposts" of modern life, *e.g.* Spitzbergen, Vorkuta in Russia, and Port Radium in Canada.

EXERCISES

1. What difficulties impede the exploitation of Equatorial lands? (*Royal Society of Arts.*)

2. Outline the climatic conditions found in the Tropical Grasslands and elaborate for one of the main areas the probable or possible lines of economic development.

3. Write an account of the "Cotton Belt" of the United States of America, describing its position, climate, farming, and industries. (*Chartered Institute of Secretaries.*)

4. What are the main geographical factors which cause the Hot Deserts of the world? In what ways has man been able to develop these areas economically?

5. What is a tropical monsoon climate? How is it caused, where is it found, and what are the typical products of regions having this type of climate?

6. Describe the vegetation found in (*a*) the Amazonian basin, (*b*) Siberia? What factors handicap the fuller utilisation of these two areas by man? (*Northern Counties Technical Examinations Council.*)

7. Outline the distribution of the world's desert areas. To what extent have they been developed and of what potential value are they? (*Northern Counties Technical Examinations Council.*)

8. Describe and account for the major temperate grassland areas of the world. Assess their economic importance at the present time.

9. What conditions characterise a Mediterranean climate? Where would you expect to find such conditions and why? (*Institute of Transport.*)

10. Compare the economic development of the three main Equatorial regions of the world. Try to account for the variations noted.

11. Give an account of the coniferous forest regions of the world under the following headings: (*a*) location, (*b*) climatic conditions, (*c*) natural vegetation, (*d*) economic usefulness and development.

12. What are the chief characteristics of the Mediterranean type of climate, in what areas is the type to be found, and what influence does it have on economic activities in these areas?

# Part Two

# FARMING AND FOODSTUFFS

## *Chapter VII*

## FARMING AND FOOD SUPPLIES

### *FEEDING THE WORLD'S PEOPLE*

As we have already seen, the world's population is growing at an alarming rate—so quickly that this growth has been termed an "explosion." Indeed, if present-day trends continue, the 3600 million people living in the world today will more than double in a hundred years. Many people believe that this very rapid growth of the world's people is *the* most important problem of our time. The problem which this growth presents resolves itself into two parts: (*a*) will there be enough living space for the population of the future, and (*b*) will the world be able to feed all the additional mouths?

The first of these problems is the least troublesome. It has been calculated that the world's people—all 3600 million of them—could be found standing room on the Isle of Wight! Clearly, then, there is still plenty of room on this planet of ours although, of course, mankind would not want to live like battery hens! And, if the worst happened, it is very likely that man would expand over the water, building vast rafts upon which he would permanently live much as the floating populations of Canton live at the present time.

The second problem is much more serious. It is true that this threat of over-population and a shortage of food has been raised in the past. Thomas Malthus writing nearly two centuries ago was convinced that world population was out-running the available food supplies. But the opening up of the "new lands" of the world and the production of grain and meat on the extensive grasslands of these "new lands" came to the rescue. Others, since Malthus' day, have raised the same warning time and again. Today, however, the threat does seem to be a very real one and eminent scholars, such as Sir John Boyd Orr, Sir John Russell, and the American, P. R. Ehrlich have warned us that we should not be complacent about it.

Accepting, for argument's sake, the seriousness of the problem, the question arises: how are all these future millions to be fed? How can the world's food supplies be expanded to meet a growth in the world's population running at over 50 million a year? The problem is, in fact, more serious than this, for already it is estimated some two-thirds of the present world population are under-fed, which means that they are suffering from either under-nourishment or malnutrition. Thus the problem is one not merely of finding additional food for unborn babies but of feeding adequately some 2000 million who are alive today.

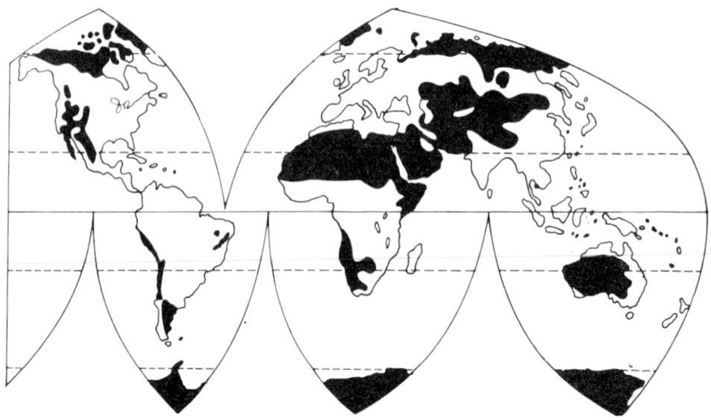

FIG. 25.—ARID AREAS

This map shows the areas of the world where the annual precipitation is too small to permit close agricultural settlement without irrigation. Roughly speaking, the areas in black have less than 10 in. (254 mm) of precipitation in middle or high latitudes and less than 20 in. (508 mm) in the tropics. Approximately one-third of the earth's surface is too dry for agriculture; a large proportion of this third is even too dry for pastoralism.

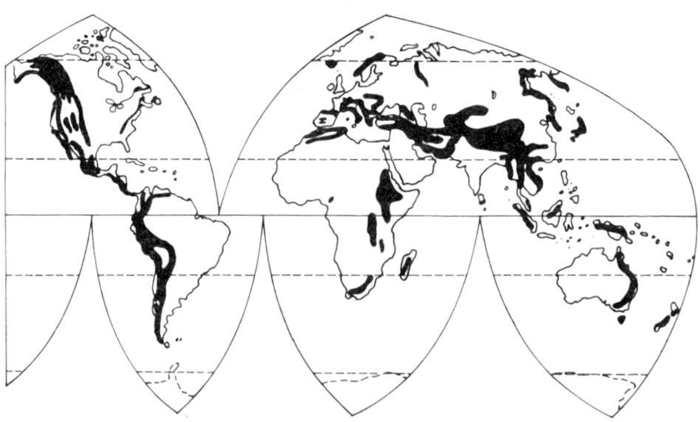

FIG. 26.—MOUNTAINOUS AREAS

The areas shown in black are the chief mountainous regions of the earth's surface. The rugged relief or the extremely elevated character of these areas prohibits close settlement. Some of these areas are in fact very thinly peopled. Areas above the snow-line are, of course, uninhabited.

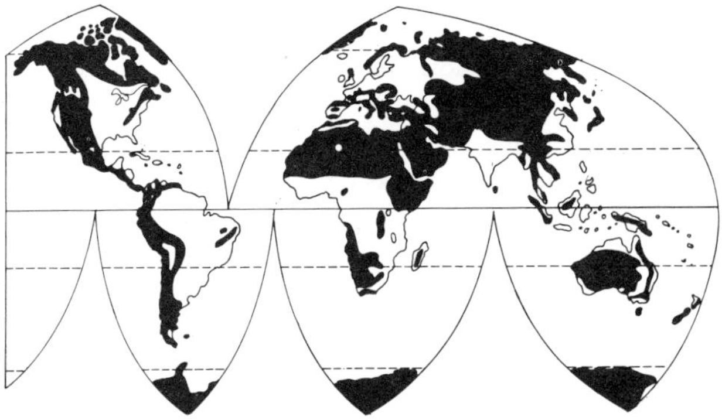

FIG. 27.—TOTAL NEGATIVE AREAS

This map combines Figs. 25 and 26 and also takes into account additional areas which are too cold for cultivation. Those parts of the earth's surface which through cold, aridity, or mountainous character make close settlement and agricultural development impossible may be described as "negative" areas. Note that the map does not take into account irrigated areas where, as for example, in Egypt, Iraq, Peru, and parts of western U.S.A., the negative nature of the land has been counteracted.

The blank areas may be termed the "positive" areas, *i.e.* those which are cultivated or are potentially cultivable. It is clear from the map that there are extensive areas in Africa and South America which at present are under tropical forest or grassland and very scantily settled but which one presumes are capable of being developed, although in many areas at considerable effort, difficulty, and cost. Much of northern Canada and the Soviet Union faces kindred problems.

Out of the earth's total habitable land area of 52 million square miles (130 million km$^2$), it has been estimated that only some 16 million square miles (60 million km$^2$) are cultivable. Of this area which is capable of being cultivated, a mere $5\frac{1}{2}$ million square miles (13 million km$^2$) are actually cultivated. It has been calculated that $2\frac{1}{2}$ acres (1 ha) of cultivated land are required to support each individual with an adequate diet. The present $5\frac{1}{2}$ million square miles (13 million km$^2$), however, gives only $1\frac{1}{2}$ acres (0·6 ha) per head—clearly an inadequate amount of land to provide even a minimum adequate diet. Were all the cultivable land used, there would be rather more than 4 acres (1·6 ha) per head of the world's total population at present. This suggests that there is more than enough land available to meet the food demands of the world's present population if the land was fully and properly used. We have to bear in mind, however, that the world's population is increasing very rapidly and that by A.D. 2000 the present-day numbers may have doubled. By A.D. 2000, therefore, the amount of cultivated land will have dropped below the necessary $2\frac{1}{2}$ acres (1 ha) per head which we have said is necessary to support each individual adequately.

In an attempt to handle this problem a World Freedom from Hunger Campaign was introduced by the United Nations Food and Agricultural Organisation (F.A.O. for short) in 1960 which ran for five years. Its objects were to raise enough money to pay for the extension of the world's food-producing areas, to teach farmers how to improve the yields of crops and animal products on existing agricultural land, and to educate backward people in the preparation, preservation, and value of different kinds of food.

F.A.O. experts have been guiding and helping farmers and fishers in Central and South America, in Africa, and in Monsoon Asia. Much valuable work has been done, but the problem has not been solved; it has merely begun to be tackled. There is much more to be done before we can even claim to have got a measure of the situation.

## *LAND AVAILABILITY AND USE*

Only 29% of the earth's surface is dry land; the remaining 71% is taken up by water surfaces which are of limited use to man, especially from the point of view of food production. Unfortunately, of the world's total land area, only a very small proportion is capable of being used agriculturally. Approximately one-fifth of the land surface is under permanent ice or too cold to grow crops; another fifth is desert, which is too dry for crop-growing (Fig. 25); a further fifth is composed of highland, too elevated, too rugged, or lacking in soil cover to be used for cultivation (Fig. 26); and yet a further fifth is covered with forest and marsh. Some of the last could be reclaimed for agriculture it is true, but there are large areas which could never be used for farming.

At the present time only about one-fifth is available for agricultural use; in hard facts this means that each person in the world has to be fed from the resources of about one acre (0·4 ha) of cultivated land and two acres (0·8 ha) of pasture. Surprisingly little, then, of the world's surface is given over to agriculture and, unfortunately, in many countries —as in England—large areas of this limited agricultural land are being lost each year as towns grow, more factories are built, and improved communications materialise.

Of the one-fifth (20%) of the earth's land area that is available for agricultural use, only about half of it (*i.e.* 10% of the total land area) is actually producing food at present. By and large, the best land is already being used. Experts believe that probably as much as 25%, or even 30%, of the total land area could produce food, providing the difficulties of farming in what are now tropical forest areas and tropical savanna areas could be overcome. This frequently demands careful research and almost always very large capital expenditure—but the possibilities are there and it could be done.

### *WAYS OF INCREASING FOOD PRODUCTION*

The problem of hunger could be conquered by the extension of food-producing acreages and by increasing agricultural yields.

The world's food producing area could be increased by bringing more land into cultivation and this could be done by: (*a*) extending and improving irrigation facilities in some of the arid areas of the world; (*b*) draining swamplands and improving flood-control measures in certain wet areas; (*c*) clearing some forested areas, especially tropical forest areas, for the cultivation of crops.

Although the additional acreages resulting from such measures would be valuable, in total they would increase the cultivable area by only a slight amount. Making "two blades of grass grow where one grew before" seems to offer a better prospect for solving the problem; in other words, increased yields rather than increased areas would appear to produce the more significant results. Increased production along these lines could be achieved by: (*a*) increasing the output and use of fertilisers; (*b*) using better seeds and improving animal strains; (*c*) introducing and improving crop rotation practices; (*d*) using more efficient farm tools and machinery; (*e*) extending the use of insecticides to control pests; (*f*) controlling soil erosion and adopting soil conservation methods; (*g*) educational training aimed at farming improvement; (*h*) improving health and thereby labour efficiency.

There is much room for better farming on land that is already in use. Over vast areas, especially in the tropical regions, farming is still carried on by backward and wasteful methods. The great majority of the world's farmers are subsistence farmers merely growing sufficient (but sometimes even less than enough) for their own needs and living almost entirely on what they produce. They need to learn how to use their land to the best advantage; they need instruction in new and better farming techniques; they require training in the use of fertilisers; they need guidance on storage and marketing; and they need better seeds and more efficient agricultural implements. Training for improvement is clearly necessary. Farm schools, agricultural colleges, demonstration farms, research centres, pilot projects of all kinds are needed in increased numbers. It has been shown that training centres and experimental centres of these kinds can bring about remarkable increases in production, but there are still far too few of them. All this, of course, costs money and many countries in the world are too poor to finance projects of this kind; this means that the richer countries must be prepared to play the role of the benevolent uncle.

## *NEW SOURCES OF FOOD*

In addition to the much more effective use of the land, greater use could be made of the sea and, also, of inland waters. Although fishing has long taken place in many areas and considerable quantities of fish are caught each year, the sea remains a much neglected resource. Man still hunts for fish instead of "farming" fish. Some peoples in the East, notably the Chinese, have long farmed fish, *i.e.* they breed fish in ponds, lakes, canals, and even in the paddy fields. Clearly, this technique of fish breeding could be extended to the sea, or at least certain parts of it; such large-scale fish farming could increase enormously the output of food. It is both interesting and heartening to read that experimental work of this kind is already being undertaken. Developments along these lines have already taken place in Africa. Fish have been introduced into Lake Victoria and into the new artificial lake created by the building of the Kariba Dam. Both these great lakes are now being developed as inland fisheries and the fish catch will provide a valuable adjunct to the diet of the local peoples.

Another possible source of food is cropping wild animals. The vast savanna lands of Africa are the haunt of large numbers of herbivorous creatures, antelope, gnu, zebra, and the like, and a scheme is being studied whereby the more thorough protection of these animals could be combined with their provision of additional supplies of food. There seems to be no reason why the conservation of wild life should not be linked with food production.

There are many plants in the plant kingdom which, in all likelihood, could be improved, developed and used as sources of food supply. But, it is also possible to process vegetable materials so that they could be made edible. For example, we use many plants as sources of vegetable oil: the seeds or fruits of the sunflower, cotton bush, groundnut and soya plants all yield valuable oil. The vegetable residue left after the oil has been expressed is then used either for making animal feedstuffs or organic fertiliser or is simply discarded as waste. This residue, however, is rich in protein and is potentially valuable as a human foodstuff. At present this residual vegetable material is frequently rendered unfit for human consumption through overheating during the soil expressing process, or because it is contaminated. But it could be made edible and it has been estimated that this residual matter would yield twenty million tons of protein or roughly double the world's present estimated deficit of protein. Experiments to turn this vegetable matter into edible, palatable and acceptable food are already well under way and, in fact, some processed vegetable food of this kind has already been marketed.

## *THE PROBLEM OF DIET AND THE DISTRIBUTION OF FOOD*

In some parts of the world malnutrition is due rather to unbalanced diets and badly prepared food than to actual shortages. Because the value of proteins is not understood by backward peoples their diet is frequently grossly unbalanced. As a result they suffer from deficiency ailments and diseases and their general health is impaired. And this, in turn, reacts upon their ability to work and to work efficiently. The whole thing is a vicious circle: because the people are under-fed or ill-fed, they have no reserves of physical strength; so they are more easily weakened by disease and become subject to chronic lassitude; lacking energy, they cannot labour efficiently and agricultural productivity suffers; inadequate production completes the circle for the people must then go hungry. Breaking the vicious circle is no easy task, but it must be done, for it is precisely in those parts of the world, *e.g.* in Africa, South-west Asia, Monsoon Asia, and Latin America, where people are under-fed, that increased food production is failing to keep up with the annual increase in population. Unless something can be done to change these conditions famine and disaster are likely to be the ultimate result.

Although millions of people in different parts of the world are crying out for food, we frequently read of huge surpluses elsewhere, perhaps of grain in North America, or coffee in Brazil, or even of fish in England. The question inevitably raises itself: why cannot these surpluses be used to feed the hungry? They could; but often there are certain difficulties in the way. Although a particular country might be generous and give some of its surplus away—as the United States has often done—a country, like an individual, must earn its keep, and it does this by selling its goods. Poor countries often cannot afford to buy such goods, even if they are desperately in need of food. Gluts, such as those mentioned above, arise as a result of gross over-production and because producers cannot sell their products at a satisfactory price. While it may pay some countries to give away part of their produce (in order to stimulate continued production in that country) there is, clearly, a limit to which this sort of thing can be carried on.

There is another aspect of this over-production to which we may draw attention. Although these surpluses may appear huge, measured against the total need the quantity is almost negligible. Again, the world surpluses are usually of cereals, *e.g.* wheat, maize, and it is not grain that is needed, but rather foodstuffs such as milk, meat, and fish which are rich in protein. The problem is also further aggravated by the fact that grains such as wheat and maize are not those to which the under-fed are accustomed, and food habits are not easy to change, even when people are hungry.

Let us conclude this brief account of the problem of population and food supplies by saying that the question of feeding coming generations satisfactorily is not an intractable one, even though it may present many difficulties. If, for example, all the world's farmers were able to raise their standards to those reached in parts of England or Holland or Denmark, the present world area which is in agricultural occupation could support many times its present population. Progress is now being made on many fronts and great efforts are being made by many peoples but a quickening in the rate of progress and even greater efforts are urgently needed.

## EXERCISES

1. The world's population is growing at the rate of about 150,000 a day. But the increase is very unevenly distributed over the continents: what are the major problems of population and food supply facing the world today?
2. In view of the rapid growth in world population and the need to feed the extra numbers, what ways and means are there of increasing the world output of food?
3. Write a geographical essay on Food and Population in the Monsoon Lands of Asia.
4. "Some two-thirds of the present world population are under-fed which means that they are suffering from either under-nourishment or malnutrition." Elaborate upon this statement.
5. Write explanatory notes on the following: Thomas Malthus; F.A.O.; fish-farming; population explosion.

# Chapter VIII

# FACTORS AFFECTING FARMING

SOMETHING like three-quarters of the world's population are engaged in farming, either the growing of crops or the rearing of animals. Agriculture is still by far the most important occupation in the world as a whole. Although here and there mining, manufacturing, lumbering, fishing, or some other activity may be more significant and more important, by and large farming transcends all other activities. Even in countries such as England, which are predominantly industrial, farming is a major activity; and even in Norway, where conditions for agriculture are most unpromising and where the fishing and lumbering industries loom large, farming is the most important single activity. Because farming is of such importance in the world's economy a considerable section of this book will be devoted to it—to the factors which affect agriculture, to the wide variety of crops that are grown, and to the animals that are reared and their products.

If a number of photographs of non-industrial occupied areas are studied, such as those in Figs. 36 and 58, the most noticeable features of the landscape are those pertaining to man's agricultural activities, *e.g.* the field patterns, the walls and hedges, the isolated farmsteads, the signs of tillage, the growing crops, the domesticated animals, etc. The way in which the land is used, *i.e.* the agricultural land-use, varies from country to country, from region to region, and maybe from district to district. One of the most interesting aspects of field study work is to find out why a particular kind of farming is carried on in a particular area or why a particular crop is grown. For instance, one of the most interesting questions the writer knows is: why is something like 50% of all the rhubarb grown in Britain cultivated within a very small and restricted area within the triangle made by Bradford, Leeds, and Wakefield? This is a problem you might try to solve for yourselves. Sometimes a particular type of soil, such as a heavy, moist clay or a light, porous sandy soil, may produce the answer as to why a given crop is grown; sometimes an economic, rather than a strictly physical, factor, such as a local demand from a centre of population or a factory, may be responsible.

### *AGRICULTURAL INFLUENCES*

Agriculture in any area—the degree to which it is developed, the type of farming pursued, the crops that are cultivated, etc.—is dependent upon a combination of factors and conditions; the most important of these are enumerated below.

1. *Climate*. Temperature, the occurrence of frosts, rainfall, and sunshine may be said to be fundamental conditions controlling farming,

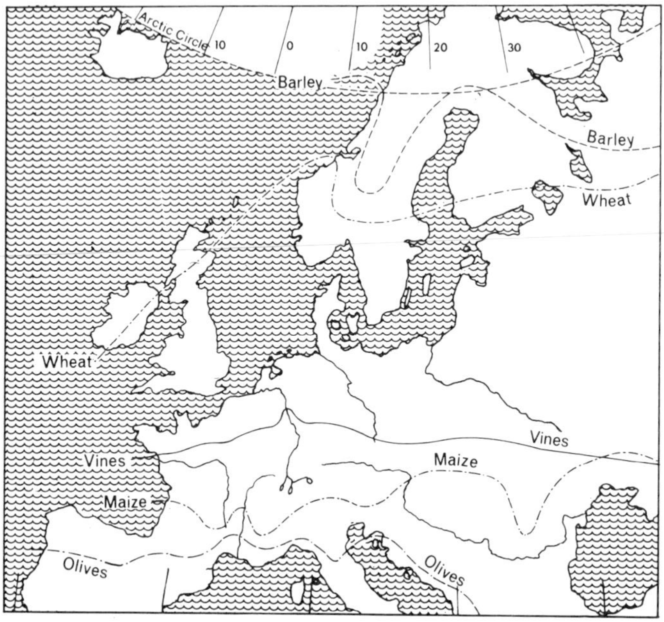


FIG. 28.—CROP LIMITS, EUROPE

The July isotherm of 10° C (50° F) may be regarded as marking the northern limit of possible cultivation. Almost the whole of Europe lies within this critical limit. But different crops require different conditions of temperature and rainfall for their growth. The lines on the map indicate the poleward limits of selected crops of importance. Man can only overcome these natural limits in two ways: (*a*) by developing plant strains which will grow quicker, thus enabling him to push the cultivation of specific crops northwards, though of course only slightly polewards; and (*b*) by growing crops under artificial conditions such as in greenhouses; but while grapes, tomatoes, and salad crops may be cultivated under such specialised conditions (as in fact they are in Iceland), clearly cereals are incapable of being grown in this way.

especially the growing of particular crops. Temperature sets the northern limit of cultivation of the various cereals, the occurrence of frosts restricts the growth of many crops such as fruits, coffee, and cotton, while the amount of rainfall will decide whether cereal cultivation or grass and fodder crops will be predominant.

[*Courtesy: French Embassy.*

FIG. 29.—TERRACING

Typical Mediterranean coastland: the steep hill slopes have been terraced to make them cultivable and help preserve the thin soil from erosion by heavy rains. (*See* Figs. 7 and 41.)

2. *Soil.* Soils vary widely in their structure, texture, workability and fertility. Heavy wet clays are not suited to arable farming for they are difficult to plough and often sour; hence they are commonly laid down to grass. Light, porous sandy soils which warm up easily are well suited to, and are likely to encourage, market-gardening. Some crops demand or prefer specific soil conditions, such as iron or lime, good drainage or waterlogged soils.

3. *Topography.* Surface relief exerts an obvious influence: large-scale cultivation is practically confined to lowland areas since upland areas are normally unsuited to large-scale tillage. In hilly areas the land, if it is to be used, may have to be terraced (*see* Fig. 29). Hilly country, moreover, makes the use of machinery impracticable and hence it is often necessary to resort to hand tillage. Mountainous terrain favours the practice of transhumance.

4. *Methods of cultivation.* Farming methods and techniques of cultivation vary widely throughout the world, ranging from primitive shifting methods to highly intensive scientific methods. Differences between primitive and advanced farming techniques are related mainly to crop rotation systems, the use of manures and fertilisers, the degree of mechanisation, the use of irrigation, and the selective breeding of animals. Methods of cultivation are closely related to productivity, *i.e.* yield of crops or produce.

5. *Organisation of labour.* By this is meant the manner in which the land is held and the way in which the available labour supply is organised. The peasant small-holding, the individually owned farm, the *metayage* or share-cropping system, the co-operative farming system, the great estate employing hired labour, and the communist collective farm are all variants of the organisational factor. The type of organisation is of great significance in farming since it influences appreciably such matters as the efficiency of farming and land care and improvement.

6. *Biological factors.* Farming is more seriously affected and limited by biological factors than is commonly realised. In Britain these are seldom of great moment and it is only when there is a serious outbreak of foot-and-mouth disease or something similar that we realise the importance of the biological factor. Plant diseases and blights, such as wheat rust, sigatoka which affects bananas, swollen shoot disease which attacks the cacao tree, and pests, such as the locust, tsetse fly, phylloxera, and the Colorado beetle and boll weevil may be controlling factors.

7. *Social customs.* Social customs, including religious influences, may exercise a strong influence upon agriculture. For instance, the practice among many peoples of sub-dividing land upon the death of the owner has led to the excessive uneconomic fragmentation of the land. The

social prestige of cattle in the African savanna lands, the Hindu veneration of cattle, and the Moslem attitude towards the pig are well-known examples of the effect which social customs may have upon animal husbandry.

8. *Markets.* Unless production is purely for subsistence purposes, agriculture is closely geared to market conditions. A shortage of a commodity or, alternatively, a glut of a commodity will react upon production. The large and dependable markets for food products in Western Europe have exerted a very strong influence upon agriculture in this region. Over-production of, and dependence upon, a particular commodity, *e.g.* coffee in Brazil, may have profound effects upon the farm economy.

9. *Political factors.* Agricultural production throughout the world increasingly has come to be influenced by political factors. Governments have introduced tariffs, bounties, subsidies, etc., to persuade or compel farmers to produce commodities in accordance with national needs. During the inter-war period the growth of economic nationalism and the desire for political security led many European countries to strive for agricultural self-sufficiency.

It will be clear that there are numerous factors at work influencing agriculture. The relative importance of these factors varies from place to place but all, to a greater or lesser degree, play some part. Find out the answers to the following questions; they will help you understand better the effect of the above factors and conditions:

1. Why is rice replaced by millets in the Monsoon lands where the rainfall is below 40 in. (1016 mm)?
2. Why is maize in the United States limited on the north by the 21° C (70° F) summer isotherm?
3. Why is coffee no longer cultivated in Ceylon when, formerly, it was an important crop?
4. Why have large areas of the East African savannas not been developed as cattle areas when in many respects they are well-suited to cattle?
5. Why are the rich alluvial soils of the polder lands of Holland often devoted to cattle farming?
6. Why have the banana plantations of Central America been moved from the Caribbean to the Pacific coastlands?
7. Why have the Hindus failed to introduce restrictive breeding among the tens of millions of cattle found in India?
8. Why has the district around Sandy in Bedfordshire become an important area of market-gardening?

9. Why was the acreage devoted to sugar-beet in England substantially increased during the years 1939–45?

10. Why in recent years has the French government initiated a policy of consolidating the peasants' plots?

11. Why has a highly mechanised type of farming come to be characteristic of the steppes of the Soviet Union?

12. Why is the yield of wheat in Denmark among the highest in the world but relatively low in Spain?

13. Cotton requires an annual rainfall of between 20 and 40 in. (508–1016 mm) but is grown in large quantities in the dry Punjab—how and why?

## *LIMITS OF CROP PRODUCTION*

As we have just noted, there are many factors affecting agriculture in a given area; let us now look at the conditions which determine specific crop production in an area.

For any crop it is possible to recognise three different limits of production: (*a*) *the geographical limit*, (*b*) *the economic limit*, and (*c*) *the actual limit.*

Certain geographical conditions, the most important of which is climate, are necessary for the growth of any particular crop. Climate is usually the predominating factor, for temperature, frost, and moisture closely control plant growth. Every plant has its own specific climatic limits beyond which it cannot thrive. A case in point is cotton cultivation which, in the United States, is limited on the north by the 200 consecutive frost-free days line and by the 25° C (77° F) isotherm for the three summer months, on the west by the 23-in. (584-mm) annual rainfall isohyet, and on the south and east by the 45-in. (1143-mm) rainfall line, for amounts in excess of this figure can ruin the crop. These may be termed the absolute climatic limits of the crop. Sometimes crops require particular soil conditions and these may confine it to more limited areas within its possible climatic limits. The particular limits of crop production imposed by climate, soil, or terrain form the geographical limits of production.

Often, for economic reasons, it is impossible to grow a crop (or to rear animals) throughout an area which may possess suitable geographical conditions. For example, the plant yield may be insufficiently great to justify its cultivation; the type of labour required may not be available; adequate transport facilities may be lacking; competition from another crop, perhaps of greater value, may be a limiting factor; and so on. An illustration of this economic factor is provided by the Ford rubber plantation in Amazonia; here a planned output demanded a labour force of 80,000 but the company could never muster more than about 25,000 workers. This inability to secure a sufficient and adequate labour supply was one of the factors contributing to the collapse of the Ford venture. It is important to note that the economic limit of production may change in response to changing economic conditions. Increased demand may lead to the expansion of the economic frontier much as the exigencies of the Second World War led to much marginal land in England being brought into use. Economic factors, then, set the limits within which production becomes a feasible economic proposition at a particular time.

Finally, we have to note that, though the geographical conditions may be suitable and though production may be economically feasible, the actual limits of production may be more narrowly demarcated. This may be illustrated by the Nigerian cacao industry. There is a wide belt of equatorial rain forest in southern Nigeria with 45–80 in. (1143–2032 mm) of rainfall and high temperatures which permit the cultivation of the cacao tree while economic production is feasible throughout much of this zone, yet the actual present-day limits of production are relatively small, the cacao-growing area being centred upon the city of Ibadan. Hence, in considering the distribution of cultivated plants we should bear in mind these three possible limits of production: the geographical, economic, and actual limits.

## *IRRIGATION*

Crop cultivation is closely limited by the natural geographical conditions, more especially by climatic conditions. Man has attempted to overcome the restrictions imposed by the geographical environment by creating "artificial soils," *i.e.* by mixing soil and adding fertilisers, and creating "artificial climates," *i.e.* by using glasshouses and by watering the land. He has achieved his greatest success in overcoming the deficiencies of the natural rainfall, although he cannot claim to have overcome this problem of shortage of moisture except in a limited way.

The application of artificial water supplies to the land to remedy the deficiency of the rainfall is known as irrigation. It has been practised for thousands of years but the great developments have occurred during the past hundred years and especially during the past twenty years. Irrigation these days is practised under three differing conditions:

1. In arid areas where there is an absolute deficiency of rainfall, *e.g.* in Egypt and coastal Peru, and where cultivation would be impossible but for irrigation.
2. In areas where the rainfall is fickle or variable both in amount and incidence as in the Monsoon Lands or seasonally short as in the Mediterranean Lands.

3. In areas where there is no real shortage of rainfall but where additional supplies of water are likely to increase both the quality and the yields of crops, as in eastern United States and south-eastern England.

During recent years perhaps the most notable development has been in the third category, although the arid and semi-arid areas served by irrigation have increased enormously.

The value and importance of irrigation becomes obvious when it is realised that fully a quarter of the earth's surface receives under 10 in. (254 mm) of rainfall annually, and a further third receives between 10 and 20 in. (254–508 mm). Where the precipitation is below 10 in. (254 mm) annually, irrigation is a necessity for the successful growing of crops, unless dry-farming methods are adopted. Even where the rainfall is as much as 20 in. (508 mm) or still higher, irrigation may be necessary if the rainfall is chancy or if temperatures are high and there is a high evaporation rate.

Let us now list, briefly, the main advantages and disadvantages of irrigation.

## ADVANTAGES

Due to the dry conditions the soil is often rich in plant foods since the minerals have not been leached or washed out of the soil.

If irrigation is by natural flood-water, the soil is constantly being re-fertilised by the deposited mud which brings new mineral constituents to the land.

The supply of water is assured and under complete control; hence the cultivator can apply water to his land when he likes and in the amounts he desires.

There is no interference with the supply of sunshine; so plant growth is seldom hindered by cloudy weather.

If temperatures are high enough, cultivation becomes possible all the year round, enabling two, three or four crops to be grown in succession.

When carefully carried out, irrigation helps to give a greater measure of freedom from plant diseases and insect pests which molest crops.

While the practice of irrigation brings many advantages to cultivation, it may also result in a number of disadvantages.

## DISADVANTAGES

Unless water is applied with discretion the soil is likely to become water-logged and sour (as has happened in parts of the Punjab and in the Nile delta).

Undesirable mineral salts are apt to accumulate in the upper layers of the soil in dry regions and this may lead eventually to the formation of saline encrustations.

Since plants secure their water supply easily, root development may be retarded and, in consequence, plant growth may be checked.

Unless due care is taken and adequate drainage provided, stagnant surface water may provide a breeding ground for snails and insect pests which may serve as hosts for disease-spreading parasites.

## *METHODS OF IRRIGATION*

There are two principal methods of irrigation: annual inundation by natural flooding and perennial irrigation using stored water.

### ANNUAL INUNDATION

This is the older, more primitive, and simpler method of irrigation. It has been practised in Egypt, where it is known as "basin" irrigation, for several thousand years and was common in many other parts of the world; it was a method employed by the ancient peoples of Mesopotamia, by the Hindus, the Chinese, and the Indians of the dry lands of North America. Annual inundation merely involves the construction of earthen embankments along the river floodplain which are broken to receive the flood-water when the river overflows its banks in flood time, and plugged to entrap the water when the river flood begins to recede. This was the traditional method of irrigation followed in Egypt and it is still carried on in parts of Upper Egypt. Annual inundation has one great advantage over modern perennial irrigation: it allows the river to bring and deposit its suspended silt which annually enriches the soil. The chief disadvantage of this traditional method of irrigation is that it permits only one crop a year to be grown.

### PERENNIAL IRRIGATION

The term perennial implies that water is available for use throughout the year. The presence of wells may provide a continuous water supply but usually we associate perennial irrigation with dams and stored water. Most modern schemes involve the impounding and storage of natural flood-water behind barrages and dams. This stored water can be run off and distributed by means of canals and ditches whenever it is required and it is available the year round. This, in turn, enables crop-growing to be carried on throughout the year. Perennial irrigation has another great advantage: unlike basin irrigation, the cultivable area is not limited to the floodplain of the river. But perennial irrigation has two serious drawbacks: first, the supply of fertilising silt carried by the river flood-waters is withheld from the land and, secondly, dams, barrages, etc., are very costly to construct and maintain.

## THE APPLICATION OF WATER

Water may be applied to the land in three main ways: (*a*) by "flush" methods, *i.e.* allowing the natural flood-waters of the river to spill over on to the land; (*b*) by gravity flow, *i.e.* diverting stream-water on to terraces or sloping land which trickles down slope through the pull of gravity; and (*c*) by lifting water on to the land either by muscle-power or mechanical pump when the water surface lies below land level.

In the latter case, mechanical contrivances are normally used these days although in some parts of the world, notably in the under-developed countries, primitive water-raising techniques are still employed. In Egypt and the Middle East generally and also in the Indian sub-continent age-old methods of water-lifting are by *shaduf*, *tambour*, and *sakia*. The shaduf is a bucket suspended from a mounted spar which can be swivelled round; the tambour or Archimedean screw, so called after its inventor, Archimedes of Syracuse, consists of a revolving spiral in a cylindrical case; while the sakia, known as a *noria* in Syria, makes use of the principle of the water-wheel. In China, widespread use is made of the treadmill worked by man-power and of water-wheels turned by buffaloes.

In the United States and Western Europe the overhead movable sprinkler system is widely used, especially in those areas where irrigation is primarily of an auxiliary character.

## *IRRIGATED AREAS*

Irrigation is of greatest significance in the arid lands of the Near and Middle East. Egypt, for instance, relies almost entirely upon irrigation for its crops; the irrigated area of 9678 square miles (24,197 km$^2$) is the same as the cultivated area. The Aswan High Dam, now under construction, will add another two million acres (80,000 ha) to Egypt's cultivated area. The Sennar Dam, on the Blue Nile, provides water for the Gezira Plain of Sudan, which is a notable cotton-growing area. Irrigation is important in Israel, Lebanon, and Syria; noteworthy is the irrigated Ghuta or Plain of Damascus, covering approximately 150 square miles (375 km$^2$), which is a fertile garden based upon the waters of the Barada and its six tributaries. In the Tigris–Euphrates lowlands of Iraq modern schemes with large barrages at Hindiya, Daghgara, and Kut have been developed to expand the cultivated area in the fertile, but semi-arid, Mesopotamian alluvial plain.

Throughout the Mediterranean coastlands, where the rainfall comes in winter, *i.e.* the non-growing season for many crops, irrigation is widely practised. Except on the desert margins, where there is practically no

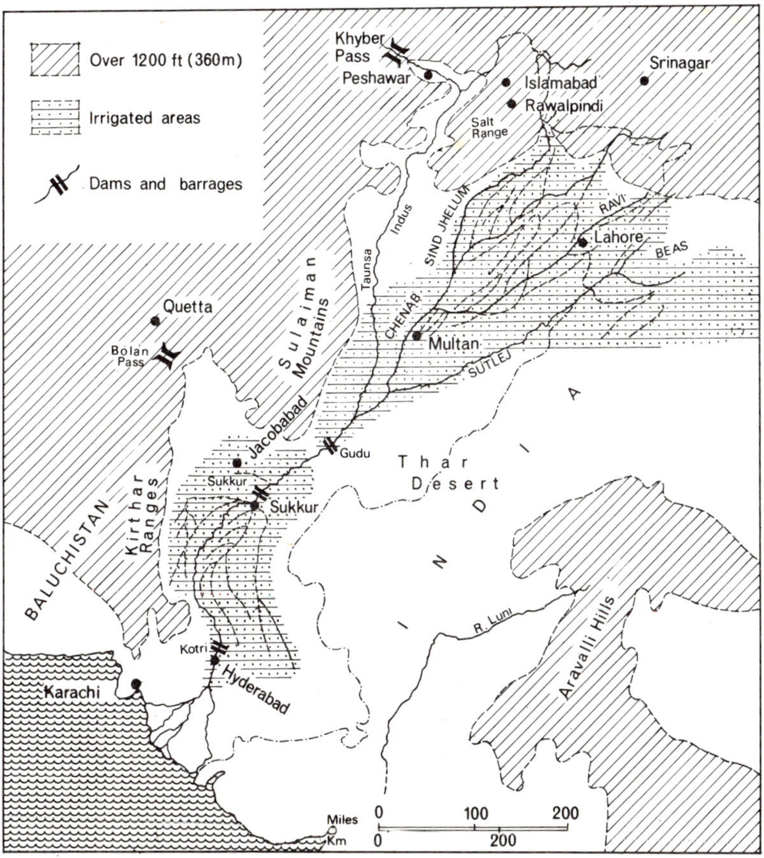


FIG. 30.—INDUS VALLEY IRRIGATION

Pakistan is a semi-arid land. It is dependent, like Egypt, upon a great river for its life and prosperity. Altogether Pakistan has 23 million acres (9·2 million ha) of irrigated land. A great network of canals feed the doabs (the interriverine areas) between the "five rivers"—the Jhelum, Chenab, Ravi, Beas, and Sutlej. Sind, occupying the lower, drier Indus Basin, is irrigated by the world-famous Sukkur Barrage which alone commands 7 million acres (280,000 ha). Among the major projects carried out so far are the Ghulam Mohammed Barrage, near Kotri, which irrigates $2\frac{3}{4}$ million acres (1·1 million ha) of virgin land; the Gudu Barrage, completed in 1963, which is designed to turn about $2\frac{3}{4}$ million arid acres (1·1 million ha) a fertile green; the Taunsa Barrage providing irrigation for about 1·4 million acres (0·5 million ha) of existing inundated land; the Rawal Dam providing 14 million gallons (63 million litres) of water daily to Rawalpindi (the former capital) and Islamabad (the new capital) and water for irrigating 8000 acres (3200 ha); and the Warsak Multi-Purpose Project which provides irrigation for 120,000 acres (48,000 ha) and generates 160,000 kW of electricity.

rainfall and cultivation is almost completely dependent upon artificial water supplies, irrigation in the Mediterranean region is mostly an adjunct

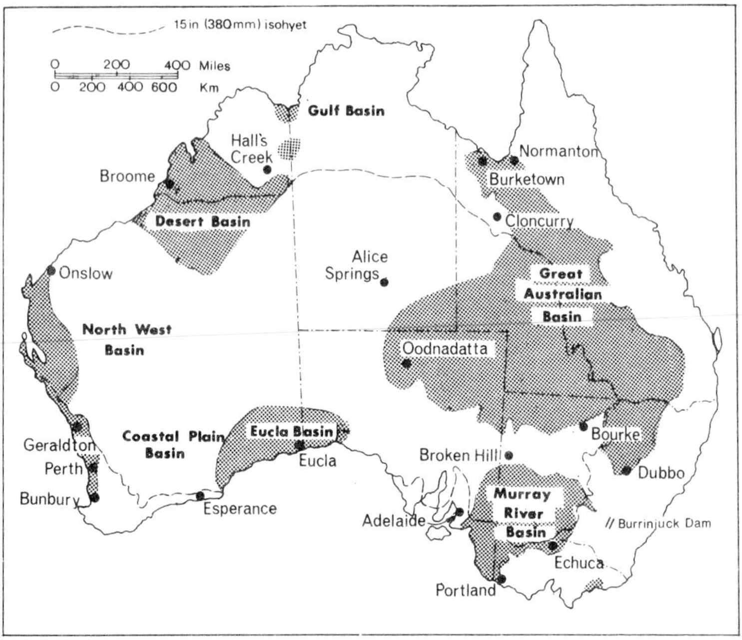


FIG. 31.—ARTESIAN BASINS, AUSTRALIA

An artesian basin (the names derives from the province of Artois in France) is an underground syncline in which ground water collects. Artesian water may collect, however, in many forms of "traps" in the rocks. Australia, "the dry continent" is fortunate in having below much of its area where the rainfall is short or marginal—the 15-in. (381-mm) isohyet suggests the area in which surface water is deficient—supplies of artesian water (underground water that gushes to the surface freely) and sub-artesian water (water that has to be pumped up). With the help of artesian water supplies many areas have been opened up for grazing. Thousands of wells have been bored. However, we should note that artesian water has its limitations: it is expensive to tap; it is often slightly saline and so unsuitable for irrigation; and frequently the quantity of water is limited and the output declines with use.

to the rainfall although cultivation on the Mediterranean coastlands of Spain, in Aragon, in the Plain of Lombardy, in parts of peninsular Italy, and in much of Greece and parts of the Aegean coastlands of Turkey leans heavily upon irrigation.

India and Pakistan are the countries par excellence of irrigation (*see* Fig. 30). China has a very large area under irrigation; much of it is served by traditional systems, but recently, more especially in the northern parts of the country, huge works have been undertaken. The great Hwang-ho scheme, which aims at taming completely this mighty river, will provide water for many additional millions of acres as well as yielding vast amounts of hydro-electric power. Already China claims to have the greatest irrigated area in the world. In interior Asia, both in Chinese and Soviet Turkestan, considerable developments have taken, and are taking, place; large quantities of irrigated crops such as cotton and fruit are produced.

Australia is "the dry continent" and cries out for irrigation; unfortunately, only a small fraction of it can ever hope to be irrigated for there is an acute overall shortage of water. In the Murray–Darling Basin the rivers supply water for the irrigation of wheat, fruit, and fodder crops. The Snowy River Scheme, which has involved the re-direction of the River, has been undertaken to augment the water supplies of the upper Murray River. In Queensland much reliance is placed on artesian and sub-artesian supplies of water; this water is useful for watering animals but unsuitable for cultivation. For the most part New Zealand is more than adequately watered, but in one or two of the drier valleys of South Island fruit is grown by irrigation.

We have already referred to irrigation in the Nile valley but there are numerous other relatively small-scale schemes dotted throughout the arid and semi-arid zones of Africa—in South Africa, at Sansanding on the Niger, in Morocco, Algeria, and Tunisia, and in the Saharan oases.

In South America there is irrigation in the Central Valley of Chile and in coastal Peru—"the land of little Egypts"—where the Andean streams can be used to provide water for the fruit, sugar, and cotton crops. The productivity of the oasis towns, *e.g.* Jujuy, Mendoza, Tucuman, of western Argentina is dependent upon the Andean streams. In the dry north-east shoulder of Brazil numerous dams provide water supplies, and in an effort to solve the tragic problem of this drought-stricken, drought-handicapped region the government is planning large-scale water conservancy schemes. Again, much of the Mexican Plateau is semi-arid and is dependent upon artificial water supplies.

In the United States irrigation is highly developed, especially in the western half of the country, and many of the world's most famous dams, *e.g.* Hoover, Shasta, Grand Coulee, are to be found in the Western states. The position is summarised in Fig. 32.

## *ANIMAL HUSBANDRY*

So far we have concerned ourselves with the cultivation of crops or arable farming. Let us now look at the other main branch of agriculture, animal husbandry, and note the chief factors affecting pastoralism.

FIG. 32.—IRRIGATION IN THE UNITED STATES

Much of the western half of the United States is dominated by semi-arid or arid conditions. The intermontane plateaus and basins of the Western Cordillera usually have less than 20 in. (508 mm) of rainfall and this may be erratic. Southern California, with its Mediterranean type of climate, has rainfall in winter but the summer growing season is dry. Parts of the south, *e.g.* Arizona, are desert. In order to overcome the great handicap of aridity, numerous great dams and irrigation projects have been completed. Already there are more than fifty projects in operation and many more are planned. Several million acres are irrigated. Many of the schemes are multi-purpose projects providing hydro-electric power in addition to water supplies for irrigation. Amongst the more noteworthy projects are the Columbia Basin project (Grand Coulee Dam) which ultimately will irrigate more than 1 million acres (400,000 ha) of unproductive land; the Minidoka project, using the water of the Snake river, which irrigates nearly 1 million acres (400,000 ha); the North Platte, Salt River, and Yakima schemes which each water over 300,000 acres (120,000 ha). Notwithstanding an impressive list of schemes and a large acreage under irrigation, it may be said that irrigation in the western United States is still in its early stages. Vast developments, one can safely prophesy, will take place in the future. Irrigation in the United States is not confined to the lands west of the 100 degree meridian; it is also carried out in the humid east, although its use here is to increase the yield and the quality of the crops.

It used to be thought that man in his cultural development had passed from the collecting and hunting stage to herding and thence to cultivation. Nowadays it is believed that man may have jumped from the hunting stage to the cultivating stage without passing through a pastoral phase. In other words, he may have discovered the art of cultivation at the same time or possibly even before he began to domesticate animals. Which came first is, so far as we are concerned here, an academic point; it matters little. The domestication of animals took place, it is thought, about seven or eight thousand years ago, and the first animals to be tamed and domesticated were the dog (the first of all, it would seem), sheep, cattle, and the pig, with the horse, camel, and elephant coming later. One of the most interesting facts is that since these early domestications man has not succeeded in domesticating any other animal of importance.

The majority of mankind lives on a vegetarian or near-vegetarian diet. In other words, man, for the most part, depends directly upon the land for his food. Primitive and backward societies, apart from a few specialised groups, such as the Eskimo, the Bedouin, the Lapps, and until recently such Asiatic peoples as the Kirghiz and Kalmuks, were basically vegetarian, eating little or no flesh; if they did eat meat or fish, it was on special occasions only. A few societies lived by herding animals and subsisted mainly upon the milk, cheese, and flesh of their animals. Quite recently in a newspaper report it was stated that one of the wealthy oil sheiks of the Persian Gulf area lived almost exclusively on camel's milk, the basic food of the Bedouin people.

Pastoral communities use plants indirectly to provide them with food; in other words, they live off their animals which in turn live off plant life. This is really an uneconomic way of using the land, and in the most populous parts of the world (the industrialised areas excepted) land cannot be spared for such uneconomic production; hence meat is a luxury for the great masses of people living in densely occupied areas. Generally speaking, dependence upon plant life enables the earth to support greater numbers than when plants are used to feed livestock, which in turn are used for food. Of course in many arid and semi-arid lands it is impossible to grow crops and, if man is to live in such areas, he must find some alternative way of supporting himself. Conditions which are unfavourable for crop cultivation may be favourable for a sparse growth of grass capable of supporting animals. Hence the raising of livestock becomes feasible and the drier areas of the earth are among the great stock-rearing regions of the world.

In general, as the standard of living rises and the purchasing power of people increases, the consumption of animal foodstuffs, such as meat and dairy produce, increases. This leads to a commercial demand for meat—

beef, mutton, pork—and dairy produce—milk, cream, butter, cheese—which stimulates the rearing of animals for flesh on the one hand and for dairy products on the other.

The peoples of Europe and those of European origin living in other parts of the world are the chief meat eaters and consume the bulk of the world's meat production; accordingly, the production of and trade in beef, mutton, lamb, and pork are determined by the requirements of these peoples.

## *FACTORS AFFECTING THE LIVESTOCK INDUSTRY*

Many factors influence the rearing of livestock and world trade in meat and dairy products. As noted above, there are many areas in the world receiving insufficient moisture for the cultivation of crops but sufficient moisture for grass which is capable of supporting animals. Indeed, it is these semi-arid lands of a marginal character, which are of considerable extent in almost every continent, which form the great stock-rearing areas of the world. The only really important exceptions to this general statement occur in Western Europe, New Zealand and the United States corn belt. In the semi-arid grassland areas, where there is plenty of room, animals are reared for beef, hides, or wool. Where the emphasis is on dairying, cooler, moister environments are needed and the dairy industry is found in wetter areas and usually in close proximity to areas of dense, urban populations, especially where milk is the chief saleable commodity.

One of the necessary conditions for successful livestock ranching is that the ranging area should be of great size. Many of the ranches are as large as an average-sized English county. This is necessary to give adequate carrying capacity, for if the pasture is poor or the animals too numerous the cattle or the land or both are likely to suffer. Insufficient and poor quality forage will lead to inferior beef while over-grazing of the pasture will be likely to induce soil erosion. Many of the tropical grasslands which would appear to offer wonderful prospects for cattle-rearing are of restricted value because of the poor quality, unnutritious grasses found there.

Much progress has been made in cattle-rearing due to breeding and pest control. In Brazil, for instance, the Santa Gertrudis breed has been developed by crossing English Shorthorn with Indian Brahman cattle. This was an attempt to produce an animal suited to the Brazilian campos. "Years of crossing, selection, and inbreeding were necessary to develop the new breed which combines the best qualities of the Shorthorn and the Brahman. The Brahman's high tolerance of heat, humidity, poor pastures, and resistance to insects and disease is combined with the Shorthorn's fine beef-producing quality and quiet temperament."* The control and elimination of insect pests have done much to help the industry, but the tsetse fly is still a serious handicap in Africa. Many parts of the African savanna lands could be devoted to cattle raising or more productive cattle raising but for the ravages of the tsetse fly.

A number of technical developments over the past hundred or so years have materially affected the livestock industry. The introduction of barbed wire enabled the land to be fenced off and made it possible for the animals to be kept under control. A century ago cattle were reared chiefly for their hides and tallow, since means of preserving flesh other than by drying or salting had not been developed. Then a German scientist, Baron Liebig, discovered a process whereby meat extracts could be produced. But the two most significant developments were the developments in the canning of meat and the invention of the refrigeration process; these revolutionised the industry and made possible the large export trade that is now carried on.

Social factors may encourage or inhibit animal husbandry. High standards of living, as we have already noted, create a demand for meat and dairy products. On the other hand, religious proscriptions about eating flesh may militate against the keeping of animals. Buddhism, for example, advocates vegetarianism and this helps to explain why in much of China, Korea, and Japan domestic animals are few, although the scarcity of arable land is another important contributing factor. A good example of direct religious influence affecting livestock farming is to be found in the reluctance of the orthodox Hindu to control cattle in any way. As a result, India is plagued with scores of millions of sacred but useless cattle for they provide no meat and little milk, consume large quantities of food which is already scarce, and help to spread disease. How beneficial to the Hindus the cow might be if its religious significance could be discarded, its numbers reduced by three-quarters, and the stock improved by breeding and inoculation!

The pig, one of the most useful of animals, is subject to religious stricture. Requiring little care, pigs are a useful animal for the small farmer, producing good food from waste and fitting well into the system of general farming. For these reasons they are reared in many countries, notably in China, the United States and Germany. Yet among Moslems, Hindus, and Jews pork as a food is forbidden; to the Moslem the pig is an "unclean" animal. Thus in many parts of the world pig-rearing is prohibited on religious grounds.

Enough, perhaps, has been said to show how livestock farming as a

* Shaw, E. B., *World Economic Geography*, John Wiley & Sons, Ltd., 1955, p. 261.

branch of agriculture is influenced, like arable farming, by a variety of factors, geographical, scientific, technical, biological, and social.

## *NOMADISM*

Animal farming is frequently characterised by the seasonal movement of the animals which occurs in two forms: as nomadism in semi-arid areas and as transhumance in mountain regions. These movements deserve and require some consideration.

Nomadism may be defined as the organised seasonal movement of man and his animals in search of subsistence. A more elaborate and precise definition is as follows: "Nomadism involves the repeated shifting of the habitat of a people in its search for subsistence. It does not consist of unrestricted and undirected wandering, but is focused around temporary centres of operation, the stability of which is dependent upon the food supply available and the state of technical advance."*

Pastoralism is a feature of the natural grasslands and semi-arid to arid areas of the world. In these regions rainfall is seasonal in its occurrence and the vegetation, accordingly, flourishes with the onset of the rains and withers and dies away during the dry season. During rainy periods germination is rapid and often quite a luxurious growth of grasses and small herbaceous plants spring up. With the onset of drought most of the grasses wither and die away, and only the tougher, coarser, hardier, drought-withstanding plants remain to provide a thin and scattered herbage. When, in early autumn, the grasses wither out on the windswept plains, the vegetation often remains green along the fringes of the plains where they abut on adjacent hill country or in sheltered valleys where streams may persist and provide patches of richer vegetation. Consequently, people depending for their livelihood upon animal grazing find it profitable to migrate with the pasture; in spring and summer they may roam the plains with their animals and seek out the shelter and better pasture of the valleys and foothill regions in autumn and winter.

A wandering or nomadic pastoral life is traditional in the dry steppe country of the interior of Asia and for thousands of years man has moved hither and thither, usually according to some definite plan, with his horses, asses, camels, sheep, and goats. Migrations are normally confined to the traditional territory of the family group, and may take place as often as forty times a year, and involve treks of 200 miles (320 km). Similarly, in the desert and semi-desert zone of North Africa and South-western Asia the nomadic peoples, mainly sheep and camel herders, follow regular routes between summer and winter pastures, calling at oases as they wander.

* THURNWALD, R., in *Encyclopedia of the Social Sciences*, New York, Macmillan, 1933, Vol. XI.

During recent decades there have been many changes in the traditional nomadic way of life. Some have been due to political, some to economic, and some to social factors. In the Soviet Union herding has been collectivised and the traditional grazing lands divided up into great farms. In the Republic of Mongolia, which is also a communist state, the nomadic way of life still persists though even here there have been changes. Grazing is being more carefully planned in advance so that long treks are no longer essential; seldom must more than 20 or 30 miles (32 or 48 km) be covered and fewer journeys need be undertaken these days since wells are being dug, hay is beginning to be collected, and shelters are being erected for wintering the animals. Moreover, breeding and pasture research stations are being set up while many of the nomads are coming into closer contact with the sedentary populations of the towns and are obtaining flour and other food commodities. All these things are gradually undermining and changing the old way of life. Similarly, in the desert belt of North Africa and South-west Asia, the discovery and exploitation of oil and other kinds of mineral wealth are having a great impact upon the traditional mode of life. It seems very likely that the days of true nomadism are numbered.

## *TRANSHUMANCE*

Transhumance is the vertical movement of flocks and herds up and down mountains according to the season in search of fresh pasture. It is, in a sense, a kind of nomadism. The practice is essentially a response to: (*a*) a shortage of arable land, and (*b*) a shortage of pasture. Transhumance is a feature of, and is, moreover, dependent upon, mountain country. A traditional activity, it is still widely practised in the Alps, the Pyrenees, the Balkans, in Scandinavia, in the Atlas Lands, in various parts of the Near and Middle East and elsewhere in the world, including Wales.

Transhumance provides an interesting illustration of human adjustment to environmental conditions—to relief, to climate, and to vegetation. The practice can best be described by taking a specific example: let us take the Swiss Alps. During the winter months cattle are kept in the valleys, for the uplands are under snow; apart from a few favoured areas where outdoor grazing is possible, the animals are housed and stall-fed. In spring, when the snows begin to disappear, the cattle are driven up the mountain slopes stage by stage or from alp to alp as each becomes cleared of snow and the grass grows. In some places the cattle are moved directly from the valley bottoms to the upper mountain pastures, the intermediate mountain shelves being reserved for hay production. Meanwhile, during the summer months, the valley bottoms are given over to the cultivation of lucerne and other fodder crops, and the lower valley slopes are left for hay. By September the animals begin their downward trek and at the end of October they are all, once again, back in the valleys and ready to take up their winter quarters.

[*Courtesy: Marcel Coen.*

FIG. 33.—TRANSHUMANCE

Transhumance in the Southern Alps, France. Upwards of 500,000 sheep are moved up the mountains—Alps, Pyrenees, and Central Massif—in summer and brought down before the snows of winter. They are carried as far as possible by railways or in lorries and then driven, as shown, to the new pastures. Where else in Europe is transhumance carried on?

A similar seasonal movement characterises much of the Mediterranean coastlands which are backed by mountains. Here, under Mediterranean climatic conditions, natural herbage is available in the lowlands only during the periods of winter rains. After March the ground vegetation quickly withers and disappears under the heat and drought of summer, and even animals such as sheep and goats which can browse on the poorest of pastures, find it difficult to subsist on the scanty, dried-up herbage. On the higher mountain slopes and summits, however, good summer pasture is available due to the effect which altitude has on rainfall and temperature and the melting of the winter snows. There is, therefore, every inducement for a seasonal movement of sheep and goats from the lowlands to the highlands. Transhumance "has been practised in the Mediterranean Lands since the earliest times. It survives still as a feature of the economy in most countries, but the increased use of lowland pastures for the growing of crops and the introduction of artificial feeding stuffs have tended to militate against the continuance of this age-old practice. . . . Nevertheless, amongst the poorer peasantry who cannot afford to buy feeding stuffs and who live in or near to suitable mountain pastures, transhumance is still very much in evidence."*

## *SYSTEMS OF FARMING*

The methods by which man organises his agriculture gives rise to the different systems of farming. This organisation is closely linked with the methods of land tenure, *i.e.* the ways in which man owns and works the land. The chief systems are as follows:

1. Private and individual ownership.
2. Peasant farming of rented plots.
3. The share-cropping or *metayage* system.
4. The great estate using hired labour.
5. The co-operative farming system.
6. The communist collective farming system.

Most private, individually owned farms are small in size. Usually they are managed and worked by the farmer and his family, although he may employ additional labour, two, three, four, or more hands according to the size of the farm and the nature of the agriculture. The profit he secures from his farm depends to a very great extent upon the work he puts into his farming and the efficiency with which he manages his farm. The farmer has a real interest in the soil and his animals and it pays him to take care of his land and his property. It is common knowledge that anyone who owns his own business is willing to devote longer hours, greater effort, and more labour to it than would an employed person; this applies to any kind of enterprise. Private enterprise farming is typically relatively small-scale, satisfying, and rewarding.

The peasant farmer who rents a plot may love the soil, like the French or Japanese peasant does, but, because he rents his land, he is less concerned with maintaining and improving its fertility. The peasant, too, often works a plot (or scattered plots) which is too small to give an economic

* Robinson, H., *Mediterranean Lands*, U.T.P., 2nd edition, p. 93.

return. Peasant farming of this kind is a common feature of much of Europe outside the Iron Curtain and of Monsoon Asia.

A system of farming once of much significance but quickly disappearing is that of share-cropping. It is a system whereby the farmer paid a proportion, often as much as half, of his produce to the landowner in lieu of rent and in which the owner furnished some or all of the seed, stock, farm implements, etc. As a system of land tenure it was most characteristic of Western Europe but was also found in the United States, especially in the cotton belt.

The great estate using hired labour is a relic of feudal medieval times. Known as *latifundia* in Southern Europe, they are still very common in Spain and southern Italy. Elsewhere in Europe they have been very largely broken up and, indeed, are slowly disappearing in Mediterranean Europe. The countries of Latin America which historically formed Spanish or Portuguese colonial territory inherited many Iberian cultural features, among them the traditional agrarian system. Thus in many parts of Latin America the land continues to be held in large properties as, for example, in Argentina with its *estancias*, in Chile with its *haciendas*, and in Brazil with its *fazendas*.

Co-operative farming, pioneered in Denmark, has spread to many parts of the world, *e.g.* to Eire and New Zealand. In co-operative farming the producers join together into co-operative organisations and by this means the farmers are able to take advantage of the best and most efficient methods of handling, processing, and marketing their produce. Co-operation also extends to the purchasing of seed, feeding stuffs, fertiliser, farm machinery, and equipment, etc. By cutting out the middleman, the farmers can both buy and sell more profitably.

Collective farming carries co-operative farming a stage further. It is the kind of farm organisation found in communist countries such as the Soviet Union and China. The original farms and small-holdings of the peasants are compulsorily merged into a single large unit. The consolidated land is then placed under one management and worked as a unit. Usually each farm has a manager who is in charge, but there is also a committee of management who help to organise the work and run the farm. Each member of the collective farm has a given job to do and must put in a certain number of days' work each year. He is usually paid a wage but may be paid in part in kind. In the Soviet Union the collective farms are known as *kolkhozes*. In addition to these, there are very large state farms, known as *sovkhozes*; these are run directly by the state.

The type of farm organisation in agriculture is an important factor since it affects such things as land improvement and the efficiency with which farming is carried out. Generally speaking, where the land is owned by the farmer, it will be better managed and kept in better condition with, in consequence, a greater yield in output whether of crops or animals. Large farm units, especially those worked by hired labour, tend (though this is by no means always the case) to be wasteful of the land and sometimes inefficient.

### EXERCISES

1. What factors and conditions affect the kind of agriculture carried on by man? Demonstrate how these apply to any region you have studied.
2. Explain the need for irrigation and describe the principal methods used, illustrating your answer by reference to specific regions.
3. Explain how *three* of the following may affect the efficiency and prosperity of a farming community: (*a*) modes of land tenure; (*b*) farmers' co-operatives; (*c*) the work of F.A.O.; (*d*) size and distribution of land holdings; (*e*) government interference in agriculture.
4. Examine the importance of irrigation in the agricultural development of the Indian sub-continent, and indicate some of the problems caused in this connection by political and social issues. (*Union of Educational Institutions.*)
5. What factors of a climatic, biological, and social nature influence the livestock industry?
6. Distinguish between the practices of nomadism and transhumance. Give regional examples of where these practices still persist.

# *Chapter IX*

# TYPES OF FARMING

## *FARMING METHOD*

In general, agriculture is carried on in two main ways; these are usually classified as intensive and extensive methods. These methods are not necessarily related, though they may be, to the systems of farming described in the previous chapter. However, let us be clear, first of all, as to the meaning of the terms intensive and extensive.

### INTENSIVE FARMING

This method is, perhaps, best defined as farming in which much capital is expended or much labour applied to a given area of land in order to increase its productiveness. The idea is to get the maximum possible yield from the land. This may be achieved in various ways: by the careful tillage of the soil, by the application of manure and fertiliser, by the careful rotation of crops, by cutting out fallow, by using every square inch of soil, by using high quality seed and stock, by applying scientific methods to farming, and by the use of mechanical aids. In intensive farming several or all of these methods are employed. Intensive farming is characterised by high yields per unit of land. Western European countries, such as Belgium, Holland, and Denmark, grow nearly twice as much wheat per acre as do Canada and the United States; this greater yield, however, is achieved only by the greater expenditure of human energy, while the cost per bushel of producing the wheat is much higher than in North America where it is grown purely by mechanical methods. Another feature of intensive farming is that it is practised in countries such as Holland, Denmark, and Japan, where land is relatively scarce. Shortage of land implies the greatest possible use of what land there is; this, in turn, leads, where climatic conditions render it feasible, to inter-cropping (the growing of one crop between the rows of another) and multiple-cropping (the growing of several crops in succession upon the same piece of land throughout the year). Finally, intensive agriculture is characteristic of densely peopled lands since there is an abundance of labour available for the care and effort demanded of this type of farming.

### EXTENSIVE FARMING

In contrast to intensive farming where the aim is to get the maximum return per unit of land area, extensive farming aims at producing the maximum amount of a commodity per unit of man-power. It is a method of farming in which the amount of capital and labour applied to a given area is relatively small. It is a method by which a relatively small crop is secured from a large area, though this crop is achieved with the minimum of attention and expense. One point needs some elucidation; we have just said that extensive farming is characterised by a low capital outlay and yet this type of farming makes use of much mechanisation, *e.g.* huge combine harvesters which, manifestly, are extremely costly; how can these two statements be reconciled? The point is that the land area farmed is large so that the capital expenditure, though perhaps great, is relatively small per unit of land. The great wheat farms of the Canadian prairies, which are highly mechanised but very economic of man-power, producing grain at a low yield per acre, provide good examples of the extensive system. Extensive farming "may lead to somewhat careless methods of farming . . . but by the use of power equipment, crops can be produced so cheaply that the cost of a food product may actually be less in a land of high wages where machinery is used than in a land of low wages where hand methods prevail. Thus rice is grown in California, Texas, Louisiana, and Arkansas at less cost per bushel than in China or Japan, yet wages in America are many times those paid in the Orient. Extensive agriculture cannot be carried out in all locations. It is best adapted to open plains or gently rolling topography, lacking in dense populations."*

Having clarified the meanings of intensive and extensive as applied to farming, we can now look at the different types of farming which are carried on in different parts of the world.

## *TYPES OF FARMING*

The way in which agriculture is carried on varies widely from area to area. There are differences in the size of farms, their organisation, the way they are worked, the degree of mechanisation used, and the types of crops grown. As a result of these differences, it is possible to distinguish a number of "types of farming," the chief of which may be summarised under the headings given below.

### TROPICAL SUBSISTENCE AGRICULTURE

Within the tropics cultivation is characteristically of the subsistence type; in other words, the native cultivator is concerned almost purely with the growing of crops (*e.g.* yams, manioc, millet, bananas, sweet

* Freeman, O. W., and Raup, H. F., *Essentials of World Geography*, McGraw-Hill, 1949, p. 301.

potatoes) to provide sufficient food for himself and his family. Usually there is little surplus, for he cultivates no more land than is necessary to provide the essentials for existence. His wants are relatively few and simple and what he does not grow he can usually obtain from the natural surroundings. Such a way of life is known as a subsistence economy.

In the hot, wet forested lands, where the soils are poor and infertile and lacking in plant foods, the earth becomes easily exhausted. In such areas the practice of shifting cultivation is carried on. The trees are cut or ring-barked and then fired to make a clearing. The wood ash provides a dressing of fertiliser. Seeds or cuttings are then dibbed in with simple instruments such as a digging-stick or a primitive hoe. Since the soil quickly loses its fertility, after one or two seasons the cultivator is compelled to move on and clear a fresh patch in the forest. Hence the native farmer may be described as a nomadic cultivator.

In some parts of the tropics, especially where land is scarcer, periodic movement to new patches of land is impossible; under such circumstances sedentary agriculture, *i.e.* settled tillage, is usual. In this case cash crops are sometimes grown in addition to the staple food crops. Subsistence agriculture, both shifting and sedentary, is found in the upland areas of Central and South America, throughout much of tropical Africa, in India, and in parts of South-east Asia.

## TROPICAL PLANTATION AGRICULTURE

Although some plantations were introduced into tropical regions by Europeans three or four centuries ago, this specialised type of farming is mainly a development of relatively recent times. Plantation agriculture, which occurs in selected areas, was made possible by the capital resources and organisational capacity of Europeans on the one hand and the labour supplied by native peoples on the other. It is important to note that it is a system suited to the production of crops which do not give an immediate yield; coffee and rubber trees, for example, do not begin to give returns until a period of several years has elapsed. Clearly a native farmer could not possibly afford to wait so long before he got a return from his land. Moreover, it is a system suited to the production of crops demanding processing and careful handling, *e.g.* rubber, sugar-cane, tea. Farming methods are far more efficient and scientific than those practised by native farmers, although some of the latter now run "small-scale plantations" on their own land, *e.g.* the cacao growers of Ghana, the rubber growers of Malaya.

Tropical plantation agriculture is characterised by certain features: (*a*) they are usually large-scale holdings, which draw most of their labour supply from the local inhabitants, as in Cuba, or which rely upon immigrant labour, as in Malaya; (*b*) they concentrate upon the production of a single crop, a practice known as monoculture, although such reliance means that plantation agriculture is very much at the mercy of market demand and price fluctuations; (*c*) they commonly require heavy investment in the form of processing plant, light railways, shipment facilities, etc., and for this reason are frequently located in coastal localities for ease and economy of export.

## ORIENTAL INTENSIVE AGRICULTURE

This is a subsistence peasant farming of a more advanced type which is practised in the monsoon lands of Eastern and South-east Asia where rice is, typically, the dominant crop. It differs from the extensive subsistence type of agriculture in the skill with which the land is worked, in the intensity of the cultivation, and in the higher crop yields secured.

The chief features of Oriental farming are: the small, often diminutive, size of the farms and the scattered nature of the plots; the practice of hand tillage by spade and hoe and the use of the water-buffalo for ploughing; the cultivation of rice as a staple and of a second (dry season) crop, such as barley, millet, beans or vegetables; the maintenance of soil fertility by the careful and constant application of animal manures, human excreta, vegetable compost, etc.; the use of a system of crop rotation to help keep the soil from exhaustion; the practice of double-cropping and inter-cropping, that is the growing of one crop between another; the widespread terracing of hill slopes; and the frequent and widespread use of irrigation (which is often needed in rice cultivation).

Oriental agriculture is very intensive in its nature and high yields are secured from small areas of land but at the cost of much human toil. Because land is scarce and the pressure of population is great, the soil is made to give its maximum yield; moreover, because of these circumstances, there is no room in the farming system for animals—land simply cannot be spared for pasture—other than pigs and poultry which are useful scavenging animals and can forage for themselves, turning waste material into good food.

## OASIS AND MEDITERRANEAN AGRICULTURE

Though sometimes treated as two separate types of farming, they may be conveniently dealt with together for present purposes. As farming types they have much in common: in both cases the major problem is shortage of water and irrigation is much resorted to; also, they grow the same kinds of crops, mainly cereals, vegetables, and tree crops—especially fruits.

Oasis cultivation is confined to localities in desert areas where supplies of water are available either from rivers or from underground sources. The chief crops are cereals, dates, vegetables, sugar, and cotton. Oasis agriculture is usually of the subsistence kind, but in some cases commercial

[*Courtesy: Japanese Embassy Library.*

FIG. 34.—INTENSIVE AGRICULTURE, JAPAN

About 20% of the people of Japan are farmers. This is a high proportion considering a mere seventh of the area can be farmed because of the ruggedness and steep slopes of the mountains. Lowland plain areas are very limited: the largest is the Kwanto Plain around Tokyo. Due to the small area of cultivable land, farming must be intensive. Japanese farms average only $2\frac{1}{4}$ acres (0·9 ha). They are worked by laborious hand methods and carefully fertilised to produce the maximum crop yields possible. The photograph illustrates inter-cropping—the growing of one crop between the rows of another different crop. Vinyl sheets are used to protect the young seedlings during the early spring planting season.

crops have been developed and with improved means of transport commercial cropping, especially of such things as high-priced vegetables and fruits, is likely to grow.

Mediterranean agriculture, as the name implies, is carried on in areas having a Mediterranean type of climate—one of hot, dry, sunny summers and warm, moist winters. It will be noted that in summer, when crops mostly grow and are most in need of water, rainfall is lacking; hence the need for irrigation. Cereals, however, generally unirrigated, are winter crops. Soils, especially in the lands surrounding the Mediterranean Sea, are thin and stony, being largely derived from limestone but small deltaic and alluvial plains are fertile and these, accordingly, are very intensively cultivated. Terracing is also a feature of Mediterranean cultivation. Subsistence farming is mainly concerned with cereals but commercial farming emphasises fruit, *e.g.* citrus fruits, vines, peaches, figs, and market-garden produce. Because of the shortage of pasture cattle-rearing is difficult and animal husbandry is mainly concerned with sheep and goats. Pastoralism is assisted by the practice of transhumance made possible by the presence of mountains.

## EXTENSIVE GRAIN FARMING

This type of farming is typical of the "new lands" of the world, especially of the interior plains of those lands—Canada, the United States, Argentina, and Australia, to which we may add the Soviet Union—which form the great granaries. The development of large-scale commercial grain-farming was essentially an outcome of the Industrial Revolution for not only did that Revolution give rise to the growing demand for foodstuffs, especially cereals, but the very use of these grassy plains was dependent upon the invention of steel ploughs, capable of turning the tough prairie sod, and of harvesting machinery.

The areas now devoted to commercial grain-farming were previously used by nomadic herders or livestock ranchers for as natural grasslands they were the natural homelands of herbivorous animals, *e.g.* horses, sheep, bison, guanaco. Because of their interior location they usually have cold winters, warm but short summers, and light rainfall. The shortness of the growing season is a limiting factor and has been responsible for the development by man of quick-maturing varieties of grain. The light and often chancy rainfall, especially on the drier margins of these areas, has led in some cases to soil erosion.

Commercial grain-farming is characterised by monoculture, the cultivation in the main of one crop, usually wheat; by the large size of the fields and farm units; by the large-scale use of machinery; by the small amount of labour used; and by the low yields per acre. In contrast to those regions where intensive tillage is practised and there is a high yield per acre due to a high labour ratio per acre and a high degree of fertilisation, extensive grain-farming uses little, if any, fertiliser and has a low man-land ratio. The real difference between these two very different and directly opposite types of farming is best expressed by saying that the one has a high output *per acre* (intensive farming) while the other has a high output *per capita* (extensive farming).

## MIXED FARMING

One of the fundamental weaknesses of extensive grain farming is the dependence upon one crop. The farmer is completely at the mercy of the current world price for his commodity. Mixed farming is in a sense the antithesis of extensive grain farming. It is a type of agriculture devoted to the growing of a variety of crops, more than one of which is usually grown for sale, or more usually to the rearing of animals as well as crops. Clearly the mixed farmer will enjoy a greater security than the farmer who specialises in the production of a single crop because of his more varied sources of income—perhaps from several, or even all, of the following: grain, root crops, fruit, hay, meat, milk, and eggs. Another characterising feature of the mixed farming system is that the farmer is practically self-sufficient in food supplies.

The size of farm under this system varies but typically it is of moderate or small size and is run by a family. Other significant features of mixed farming are: (*a*) well-developed systems of crop rotation; (*b*) the keeping of some animals; (*c*) the use of green-manure, animal manure, and chemical fertiliser; and (*d*) the relatively high degree of capital investment, *e.g.* in tractors, machines, and other farm equipment.

Mixed farming is a response to a variety of environmental, economic, social, and historical conditions. Obviously rather special environmental conditions (*e.g.* of soil and climate) are needed which suit both crop-growing and animal husbandry. Hence the practice of mixed farming is subject to fairly strict environmental limitations. Since, too, the farmer produces commodities mainly for the market, he must be relatively near consuming centres and have relatively easy and cheap transport available to help him dispose of his produce, which is often perishable. Thus mixed farming is also subject to certain economic limitations. Finally, there are cultural limitations, for this type of farming demands intelligence, skill, foresight, and adaptability on the part of the farmer.

Mixed farming is a type of farming characteristic of, and largely confined to, Western and Central Europe and the north-eastern part of the United States.

## COMMERCIAL DAIRY FARMING

The demands of large cities for supplies of fresh milk have led to the growth of intensive dairying in their vicinities, *e.g.* around London and Buenos Aires, but commercial dairy farming is characteristically a development of lands where summers are too cool and wet for cereal farming but are well-suited to the growth of grass and the production of hay, *e.g.* much of north-western Europe, parts of south-eastern Canada and north-eastern United States, and New Zealand. Some countries, notably England, Eire, Holland, Denmark, and New Zealand, have become highly specialised producers.

Natural pasture normally plays an important part in dairy farming but in some cases, *e.g.* in Denmark, shortage of land has led to a great reliance upon feedstuffs such as grain and root crops. Generally speaking, in all dairying countries grass is supplemented by leguminous forage, root crops, and grain.

Commercial dairying is dependent, directly or indirectly, upon urban markets where the population has a high living standard. Dairying is concerned with the production of fresh milk and cream or their derivatives, butter and cheese. Fresh milk, obviously, must be produced near to the consuming areas for milk is a highly perishable commodity. Milk cannot normally be carried much farther than a distance of some 200 miles (320 km). Butter and cheese, especially in refrigeration, can be carried over long distances. Dairy regions tend to specialise; for example, Denmark and New Zealand concentrate upon butter production, Holland and Switzerland on cheese. A notable feature of most dairying countries is the associated pig-rearing industry, for the skimmed milk is a valuable pig food.

## SPECIALISED HORTICULTURE

Horticulture is a type of intensive highly specialised commercial farming frequently, though not always, carried out on a small scale. The term horticulture originally meant the cultivation of a garden but in its modern and broadest sense it includes: (*a*) market-gardening; (*b*) glasshouse cultivation; (*c*) flower culture; (*d*) plant nurseries; and (*e*) commercial orchards. As generally understood, horticulture means the cultivation of vegetable, salad, fruit and flower crops, usually on small plots and with a much higher degree of intensiveness than in field cultivation, which are specially produced for human consumption and for urban markets.

The development of horticulture is very largely a response to two factors or conditions: (*a*) the growth of large urban centres demanding fresh vegetables, fruit, and flowers; and (*b*) high standards of living which make it possible for people to afford high-priced vegetables, fruit, and flowers. For these reasons horticulture is mainly to be found in Western Europe, North America, and in other areas of white settlement. It is carried on either in the vicinity of large towns, *e.g.* around London and Paris, or in areas remote from markets where "early" crops (*primeurs*) can be produced, *e.g.* in Cornwall, in southern France, and in Florida. The latter areas are usually served with good communications and make use of special trains, refrigeration equipment, and even air transport.

FIG. 35.—BRITISH AGRICULTURE

TABLE III

*The Farm Budget for the United Kingdom—1969–70*

| *Sales in £ million/percentage of total* | | | *Expenses in £ million/percentage of total* | | |
|---|---|---|---|---|---|
| Milk and milk products | 458·0 | 21·7 | Labour . . . | 333·5 | 18·8 |
| Fatstock . . . . | 657·0 | 31·1 | Rent and interest . | 198·0 | 11·2 |
| Eggs and poultry . | 322·0 | 15·2 | Machinery . . | 284·0 | 16·0 |
| Farm crops . . . | 403·0 | 19·0 | Feeding stuffs . . | 524·0 | 29·5 |
| Horticultural products . | 226·0 | 10·7 | Fertilisers . . . | 150·5 | 8·5 |
| Others . . . . | 48·5 | 2·3 | Others . . . | 285·0 | 16·0 |
| Total . . . | 2114·5 | 100·0 | Total . . | 1775·5 | 100·0 |

Net income £535·5

FIG. 35 (*facing page*).—BRITISH AGRICULTURE

Farming is one of Britain's major industries. In June 1970, there were 274,000 full-time farm workers, two-fifths of the number in 1946. They produce over half of Britain's food supplies. Farm products, however, furnish only a minute fraction of British exports. Farming is mainly of the mixed type. Out of the 78 million acres (31 million ha), 64 (25) are used: roughly a quarter is under crops and a quarter under grass. Most of the arable and grassland is in lowland Britain. Rough pasture, covering another quarter, occurs mostly in Scotland (13 million acres; 5·2 million ha). The remaining quarter comprises waste and built-up land.

Cereals, mainly wheat and barley, are largely confined to the warmer, drier, sunnier eastern side of Britain, especially East Anglia, the east Midlands, south Lincolnshire, Holderness, and the eastern coastal lowlands of Scotland. Oats, of lesser importance, are more widely spread. Yields of cereals per acre are high. Annual production of barley, wheat, and oats is about 8, 3½, and 1 million tons respectively. Root crops are widely grown but mostly for animal fodder. Sugar-beet (7 million tons) is closely confined to south Lincolnshire, Norfolk, Suffolk, and Essex in England and to Fifeshire in Scotland. Potatoes are important in the low-lying peaty soils of south Yorkshire, south Lancashire and the Fens, and in Lincolnshire, Norfolk, and Essex. Seed potatoes are a speciality of Scotland, notably in Ayrshire, Midlothian, and Fifeshire.

Crops are usually grown on a rotation basis. In England, a typical six-year rotation is sugar-beet or potatoes, cereals followed by a second cereal, clover or lucerne, and grass. The growing of grass for a short period, within a cycle of rotation, to provide temporary pastures, is known as a ley.

Market-gardening, covering a mere 2½% of the arable area, is very important, giving more valuable returns than any other branch of agriculture. Over 400,000 acres (160,000 ha) are under vegetables, 300,000 acres (120,000 ha) under fruit, and over 4000 acres (1600 ha) under glass. Light, warm, sandy soils are especially suitable but proximity to markets is the prime consideration, for the produce must be fresh. Thus market-gardening often occupies costly land on the fringes of urban areas. Early produce (fruit, flowers, salad crops, early potatoes) is a speciality in Pembrokeshire, Cornwall, the Scilly Isles and the Channel Islands, where the winters are mild and late frosts unlikely. Fruit is mostly grown in the warm, sunny, sheltered and drier areas, notably in Norfolk, the Fenlands, and the Isle of Ely (soft fruits), Kent (hops, cherries, soft fruits), Herefordshire and Worcestershire (hops, stone fruits, hard fruits), and Strathmore and the Carse of Gowrie in Scotland (soft fruits). Damsons are a speciality of Cheshire. Apples are important in Somerset and Devon for cider-making.

Britain has 12 million cattle: about 8 million are reared for beef, mostly in the drier, eastern areas, and about 4 million for milk, mostly in the milder, damper western districts where the rainfall is 30 in. (762 mm) or over. Some dairy herds occur in drier districts and are fed in part on imported fodder: such areas are near large towns which make heavy demands upon fresh milk. Pigs and poultry are closely associated with dairy farming. The Fylde and Craven districts are important for poultry, while East Anglia is noted for turkeys.

Some 29 million sheep, comprising over thirty different breeds, producing wool and meat, are reared, mainly on the high moorlands and hill pastures of the Southern Uplands, Lake District, Pennines, Welsh Hills, and on the dry chalk and limestone pasture lands of the Cotswolds, Northampton Uplands, Lincolnshire Wolds, and North and South Downs.

For example, before 1939 "rhubarb specials" used to run from the West Riding rhubarb triangle to London, high-speed refrigerated trains run from Florida to the industrial north-east of the United States, while fresh cut flowers, which can stand the high transport costs, are flown from the Riviera to the Paris and London markets.

[*Courtesy: U.S. Information Service.*

FIG. 36.—INTENSIVE HORTICULTURE

An irrigated farm in the Imperial Valley, California. Fruits and vegetables are the main crops of these orderly, well laid-out fields. High-value market-garden produce, or truck-cropping to use the American term, is mostly grown near to consuming centres; if distant from the latter, there must be good and speedy transport services for the produce needs to be marketed fresh.

Typically, the land is very intensively cultivated, fertilisers are heavily used and much highly skilled labour is employed. Specialised packing is often necessary and much use is made of boxes, containers, etc. In view of all this, production costs are high; on the other hand, the commodities usually fetch high prices on the market.

## NOMADIC HERDING

The herding of livestock is an old and simple way of life. In its present form of livestock ranching it is still an important type of farming; but the traditional activity of wandering with flocks and herds is slowly but surely disappearing. Nomadic herding is largely confined to those areas of the earth's surface that are too dry or too cold for crop growing. Fundamentally, this migratory way of life is related to two natural conditions: the presence of pasture and the occurrence of water supplies, of which the latter is more important because it is scarcer.

The livestock reared under this system include cattle, sheep, goats, camels, and reindeer. Along with them, a few work animals are usually kept, *e.g.* horses, asses, to assist in herding, mustering, and the carrying of camp equipment. Nomadic herding, which is so closely bound up with the natural environment, is subject to recurring disasters, such as the failure of pasture due to variations in the rainfall, the drying up of water-holes due to lack of rainfall, locust plagues which eat up every green thing, the burning of moss and lichen pastures which are very slow to recover, and animal diseases.

The nomadic way of life is still followed by the cattle herders of the Sudan and East Africa, by the camel, sheep, and goat herders of the arid zone, by the cattle, sheep and horse herders of Mongolia, and by the Lapp and Eskimo herders of reindeer. The Kirghiz and other nomad peoples of Central Asia have now become settled livestock rearers following upon the establishment by the Soviet government of either collective or state stock farms.

## COMMERCIAL LIVESTOCK RANCHING

In contrast to nomadic herding which is migratory, which provides a largely self-sufficient existence, and which yields few products for commercial exchange, commercial livestock ranching is geared to production for sale, either of meat, hides, hair, or wool. Cattle-rearing for beef and sheep-raising for wool are the dominant activities. They are characteristic of the world's natural grasslands, especially the drier portions (since the moister parts have usually been taken over by the cereal farmer). Increasingly, commercial ranching has encroached upon the territories which were formerly the preserve of the nomadic herder.

The basic differences between nomadic herding and livestock ranching are: (*a*) in the purpose for which the animals are reared (noted above); (*b*) the intensiveness of production; and (*c*) the care taken in breeding. To the nomad herders, the possession of mere numbers is often the most important thing; to the commercial livestock rearer, quality rather than quantity is usually the primary aim.

A feature of livestock farming, whether of the nomadic or ranching type, is that the ratio of animals to area is very low, *e.g.* in parts of Australia the sheep density works out at one sheep per 5 acres (2 ha) and parts of the

Rupununi grasslands of Guyana average only one beast to every 70 acres (28 ha). Densities are not always as low as these of course but, in general, the carrying capacity is low due in part to the scanty pasture and in part to the poor grass, which is not very nutritious.

## EXERCISES

1. "Plantation agriculture is a matter, not directly of physical conditions, but rather of world economic and political circumstances." Discuss.
2. Explain the basic differences between "extensive" and "intensive" methods of farming. Quote regional examples of each.
3. In what ways does agriculture, in the vicinity of large towns and industrial areas in Britain, tend to differ from the general regional pattern?
4. What is understood by a "plantation economy"? Illustrate your answer with a geographical account of a plantation crop in either South America or Asia.
5. Write explanatory notes on *three* of the following: horticulture, shifting agriculture, mixed farming, monoculture.
6. Explain the need for irrigation and describe the principal methods used, illustrating your answer by reference to specific regions.
7. What are the chief differences between nomadic herding and commercial grazing?
8. Explain why the majority of farmers in the lowland areas of north-western Europe specialise in dairy farming.

# Chapter X

## PLANT FOODSTUFFS: CEREAL CROPS

THE term "cereal," which comes from Ceres, the Roman goddess of harvest, is used for the various grain crops produced by man: the wheat, barley, oats, and rye of the cooler climates, and the rice, maize, and millets of the warmer climates. Cereals are really species of cultivated grass. Since the Neolithic or New Stone Age Revolution and the introduction of the art of cultivation, some 7000 years ago, cereals have formed the basic food of the overwhelming majority of mankind. Of all man's cultivated crops, cereals are the most important with wheat and rice contending for first place and maize following not far behind. Wheat, rice, and rye are used chiefly, indeed almost entirely, for direct human consumption, whereas maize, barley, and oats are mainly used indirectly, being fed to animals, though a little of each is eaten by man. Cereals are also used for the making of starch, in the brewing of beer, in the distilling of spirits, etc.

### *WHEAT*

Wheat is the principal bread cereal of temperate climates and the chief foodstuff of the white peoples. Wheat, however, is a fairly adaptable plant and can be grown in a wide variety of climates—for example, it is cultivated in the equable climate of England, in the extreme climate of the Russian steppes, in Mediterranean lands such as Italy, and in the hot climate of Pakistan—hence the cultivation of wheat is widely spread throughout the world and almost every month in the year sees a harvest in some part of the earth.

This widespread cultivation means that numerous varieties of wheat have been produced by man. Wheats are grouped in different ways: first into (*a*) winter wheat, *i.e.* where the winters are not too rigorous and the seed can be sown in autumn; and (*b*) spring wheat, *i.e.* where, because of the severity of the winter, the seed is not sown until spring; and, secondly, into (*a*) soft wheat, which is grown in areas having a fair degree of humidity, but which does not make good bread flour; and (*b*) hard wheat, which is grown under dry conditions, has a high protein content and makes good bread flour.

We should also note that wheat is cultivated under two different systems of farming: by intensive methods and by extensive methods. Most of the wheat that enters into world trade is grown extensively on the temperate grasslands of the world.

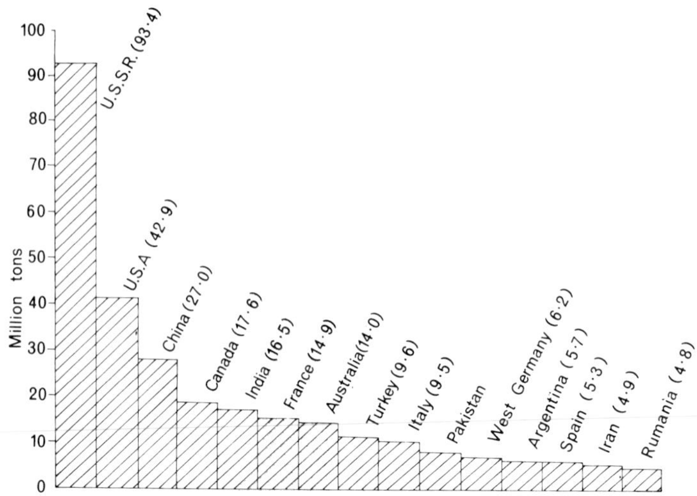


FIG. 37.—WORLD WHEAT PRODUCTION

## CONDITIONS FOR GROWTH

The conditions most suited to the cultivation of wheat are: (*a*) a moderate rainfall of about 15–30 in. (381–762 mm); (*b*) a cool and moist germination period; (*c*) a temperature of at least 16° C (60° F) and sunshine for ripening; (*d*) approximately 100 days frost-free growing period; (*e*) fertile soil, preferably a heavy loam; (*f*) level or undulating land to facilitate mechanised operations.

## AREAS OF PRODUCTION

The chief areas of production are:

*Europe.* Almost every European country grows some wheat, although in Norway, Sweden, and Finland cultivation is marginal and wheat can be grown only in the more favoured southerly areas. Ireland is rather too damp for successful wheat cultivation. In England, France, Belgium, and Western Germany considerable quantities of wheat are grown by intensive methods but production does not cover local requirements. The Mediterranean countries of Portugal, Spain, Italy, and Greece also grow appreciable quantities. The Danubian countries are important producers, but cultivation is more usually of the extensive kind.

*The U.S.S.R.* The Soviet Union is a major world producer. The rich soils (black earths) of the steppelands of the Ukraine and of the grasslands farther eastwards in Soviet Central Asia grow vast quantities under the extensive method of cultivation. Mechanisation is practised on a large scale as in the prairie lands of North America. The Soviet Union vies with the

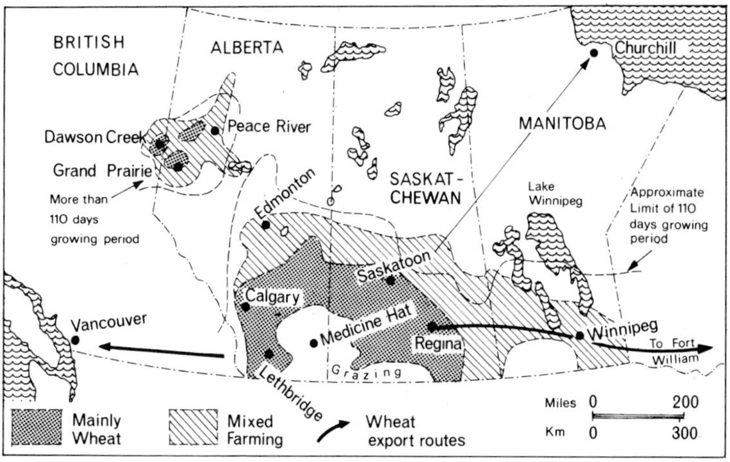


FIG. 38.—WHEAT PRODUCTION ON THE CANADIAN PRAIRIES

The prairie region of Canada forms one of the world's greatest wheat-growing areas. Over much of the true prairie country wheat occupies at least half of the land cropped. The region offers several advantages for wheat: (i) a gently undulating relief ensuring good drainage and assisting mechanised farming; (ii) a deep and naturally rich grassland soil containing much humus; (iii) cheap and abundant land which has facilitated large-scale farming and allows the practice of fallowing; (iv) a rather short growing period but with summer temperatures of around 16° C (60° F) and dry, sunny weather for harvesting; (v) a moderate rainfall of 15–20 in. (381–508 mm) some of which falls in the early growing period; (vi) a good transportation network in the priaries and between the prairies and the ports. In spite of these great advantages for large-scale, mechanised production, the Canadian prairies are subject so various natural catastrophes, such as late or early frosts, droughts, dust storms, river-floods and grasshopper plagues, which may damage or even destroy the grain crop on occasions.

About 95% of Canada's wheat comes from the prairies. Of the total production of about 18 million tons about two-thirds are exported. Canada has a large and reliable market in the United Kingdom, which usually takes about one-third of the exports. The remainder normally goes to other Western European countries, South Africa, and Japan. Most of the prairie wheat moves eastwards through Winnipeg and thence to the lakeside terminals of Fort William and Port Arthur from where it goes by boat or rail to Montreal, Buffalo, and New York. A second movement is westwards to Vancouver, which is the chief Pacific outlet, or Prince Rupert, a subsidiary outlet. A third, but much smaller movement, is to Churchill on Hudson's Bay but this route is usually open only from late July to early October.

Although wheat is the chief crop of the Canadian prairies, two points are worth emphasising: first, other cereals are cultivated besides wheat, especially oats, which is mostly grown as an animal feedstuff, and barley, mainly used for malting; secondly, during the past few decades there has been a definite trend towards agricultural diversification and non-cereal crops, such as vegetables, sugar-beet, linseed, soya-beans and sunflowers, are now being grown in some areas. Increasing use is being made of irrigation.

United States as the world's premier producer of wheat but she exports very little. Until 1914 Russia was a major exporter.

*North America.* The central plains form the world's greatest producing area accounting for about a quarter of world total output. In the Canadian Prairie Provinces and in the northern states of the United States the wheat is spring wheat because of the hard, killing frosts. The winter wheat belt stretches from northern Texas, Oklahoma, and Kansas, the three most important producing states, eastwards to Pennsylvania. A third wheat-growing area of lesser importance is the Columbia Plateau west of the Rockies.

*South America.* In the temperate region of South America three areas are devoted to wheat production: in the Argentinian–Uruguayan pampas region where extensive cultivation is carried out on a large scale; in southern Brazil, especially in the state of Rio Grande do Sul; and in central "Mediterranean" Chile. Argentina, the most important producer, grows hard wheats because the climate ensures rapid maturity.

*Australia.* Australia is an important wheat-growing and exporting country. The temperate grasslands of the Murray–Darling Basin, especially those parts receiving between 15 and 20 in. (381–508 mm) of rainfall, and the areas around Adelaide and Perth having a "Mediterranean" type of climate are the chief wheat lands. Hard wheats are grown under the extensive system.

*Monsoon Asia.* In the northern part of West Pakistan and India wheat is grown as a winter crop for although this is the cool season temperatures are high enough for wheat. In the Punjab wheat is grown mainly with the aid of irrigation. In China the principal wheat-growing area is the Hwang-ho lowlands.

## *BARLEY*

Barley was one of the first grains to be cultivated and was originally the chief breadstuff. Although it continues to be used in several countries as a human foodstuff, nowadays it is more commonly used as an animal feedstuff and for the making of alcoholic drinks, *e.g.* whisky and beer, the former being prepared by distillation, the latter by the fermentation of malted barley.

Generally speaking, barley will grow wherever wheat will grow, but it also has a wider range than wheat; indeed, barley has the greatest geographical range of all the cereals. Since it will ripen at low temperatures, barley will grow in higher latitudes than any other cereal and occasionally is to be found beyond the Arctic Circle. On the other hand, it is more tolerant of dry conditions than wheat and will flourish in hot, arid regions, and, also, on lighter more porous soils. Barley, however, is more sensitive to moisture than oats so that in damper areas oats often replace barley. Barley also gives a higher yield per acre than wheat or the other temperate cereals.

World production of barley is only about 60% of that of wheat. In 1970 total world production amounted to 130 million tons. Almost all of this was grown in the northern hemisphere; scarcely any barley is grown south of the equator. The chief producers are the Soviet Union, the United States, France, Canada, the United Kingdom, Turkey, West

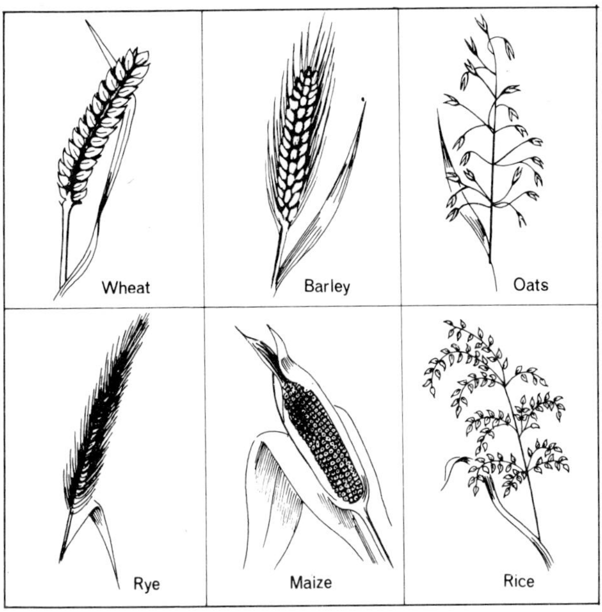


FIG. 39.—TYPES OF CEREAL

Germany, and Denmark. Very little barley enters into international trade, usually only about one-twentieth of total world production.

## *OATS*

In his famous dictionary Samuel Johnson defined oats as a grain fed to horses in England and to men in Scotland. When this definition was quoted to a Scotsman he cannily retaliated with: "Yes, and that is why you have such good horses in England and we have such fine men in Scotland!" Oats are grown both as a foodstuff and a feedstuff and are equally good for humans and animals. Unfortunately, oats cannot be made into bread though some are made into oatmeal which is used in the making of oatcakes, biscuits, and porridge. Oats are still important as a foodstuff in Scotland, Scandinavia, and Newfoundland, countries in which other cereals do not thrive very well. But the great bulk of the oats grown, some 55 million tons in 1970, mainly in the United States, Canada, and the Soviet Union, are used for feeding animals and oats rank second only to maize as a fodder crop. Both the grain and the straw can be used as feeding stuff.

Oats are the characteristic crop of the cooler, moister regions. They are more tolerant of moisture than either wheat or barley and also less exacting in their soil requirements. However, they need a slightly longer growing season than barley and so are not found in as high latitudes. This longer growing season restricts their usefulness in lands in higher latitudes just as their need for moisture places a limit upon their cultivation in warmer, drier areas. As in the case of barley, very little oats—usually less than 2 million tons—enters into world trade.

## *RYE*

Rye has long been known as the poor man's crop. From it is made the "black bread" eaten by the peasants and town people of central and eastern Europe. Over much of Europe rye is still the chief breadstuff, although it is now often mixed with wheat to make a more palatable and whiter bread. Rye, incidentally, is the only other grain besides wheat which will make good bread. Pure rye bread is very nutritious if not very appetising to look at. Considerable quantities of rye are used in the making of whisky (rye whisky) and vodka. But perhaps the chief use of rye these days is as a feedstuff for livestock, especially cattle and pigs, and an increasing proportion of the crop is being used for this purpose. Of the major cereals, however, rye is the least important and the total world production is slightly more than 30 million tons of which nearly half is grown in the Soviet Union. Poland, East Germany, and West Germany are substantial producers and small quantities are grown in the United States, Canada, Argentina, Austria, Czechoslovakia, and Turkey.

Rye is typically a crop of cool, damp climates and will thrive in poor soils, such as infertile sands. It is a hardy crop and will grow in areas which are too cold for most cereals and under a wide range of rainfall. In Europe, where about 90% of the world output is produced, rye is usually grown to the north of the wheat belt, *i.e.* in the wide zone of mainly infertile glacial sandy soils which stretches from the heathlands of east Holland across the North European Plain as far as the Ural Mountains.

## *MAIZE*

Maize or Indian corn, known by Americans simply as corn, is native to the western hemisphere. Since the discovery of America, maize has been carried throughout the world and is now an important crop in every continent; in Oceania alone it is relatively unimportant. Although maize can be ground into meal for human consumption, it does not make good bread; even so, it forms the staple foodstuff of many areas, notably in Mexico and the Central American Republics, where it is made into tortillas, flat cakes which are eaten warm, in southern Europe, notably in Italy where maize meal is made into polenta, and in South Africa where it is eaten by the Negroes. In the United States some maize is eaten as a vegetable (corn-on-the-cob), as a breakfast dish (hominy), and in its roasted form (popcorn). Some maize is ground up into corn flour used in bakery and as a constituent of blancmanges and custard powders. Some, too, is used for the manufacture of corn oil, starch, alcohol, and certain alcoholic drinks.

Maize is most important, however, as an animal feeding stuff and the grain, corn meal, and the growing plant are all used. It is the best cereal for fattening stock and is fed to millions of cattle and pigs in the United States Corn Belt; hence the saying that grain is "marketed on the hoof" and "more than one-third of the American corn crop squeals as it goes to market."

## CONDITIONS OF GROWTH

Although maize grows under a considerable range of environmental conditions, for optimum growth and production the following conditions may be said to be necessary: (*a*) a growing period of at least 140 days free from frost; (*b*) an average summer temperature of between 21° and 27° C (70° and 80° F); (*c*) warm nights with the temperature averaging about 14°C (58° F); (*d*) an annual rainfall of between 25 and 50 in. (635 and 1270 mm); (*e*) frequent showers of light rain during the growing period; (*f*) plentiful sunshine during the ripening period; (*g*) a rich, well-drained, deep soil abundant in nitrogen.

It will be clear from the foregoing requirements that maize is essentially a warm-weather plant needful of high temperatures both by day and night during the growing period. Although it will grow, it will not grow well nor will the grain mature in areas having a cool summer or where the mean night temperature falls below 13° C (55° F). But probably the crucial factor in its growth is moisture and deficiency in rainfall is the principal cause of damage to the crop.

## WORLD PRODUCTION

Although maize cultivation is widely distributed throughout the world, nine major areas of production can be distinguished: (*a*) the Middle West of the United States; (*b*) the Central American Highlands; (*c*) the south-eastern part of Brazil; (*d*) the humid pampa of Argentina; (*e*) the Danubian Basin; (*f*) the Caucasus–Black Sea borderlands of the Soviet Union; (*g*) the Punjab–Middle Ganges Lowland; (*h*) central and northern China; (*i*) the Republic of South Africa.

The total world production of maize in 1969 was 264 million tons, only a little less than the total wheat output. Present-day production is some 100% higher than in pre-war days; this increase is largely due to the development of high-yielding hybrid types of maize produced by selective cross-breeding. Thus, although there has been a general reduction in the total world area devoted to maize, especially during recent years, the actual maize yield has continued to increase.

TABLE IV

*Maize Production (thousand metric tons)*

| | *1963* | *1970** |
|---|---|---|
| U.S.A. | 103,933 | 104,135 |
| China (estimated) | 21,440 | 25,000 |
| Mexico | 6424 | 9000 |
| South Africa | 4190 | 6113 |
| U.S.S.R. | 23,461 | 9428 |
| Brazil | 7700 | 14,206 |
| Argentina | 4527 | 9360 |
| Rumania | 6023 | 6113 |
| Yugoslavia | 5380 | 6928 |
| World total | 226,200 | 260,042 |

These figures indicate the dominance of the United States as a maize producer. Note the big drop in Soviet production between 1963 and 1969.

The great bulk of the world's maize crop (some 80%) is consumed by animals, and some three-quarters of it never leaves the farm on which it is harvested. World trade in maize, which is small, is due chiefly to the demands of such countries as have intensive animal industries and require large quantities of maize as a feedstuff. The United States is the world's leading exporter although only about 3% of its output finds its way into the world market.

## *MILLETS*

The term millet covers a variety of species of cultivated grasses; these various millets form the humblest members of the grain crops. The many kinds may be grouped into either large millets such as *kaoliang* and *jowar* and small millets such as *cat-tail millet*. Millets are a low-grade food and their cultivation is an indication of a low standard of living. Generally

* Here, as in other Tables of this type, the student is recommended to discover and fill in the most recent figures as these become available. Note that here, as in all other Tables, the world total figure includes producers not listed.

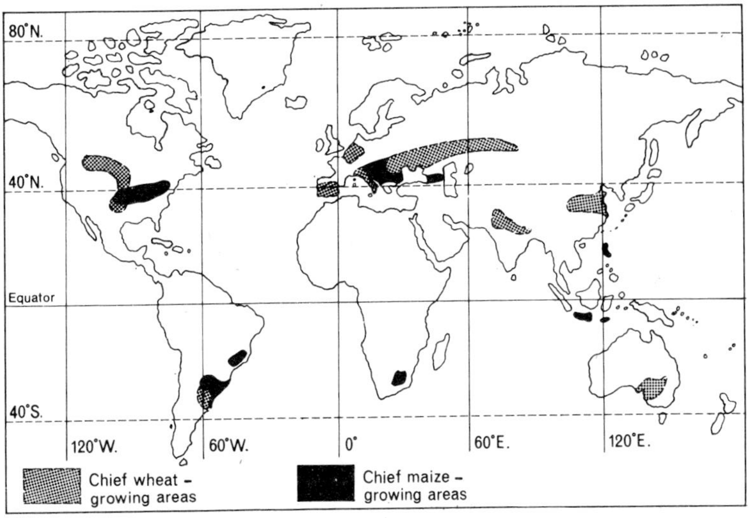


FIG. 40.—DISTRIBUTION OF WHEAT AND MAIZE

speaking, they form the staple food of peasant farmers in the hot, rather dry areas of the world. In tropical and sub-tropical regions of light rainfall, where water supplies are insufficient for the cultivation of either wheat or rice by irrigation, millet forms the chief grain crop.

Sorghum or Great Millet is much grown in Africa where it is also called *durra* and *guinea-corn*. In the Indian sub-continent, the Great Millet, known as *jowar*, and the smaller spiked millet, known as *bajra*, are important crops occupying a greater acreage of land than any other crop apart from rice. Kaoliang, a giant millet, grown in northern China and Manchuria is valuable both as an animal and human food. The millets, widely cultivated as a subsistence crop, are also grown in the United States and in southern Europe as a fodder crop. One type of millet, the so-called cat-tail millet, yields very tiny grains and is frequently sold for bird-seed.

## *RICE*

Rice is the staple foodstuff of most of the peoples of south-eastern Asia. Its high productivity together with the great demands for labour in its cultivation make it a crop well-suited to an area where some 40% of the world's people are concentrated on a mere tenth of the earth's land surface. Rice cultivation is by no means confined to the monsoon region of Asia, but the overwhelming proportion of world production comes from this area. Although rice can be ground into flour and ground rice is used in cake-mixtures, it cannot be made into bread; hence rice-eaters usually boil their rice and frequently make it more palatable by mixing it with morsels of meat, prawns, eggs, etc.

## METHODS OF CULTIVATION

The rice plant, like wheat, is a member of the grass family. There are several species of rice plant which grow under diverse conditions of climate and soil but two main classes, of which the second is easily the more

[Courtesy: *Indonesian Embassy.*

FIG. 41.—RICE FIELDS

These terraced rice fields are in West Java; they are typical of the methods of rice cultivation used in the Monsoon Lands.

important, are recognised: (*a*) upland or hill rice cultivated on hill slopes receiving plentiful and frequent rainfall; (*b*) lowland or swamp or wet rice which is grown in level flooded fields and often by irrigation.

The cultivation of wet rice differs from that of all the other cultivated grains in that it is grown in water during the earlier part of its growth period. Natural flooding or irrigation are, therefore, necessary. Usually, in Asia, the rice plants are first raised in field nurseries under water and then are transplanted by hand into the growing fields where they are still immersed for several weeks. This, and subsequent attention, calls for unremitting labour and much skill, patience, and care. Only a plentiful supply of workers could meet the heavy demands which rice growing places upon labour. Rice is a fast-growing plant and in the later stages of its growth the water is drained from the fields. The grain is hand harvested, usually being cut with a sickle, and then left to dry before threshing.

In some areas outside the monsoon region as, for example, in Louisiana in the United States and in the Rhône delta in France, rice is sometimes grown by mechanised farming methods. Recent reports from China indicate that mechanised techniques of rice-farming are being rapidly introduced there, too. Developments of this kind in the monsoon lands are likely to lead to revolutionary changes in the traditional agriculture.

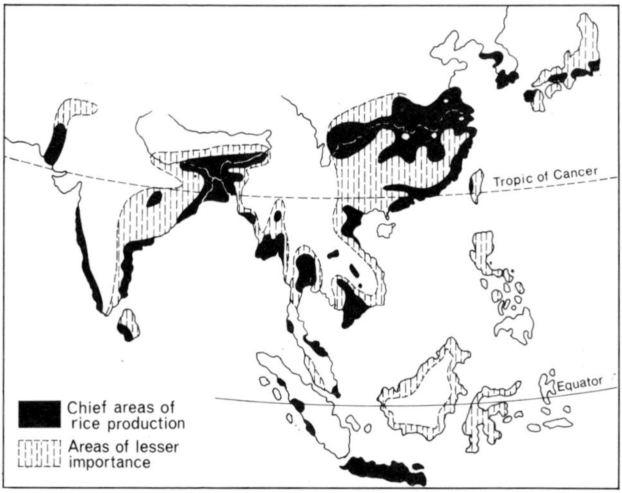


FIG. 42.— RICE-PRODUCING AREAS OF THE MONSOON LANDS

The great bulk of the world's rice production comes from the areas shown on this map. The chief areas are the level alluvial lowlands which are naturally flooded or can be easily irrigated. The great deltas of the Ganges–Brahmaputra, Irrawaddy, Menan Chao Phya, Mekong, Red, Si, and Yangtse Rivers offer perfect conditions for paddy.

## GEOGRAPHICAL REQUIREMENTS

Although rice is grown under differing climatic conditions, the primary requirements for its successful cultivation are: (*a*) level land (either naturally so, such as deltaic plains, or made so by terracing) with an impervious subsoil to hold the water; (*b*) abundant moisture during the early period of growth; this means a rainfall of about 60 in. (1524 mm) a year or compensating irrigation water; (*c*) temperatures of between 18° and 20° C (60° and 68° F) during the early growing period and about 24° C (75° F) during the ripening period; (*d*) alluvial soils are to be preferred but rice is not a soil-exhausting crop and it can be grown year after year in the same field with little ill-effects; (*e*) a plentiful supply of cheap labour since sowing, transplanting, etc., is tedious, time-consuming work.

## PRODUCTION AND TRADE

Total world production of rice in 1968 was 284 million metric tons—approximately the same as wheat. China and India are the major producers. Both these countries have something of the order of 75 million

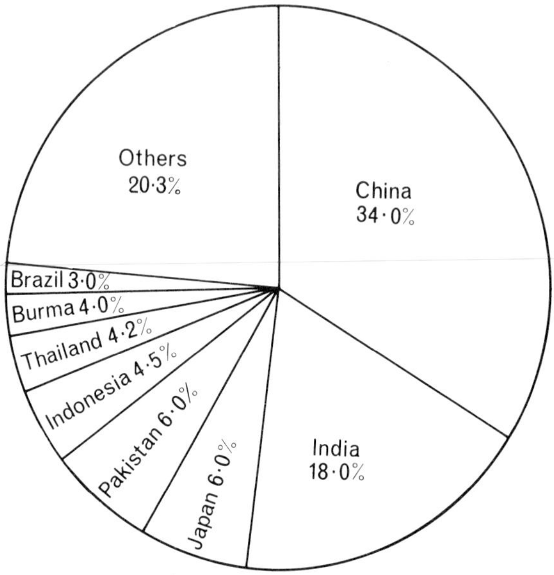


FIG. 43.—WORLD RICE PRODUCTION

This diagram is a kind of graph; geographers call it a pie-graph. It is simply a divided circle. It has the advantage of showing figures visually: one can see and estimate proportions at a glance. Pie-graphs are easy to construct. A circle is divisible into 360 degrees; hence, if we think of the whole circle as 100%, 3·6 degrees is equal to 1%. To plot a given set of figures, such as the ones for world rice production, we must reduce those figures to percentages and then multiply the percentage figures by 3·6; for example, India would be $18 \times 3{\cdot}6 = 64{\cdot}8$. In a circle, drawn to a given size, this number (remember, they are now degrees of a circle) can be marked in from a radius by using a protractor and reading off 64·8 degrees. Any other number can be dealt with in the same fashion.

acres (30 million ha) devoted to rice. Substantial areas are given over to rice production in all the monsoon lands of Asia and the East Indian region. Outside the Asian area, rice is grown in Egypt, Madagascar, West Africa, Brazil, the Guianas, and Panama, and it is also cultivated on a commercial scale in a few localities where summer temperatures are high enough and where there are facilities for irrigation, as along the Gulf Coast of the

United States, the Mediterranean coastlands of Spain, in the Rhône delta of France, in the Po valley of northern Italy, and in the Tisa valley district of Hungary.

TABLE V

*Rice (Paddy) Production (thousand metric tons)*

| | *1963* | *1970* |
|---|---|---|
| China | 85,000 | 102,000 |
| India | 54,734 | 63,672 |
| Japan | 17,157 | 16,479 |
| Pakistan | 17,724 | 8300 |
| Indonesia | 11,764 | 17,529 |
| Thailand | 10,168 | 13,270 |
| Burma | 7457 | 8162 |
| Brazil | 6200 | 7553 |
| S. Vietnam | 5327 | 5716 |
| World total | 257,500 | 307,657 |

Most of the rice which is grown is consumed in the country of its origin and there is little international trade in rice. Only about 5% of the total production enters into world trade and most of this trade is carried on between the monsoon countries of Asia, from rice-surplus to rice-deficit areas. Burma and Thailand are traditionally the most important rice exporters and account, respectively, for about 30% and 25% of the world's rice export trade; in both cases, however, they export less than a quarter of their total production. A recent trend has been the export to the Far East of rice from the western hemisphere. The appearance of the United States as an exporter (on a small scale) is a new feature of the world market.

EXERCISES

1. Locate precisely the main wheat-producing areas of the world. What factors control the successful growth of this crop? Name the main countries with a surplus for export and note their chief markets.

2. The average yield of wheat, in bushels per acre, in Denmark, Holland, and England is more than twice that in Canada and Australia. Give reasons for this difference.

3. What are the physical conditions that have helped to make Canada the greatest wheat exporting country in the world? (*Chartered Institute of Secretaries.*)

4. "Barley cultivation has the greatest geographical extension of all cereals." Discuss this statement.

5. (*a*) State precisely the conditions required for the cultivation of rice. (*b*) Explain why the minor rice producers tend to be the major exporters.

6. Write a brief account of *two* of the following: (*a*) wheat-growing in the Soviet Union; (*b*) maize cultivation in South Africa; (*c*) rice-growing in China; (*d*) rye cultivation in Europe.

7. Major cereal-producing countries are not necessarily major exporters of cereals. Discuss this statement with reference to maize and rice.

8. Give an account of the world trade in wheat, noting (*a*) the major producing countries, (*b*) the chief markets for wheat, and (*c*) the main routes by which wheat reaches its destinations.

# Chapter XI

# PLANT FOODSTUFFS: OTHER VEGETABLE PRODUCTS

## *VEGETABLE FOODSTUFFS*

EXCEPT among a limited number of people, who are basically flesh or fish eaters, various vegetable crops, after grain, form the chief article of diet. In some areas, where cereals cannot be, or are not, grown, vegetable foods take their place; this is especially true of certain tropical areas. Vegetable crops are necessary for good health, particularly the so-called green vegetables, although a diet composed solely of vegetable foods is lacking in balance and should be supplemented by other foodstuffs of a different category such as fruit, fish, meat, and milk. Here we are concerned with the chief vegetable foodstuffs consumed by man; broadly, they fall into four main groups:

1. Starchy tubers.
2. Root vegetables.
3. Pulses.
4. Green vegetables.

### STARCHY TUBERS

A number of tuberous plants (*i.e.* plants which develop root swellings ripening underground) have become important food staples in some parts of the world. In some cases they replace grain foods, in others they merely supplement cereals. Tubers are starchy foods and less nutritious than grain. They are, however, easily cultivated. The chief tubers used as food are yams, manioc, sweet potatoes, and potatoes.

*Yams.* These are the milky edible tubers of various species of *Dioscorea*, small bush-like plants which grow in the humid tropics. More than two hundred species are known. Various of them are grown in most tropical countries. They are common in tropical America, in Africa, in South-east Asia, and in Polynesia. Yams vary in size from that of a large potato to that of a marrow and the so-called "winged yam" cultivated in Polynesia may reach a length of up to 3 feet (0·9 m). Yams are generally cultivated in primitive fashion by sedentary and shifting peasant subsistence farmers. They may be eaten boiled, like potatoes, or roasted.

*Manioc.* This is another starchy tuberous plant which grows in most tropical regions but is especially important in South America where it is widely eaten, notably in Brazil; it is also cultivated in wet tropical Africa and in equatorial Asia. While it is possible to eat the roots of the sweet manioc fresh or roasted, the root of the bitter manioc contains prussic acid and this must be extracted before the tubers can be eaten. The roots are ground-up, steeped in water, and heated; the water is then squeezed out and the starchy pulp dried and then ground into flour, which is known as cassava. Tapioca, with which we are all familiar, is made from cassava. Cassava is usually eaten in the form of a porridge, often accompanied by some more highly seasoned food.

*Sweet potatoes.* The sweet potato, which is very similar in appearance and use to the potato grown in Europe, is cultivated in tropical and warm temperate lands. It is common in Central and South America, which is probably its original homeland, in the more southerly parts of Eastern Asia, where it is often grown during the dry season by irrigation, and in many of the Pacific Islands where, along with coconuts and fish, it forms the basic diet of the people.

*Potatoes.* The common potato, a native of the Andes, traditionally is said to have been introduced into Europe by Sir Walter Raleigh who brought it from Virginia to Ireland in 1586; this myth has been exploded, for Raleigh did not introduce any plant from the Americas into the British Isles at this time. The potato is now grown in practically every country in the world. Potatoes are grown in very large quantities by the white peoples but they are especially important in France, the Low Countries, West and East Germany, Czechoslovakia, Poland, and Russia. In these countries the potato is the staple food for millions of people; in other European countries it forms an important article in the diet. The potato is also widely grown in the United States and Canada, especially in those areas where the summers are inclined to be cool and moist. Potatoes form a valuable energy food, though they are inferior in their nutritive value to grain. They are a very useful crop since, under good conditions, they yield well, producing a greater quantity of food per acre than any of the cereals. Most of the potato crop is consumed as a human foodstuff but some (in parts of Europe as much as 40%) is used as animal fodder. Potatoes are also used for the production of starch, dextrose (glucose), industrial alcohol, and potable spirit.

## ROOT CROPS

Similar to the starchy tubers in that they are root swellings and grow partially underground are root crops such as carrots, parsnips, turnips, swedes (Swedish turnips), and mangolds or mangel-wurzels (the German term). All these are essentially crops of the cool temperate lands although the carrot is often cultivated in the oases of the hot deserts. Root crops play an integral part in the crop rotation systems of modern scientific farming. Carrots and parsnips excepted, root crops are mainly grown as fodder crops. They contain little nutriment and the turnip, for instance, consists of 90% water. The red carrot is a native of Britain. Mainly a human food, it may be used as an animal feeding stuff. On the Continent, a white variety is common. Beet is not commonly used as a table vegetable but one variety of beet is the sugar-beet which is important as a source of sugar (*see* p. 164). The mangold is actually a variety of beet. All root crops, whether they be used as human or animal food, are consumed locally, for they are bulky, low-value commodities incapable of standing high transport costs.

**PULSES**

This is a general term for leguminous plants including peas, beans, soya-beans, groundnuts, lentils, etc. The word pulse is usually reserved for those species of the natural order of *Leguminosae* which supply seeds or pods capable of being used for food. The food value of the pulses is mainly due to a protein (known as legumin) which is present in the seeds. Peas grow in cool climates and are mostly, though not entirely, grown in the temperate zone. Beans, of which there are many varieties, are widely grown as they flourish under a great range of climatic conditions. They may be cultivated for human food, as in China and Brazil, or as an animal feedstuff. Chick-peas, known as *gram* in India, are an important crop and article of trade in the Mediterranean lands of Europe and north Africa. They are a common item in the diet of Iberian peoples both in Europe and Latin America; also in India and Pakistan. The chick-pea is a valuable foodstuff in areas where meat is scarce as also is the soya-bean. The latter, of which there are several varieties, is indigenous to Eastern Asia and is much cultivated there, particularly in China (especially in Manchuria). Soya-bean cultivation has spread to the United States where it is grown in very large quantities but chiefly as a source of vegetable oil (*see* p. 209). Groundnuts (pea nuts, monkey nuts) are widely grown in the world in warmer latitudes, especially in China, India, Java, the Sudan zone of Africa, and in the south-eastern part of the United States. Lentils provide a very nutritious food and are much cultivated in India and in the region around the Mediterranean Sea.

**GREEN VEGETABLES**

A variety of vegetables, many of them belonging to the cabbage family (*e.g.*, Savoys, broccoli, cauliflower, Brussels sprouts), are grown for human consumption. They provide good food and green vegetables are an important item in a health-giving, balanced diet. It was the lack of fresh vegetables in the diet of the sailors of earlier times that led them to suffer from the disease of scurvy, which was caused fundamentally by the absence of certain vitamins contained in vegetables. Vegetables are grown as subsistence crops by numerous farmers the world over, especi-

[*Courtesy: Elders and Fyffes, Ltd.*

FIG. 44.—BANANA PLANTATION

Here, on a banana plantation, the "backer" is taking the weight of a bunch of bananas while the cutter prepares to sever it from the parent plant. Each plant bears only one fruiting head of bananas. Note the size of the leaves and the loose vegetable matter covering the ground. It is possible to use this "trash" for making paper, as in Ecuador which is a leading producer of bananas.

ally in garden-plots near their living quarters. Many vegetables, however, together with salad crops, such as lettuces, spring onions, etc., are intensively cultivated on small specialised plots for nearby markets. Fresh vegetables have a ready sale in urban areas.

## *FRUITS*

Fruits fall into three principal classes:

1. Tropical fruits: bananas, pineapples, dates, coconuts, and Brazil nuts.
2. Sub-tropical and warm temperate fruits: oranges, lemons, limes, grape-fruit, figs, grapes, almonds, and walnuts.
3. Deciduous fruits: apples, pears, plums, peaches, and the various "soft" fruits and "small" fruits.

### TROPICAL FRUITS

These are the fruits grown in equatorial and tropical latitudes. There are numerous types but only a few enter into foreign trade; those that do are, however, of considerable importance. There is a great trade in bananas which are grown in the West Indies, the Central American Republics, Ecuador, Brazil, and the Canary Islands in large quantities. The banana tree requires hot, moist conditions. For export, the bananas are picked green and transported in special temperature-controlled holds.

Pineapples are grown in tropical regions, especially in areas near the sea. They prefer light, rich soils. Most of the pineapple entering international trade is in the canned form, either fruit or juice. The chief producing and exporting countries are Hawaii, Malaya, the Philippines, Mexico, and Cuba. Hawaii produces about half of the annual output of pineapples amounting to 1½ million tons; production is on highly organised commercial plantations. The bulk of the international trade in the fresh fruit comes from Cuba and Mexico.

Dates come from the datepalm which likes to have "its feet in heaven (water) and its head in hell (burning hot sunshine)." The datepalm flourishes in the oases of the hot deserts. Most of the export dates come from the palmeries of Iraq, Algeria, Morocco, and Tunis. Iraq is *the* outstanding producer and exports about two-thirds of its total output.

The coconut is of little importance as a fruit but of great value for copra (*see* p. 207). Brazil nuts grow wild in Amazonia and are collected for export. The Amazon river ports of Manaus and Pará (Belém) are the great collecting and shipping centres.

## SUB-TROPICAL AND WARM TEMPERATE FRUITS

From the point of view of world commerce, these may be said to be covered by citrus fruits, grapes, and nuts.

The citrus fruits include the orange and its relations the tangerine and the mandarin orange, the lemon, the grape-fruit, and the lime. The orange came originally from China although it has come to be looked upon as a typical "Mediterranean" fruit. The orange, however, demands warm winters and water throughout the year, hence irrigation is required. The chief areas of commercial production in the Mediterranean Basin are Israel, where the famous Jaffa oranges are grown, the coast lands of South Italy and Sicily, and the coast lands of Valencia and Andalusia in Spain. Andalusia is well-known for its production of bitter Seville oranges which are exported for the making of marmalade. It is said that during harvest time one million oranges are despatched to Britain every day for the marmalade industry! Other important producing and exporting areas of oranges are California, Southern Brazil, South Africa, and South-western Australia.

Lemons are even more sensitive to cold than oranges and will thrive only in frost-free areas. The bulk of the world's lemons come from Sicily. Mandarin oranges, chiefly for canning, come from Algeria. The grape-fruit has become popular during the past few decades and is now fairly widely grown in sub-tropical areas; noteworthy producers are Florida, British Honduras, and South Africa. Limes, cultivated mainly for limejuice, come chiefly from the West Indies.

Grapes are marketed as wine, as dried fruits (currants, sultanas, raisins, and muscatels), or as table grapes. Currants, made from small seedless grapes, are a speciality of Greece and particularly of the area around Corinth; the dried grapes were known as *raisins de Corauntz*, a term which eventually became anglicised as currants. Nowadays, Australia produces large quantities of currants. Raisins are produced chiefly in southern Spain, Greece, and Turkey; the state of California in the United States is also a notable producer. Note that the dry, sunny summers of "Mediterranean" regions are ideal for fruit-drying.

Considerable quantities of nuts are produced in the Mediterranean Lands, chiefly almonds, walnuts, chestnuts, and filberts. Spain, Italy, and Turkey are the leading producers.

## DECIDUOUS FRUITS

Deciduous fruits grow in both warm temperate and cool temperate latitudes. Stone fruits, such as apricots, peaches, and plums, for example, are common to both zones.

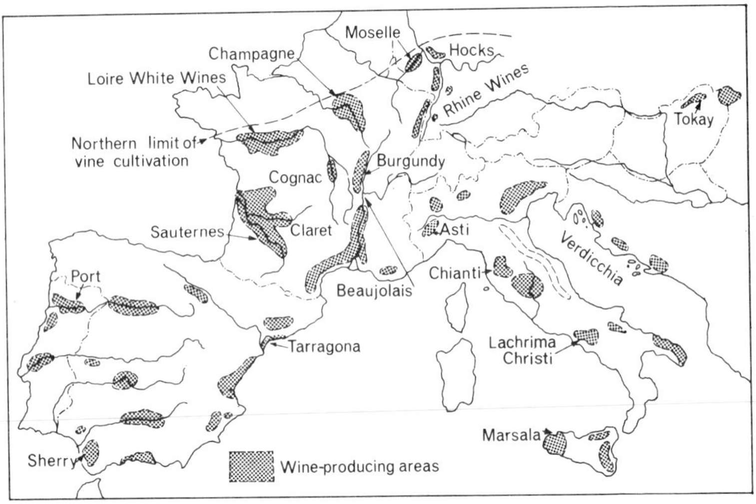


FIG. 45.—WINE PRODUCTION

Wine can be made from a variety of vegetable crops but the grape is the source of most wines. The growing of vines is known as viticulture. The production of high-quality grapes demands great care and skill. Though essentially a fruit of "Mediterranean" regions, grapes may be, and are, grown outside such climatic regions. The grape requires average summer temperatures of 18–21° C (65–70° F,) abundant sunshine, and moderate rainfall in winter. It is not very fussy about soil conditions but the soil must be well drained. It does thrive well, however, on dry chalk and limestone soils. The kind of wine depends very much upon the soil, for the same vine will produce different wines if grown in different soils. The quality of the wine depends very largely upon the weather conditions during the ripening and harvesting period—these conditions are known as the vintage—and upon the fermentation process and subsequent storage. The lands around the Mediterranean Sea are *the* great wine producers, accounting for 75% of the total world production. France, Italy, and Spain are the greatest producers. France is the leading world exporter. Spain is famous for sherry, Portugal for port. Outside Europe and Mediterranean north Africa wine is of relatively little importance although some wine is produced in the Cape Town area of South Africa, in southern Australia, in the oases of western Argentina, in central Chile, and in southern California.

| *Leading world producers of grapes (percentage of total world production)* | | *Leading world producers of wine (percentage of total world production)* | |
|---|---|---|---|
| Italy | 20 | Italy | 26 |
| France | 18 | France | 22 |
| Spain | 8 | Argentina | 10 |
| Turkey | 7 | Spain | 8 |

Apples and pears, both native fruits of Western Europe, are widely cultivated in the temperate belts of both the northern and southern hemispheres. Apples are much the more important of the two. In fact, greater quantities of apples are produced than any other fruit; world total output is about 20 million tons a year. The chief producing countries are: the United States (particularly in the north-west), Canada

[*Courtesy: High Commissioner for Canada.*

FIG. 46.—APPLE FARMING, ANNAPOLIS VALLEY

The Annapolis valley in Nova Scotia (Maritime Provinces of Canada) is a long, narrow, sheltered valley famous for apple-growing, although some pears, cherries, plums and small fruits are also grown. Much of the fruit is now converted into bottled apple juice. The Lake Peninsula of Ontario and the Okanagan Valley in British Columbia are other important Canadian apple-growing areas.

(British Columbia, the Lake Peninsula of Ontario, and Nova Scotia), the Soviet Union, France, Western Germany, and Italy. In France, about 90% of the production consists of the cider variety. The United Kingdom and Western Germany are the major importing countries.

The United States is the leading grower of both apricots and peaches. Italy and France are important producers of peaches. The international trade in fresh apricots and peaches is small because of the difficulty in transporting them, but the fruits in their dried and canned form are important commodities of trade.

Plums exist in many varieties, including damsons and greengages.

Plums are grown chiefly in Western and South-eastern Europe, the Soviet Union, and the United States (California). Italy is a leading exporter of fresh plums, Yugoslavia of dried plums (prunes). As in the case of cherries, a small proportion of the plums is used in the making of liqueurs.

The so-called "small fruits," which include strawberries, raspberries, blackberries, gooseberries, redcurrants and blackcurrants, bilberries, etc., are cultivated chiefly in cool temperate regions.

## *BEVERAGES*

Tea, coffee, cocoa, and yerba maté are the world's chief beverages. The first three are familiar, the last less so. Yerba maté is a drink, similar to tea, which is very popular in South America, especially the A.B.C. countries, *i.e.* Argentina, Brazil, and Chile. While it is usually drunk cold, tea, coffee, and cocoa provide hot drinks which are much appreciated in temperate and cold lands because they are warming, stimulating, and refreshing. However, the drinking of iced coffee, cocoa, and even tea has become something of a vogue in recent years.

All four beverages are grown in tropical or sub-tropical regions. Cocoa is essentially a product of the equatorial lands and will not grow far from the equator. Tea, coffee, and yerba maté are mostly grown in lands near the tropics or just outside. Whereas the valuable parts of the tea and yerba maté plants are their leaves, the important parts of the coffee and cacao trees are their seeds, which are known as "beans." Note that we have said cacao tree—not cocoa tree; cocoa is the processed bean of the cacao tree.

Again, all four beverages are plantation crops though yerba maté may be, indeed often is, collected wild, while tea may be grown, as it sometimes is in China and Japan, as an individual farm crop. Most of the tea and coffee are produced on large estates. Cacao may be grown on big estates but the greater proportion of it, coming from West Africa, tends to be raised on small, native-owned "plantations" or plots.

### TEA

Tea, as understood in commerce, is the dried leaf of a small evergreen shrub which is native to, and is mainly grown in, the monsoon area of South-eastern Asia. There are several varieties of tea plant but the real distinction is between the Indian and Chinese types, the former possessing thin, relatively large leaves which permit rapid transpiration and enable it to be cultivated under hot, humid conditions, and the latter having small, thick leaves which enable it to withstand a certain amount of cold and drought.

The tea-plant requires a long growing season of tropical or sub-tropical temperatures, around 24° C (75° F), but as it is a relatively hardy plant it is not injured by slight frosts. It needs abundant moisture, a minimum of 40 in. (1016 mm), with no upper limit, but it demands good drainage. Since it is intolerant of stagnant water in the soil, the tea-plant is customarily grown on hill slopes, although nowadays many tea-gardens are located on valley floors where these can be well-drained. On the whole, a better quality tea is produced on hill sites, but the estates have to face the problem

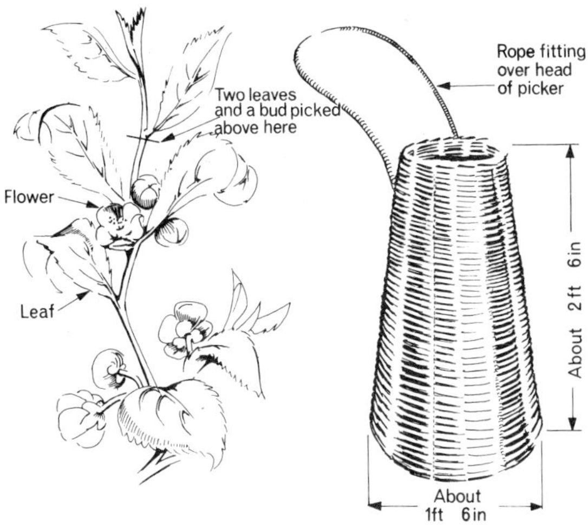


FIG. 47.—TEA-PLANT

of soil erosion; valley sites (if properly drained) are obviously easier to manage and, though the shrub produces a greater quantity of leaf, the tea is of a lower quality. The soil should be deep, fertile, and well-drained; otherwise, the tea-plant is not fussy about its soil requirements and it will thrive on a variety of soils, *e.g.* alluvial, volcanic, etc. One final point about its climatic requirements is worth noting: the seasonal distribution of temperature and rainfall is significant; whereas marked seasonal changes produce well-flavoured teas, uniform climatic conditions, such as one gets in equatorial regions—in Java, for example—give poor quality teas.

Tea-growing makes heavy demands on labour for propagation and planting, for pruning, for picking, and for processing. The tea bushes are pruned to keep them down to a height of about 4 ft (1·2 mm) and to produce a flat "top." The tea plant flushes quickly, *i.e.* the leaves grow quickly after a crop is picked, hence frequent pickings must be made, and in Ceylon as many as sixteen a year may be made. Picking is a skilled business and plenty of deft, cheap, hand-labour is needed. Mostly female labour is employed for picking. The leaf is taken to the processing plants on the estates where withering, rolling, fermentation, drying, sorting, and packing (in foil-lined chests to preserve the flavour and prevent deterioration in carriage) is undertaken.

The chief tea-producing areas in the world are: (*a*) India, in Assam in the north-east and in the Nilgiri Hills region of the southern part of the

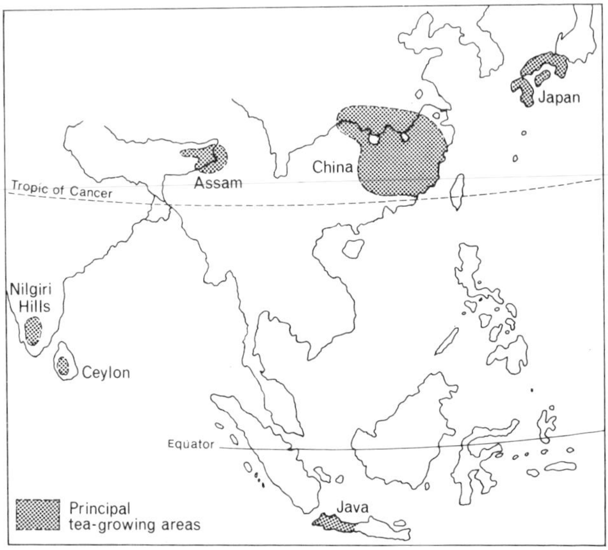


FIG. 48.—TEA-GROWING AREAS

Most of the world's tea comes from the areas depicted on the map. The Chinese and Japanese are tea-drinkers and little is exported from China and Japan. India and Ceylon are the two greatest commercial growers of tea. In the latter countries, along with Java, tea is cultivated on the plantation system, whereas in China and Japan the bush is more usually grown as part of the general farming system.

peninsula; (*b*) Ceylon, especially on the south-western slopes of the central hill country; (*c*) China, particularly in the hilly country to the south of the Yangtze, in the province of Fukien, and in the Szechuan Basin; (*d*) in southern Japan, especially in the foothills to the west of Tokyo; (*e*) in Indonesia, mainly in the western hill country of Java; and (*f*) in East Africa (Kenya, Malawi, and Natal). Some tea is also grown in Taiwan, Bangladesh, and the Soviet Union.

Tea has been grown in China from the very earliest times and until about a century ago China was the most important exporter. During the present century she has been surpassed as an exporter by both India and

Ceylon. These two countries supply a very high proportion of the tea that enters the world market. The bulk of their exports go to the United Kingdom which is the largest consumer of tea in the world. The tea is usually blended in the United Kingdom (London is the great tea market) and a proportion of it is re-exported to Ireland and some of the European

TABLE VI

*Tea: Chief Producers (thousand metric tons)*

| | *1960* | *1970* |
|---|---|---|
| India | 319·9 | 422 |
| Sri Lanka (Ceylon) | 197·2 | 212 |
| China | 158·8 | 172 |
| Japan | 78·0 | 91 |
| Indonesia | 41·5 | 67 |
| U.S.S.R. | — | 67 |
| Bangladesh | — | 31 |
| World total | 950·0 | 1346 |

countries whose peoples are also tea-drinkers. Most of the East African producers also find their market in the United Kingdom. Japan, Taiwan, and Indonesia export some of their tea to the United States (although the Americans are not great tea-drinkers) and to other countries.

## COFFEE

Coffee is a tree which attains a height of between 16 and 20 feet (4·8–6 m) and which yields a fruit, similar to the cherry, containing the coffee beans. Ethiopia is probably the home of the coffee tree. Originally grown in the western province of Kaffa, whence coffee derives its name, the plant was probably taken to Yemen, in the south-western corner of Arabia, where it was carefully cultivated to give the variety known as Mocha. From Arabia the coffee plant was introduced to Central and South America and today the majority of the world's coffee comes from the Americas. It is interesting to compare coffee with cacao for the latter, which had its origin in the Americas, is now mostly grown in Africa.

Coffee requires high, but equable, temperatures ranging from 17° C (63° F) in the coldest month to 22° C (72° F) in the warmest, and although the tree can withstand light frosts, heavy frost will quickly kill it. It needs shade from the direct rays of the sun and on this account is often grown under taller shade trees. The final requirement is a moderate rainfall of up to about 70 in. (1778 mm) with a marked summer maximum although, like tea, it must have good drainage and deep, rich soils.

Brazil, which has long been the world's premier producer, possesses in the hinterland of Rio de Janeiro ideal conditions for coffee cultivation. The coffee area lies on the gently sloping, westward slopes of the Brazilian Plateau, usually below 3000 feet (900 m) to avoid frost. The plantations, or *fazendas*, avoid the valley bottoms because of the risk of frost (cold air drains into the valley bottoms) and because the slopes are better drained. Here, the average summer temperature exceeds 21° C (70° F), the rainfall is high, about 60 in. (1524 mm), with a summer maximum and there is a relative dry season during the harvest period. Again, the prevalence of a morning mist-haze provides protection against the sun's rays. Here, too,

[*Courtesy: Brazilian Embassy.*

FIG. 49.—A COFFEE FAZENDA

A view of part of a coffee *fazenda* near São Paulo in Brazil. *Fazendas* are huge estates: many of them cover several hundred thousand acres. Parts are cleared of their tree growth and then the coffee saplings are laid out in orderly rows about 12 or 15 feet (4–5 m) apart on the gently sloping plateau surface or on the upper slopes of the river valleys cut in the plateau. The trees when grown are pruned to a height of about 8–12 ft (3–4 m) to encourage fruiting. Large numbers of workers are required on the *fazendas* with the result that communities develop, as many as 5000 strong sometimes, and a self-contained settlement with shops, schools, churches, store-houses as well as workers' dwellings grows up.

is to be found a particularly favourable soil, known as *terra roxa*, derived from weathered igneous rocks, which is deep and rich, contains much humus, and possesses iron and potash, chemicals which the coffee tree likes. When harvest time comes, the cherries are collected, spread out to dry in the sun, and then the beans are bagged. A network of railways threading the coffee area assists in the collection of the bagged beans which are taken to the ports of Santos and Rio for shipment abroad. Most of the *fazenda* workers are the descendants of Italian immigrants.

Colombia is the other important South American producer. Colombia specialises in the production of a finer, milder variety known as *café suave*. Coffee is cultivated at a greater height—in places up to an altitude of about 6000 feet (1800 m)—in Columbia because it is nearer to the equator. In central America coffee, usually of a very high quality, is

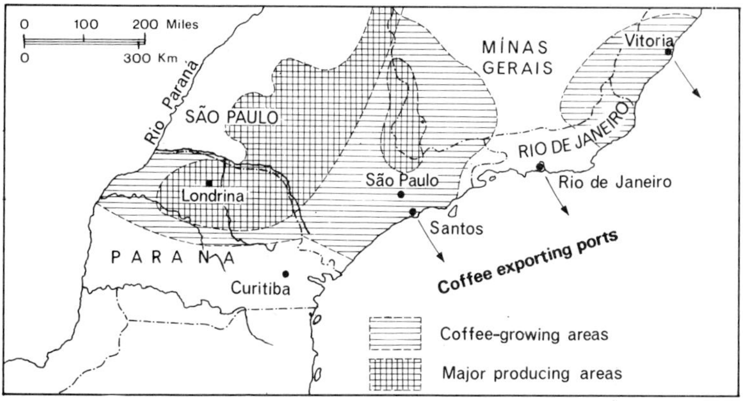


FIG. 50.—COFFEE-GROWING IN BRAZIL

The coffee region of Brazil lies in the hinterland of Rio de Janeiro approximately in the latitude of the southern tropic. The major producing area lies in the central districts of São Paulo State and northern Paraná. The great city of São Paulo is the "coffee capital." Santos is the chief coffee-exporting port although some goes via Rio and Vitoria.

grown on the western slopes of the cordillera where there is a dry season. Latin America produces over 75% of the world's coffee.

On account of the coffee plant's susceptibility to disease, and partly due to over-production, coffee has failed to compete for plantation space in Ceylon and Java, although some coffee is grown in south-eastern Java. Once cultivated in Ceylon, it has been replaced by tea. In many parts of tropical Africa, notably in Uganda, Kenya, Tanzania, and Angola,

TABLE VII

*Coffee: Chief Producers (thousand metric tons)*

| | *1960* | *1970* |
|---|---|---|
| Brazil | 1796·0 | 755 |
| Colombia | 456·0 | 570 |
| Ivory Coast | 185·1 | 240 |
| Uganda | 118·6 | 204 |
| Angola | 132·0 | 204 |
| Mexico | 123·0 | 185 |
| Ethiopia | — | 205 |
| Indonesia | — | 204 |
| El Salvador | — | 147 |
| World total | 3895·0 | 3985 |

small quantities of coffee are grown on frost-free hill slopes. Ivory Coast in West Africa is normally the largest African producer.

Of the world's production Brazil accounts for approximately 25%, Colombia 10%, Central America (chiefly El Salvador, Guatemala, Honduras) 12%, and Africa 14%. The United States is the leading world consumer absorbing some 40% of the total world production. Many continental countries, *e.g.* Holland, Germany, Switzerland, are important consumers. Since 1945 coffee consumption has shown an increase in the main coffee-consuming countries, but perhaps the most notable development has been the big increase in those countries, *e.g.* the United Kingdom, Canada, and Australia, which traditionally are tea consumers.

## CACAO

Cacao, in its native home and distribution, is almost the reverse of coffee. The cacao tree is native to Latin America. When Europeans first went to America they found that the Indians had a drink which they called *chocolatl*—our word chocolate clearly derives from this word. Cocoa (which is a corruption of cacao) was introduced to Spain and, in due course, to England. The English disliked the rather bitter taste of the cocoa and began to mix it with milk and to add sweetening to it. Although some cocoa is used as a beverage, the bulk of the cocoa produced is used in the manufacture of chocolate.

Cocoa is made from the dried beans extracted from cacao pods. The pods—the fruit of the cacao tree—grow from the trunk and main branches. Shaped something like a lemon but two or three times as large, the reddish-purple pods are cut from the tree and split open. The "beans," which are encased in a white pulp, are extracted and then allowed to ferment for several days, after which they are dried and bagged for export. In the importing countries the beans are processed: if cocoa is the desired product the beans are ground up, the fatty content (known as cocoa-butter) extracted, and the residue powdered; if chocolate is to be made, the powdered cocoa is mixed with a proportion of cocoa-butter and sugar along with milk if "milk" chocolate is desired.

The cacao tree is a plant of equatorial regions requiring high temperatures and a plentiful supply of moisture all the year round. It is grown in equatorial lowlands where temperatures of around 27° C (80° F) are maintained throughout the year, where a well-distributed annual rainfall of between 50–80 in. (1270–2032 mm) occurs, and where soils are deep and rich and, though moist, are not saturated. Shade trees are required, as in the case of coffee, and frequently the banana tree is grown to provide the necessary protection, to remove surplus moisture from the soil and also to shield the trees from wind. Optimum conditions are usually found in equatorial coastal lowlands which are a little removed from the littoral.

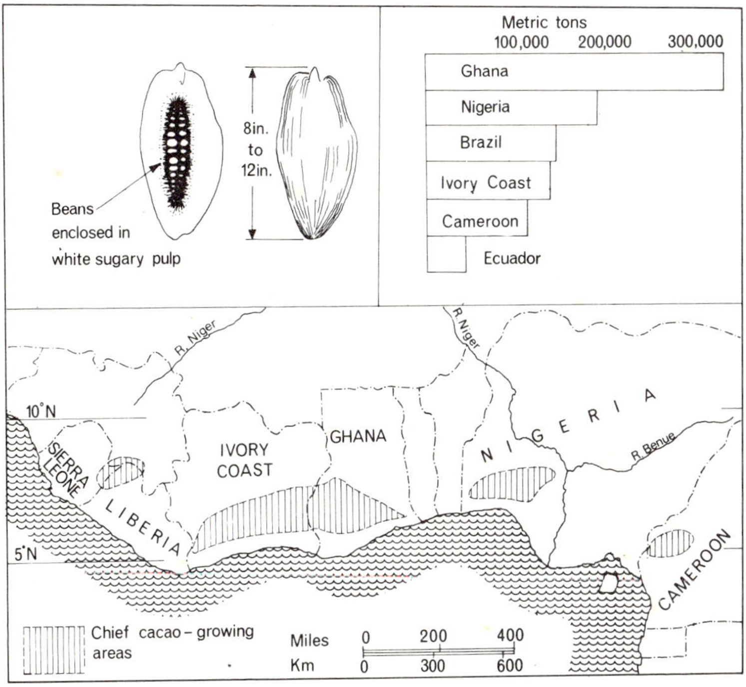


FIG. 51.—CACAO IN WEST AFRICA

About 80% of the world's total production (over 1¼ million metric tons) of cacao comes from the countries of West Africa. Cacao is a plantation crop but here production is very largely in the hands of the native inhabitants who grow the cacao trees in small plantations or on their farms—quite unlike plantation agriculture in most other areas of the world where it is usually large-scale and in the hands of Europeans.

The West African cacao growers are often quite prosperous. They depend on the sale of their crop for their money income; thus cacao is the "cash crop." The buying and selling of the cacao beans is also, to a large extent, in the hands of West African native middlemen. They, in turn, sell the crop to chocolate firms (*e.g.*, Rowntrees', Terrys', Cadbury's, and Fry's) who have representatives residing in West Africa.

Two main problems confront the native cacao farmer. The first is the occurrence of cacao tree diseases such as "swollen shoot" and "black pod." The only really effective remedy is to cut down and burn the affected trees. But the native farmer is loath to destroy what, in effect, is his livelihood, hence the diseases tend to spread instead of becoming checked. Secondly, when cacao prices are high, the cacao growers are tempted to devote too much of their land to their cash crop at the expense of their food crops. Unfortunately, if cacao prices should fall he may face starvation, partly because he has neglected to grow sufficient food for his needs and partly because he lacks the money to enable him to buy the necessary food.

Although cocoa is native to America, West Africa is now the world's dominant producing region accounting for about 80% of the total output. Ghana, Nigeria, Ivory Coast, and Cameroon are the principal West African producers. One notable point about production in West Africa is that, unlike tea and coffee which are grown on large estates often owned and managed by Europeans, the cacao is produced on native farms and small-holdings. In the Americas, Brazil is the chief grower, most of the cacao being grown on plantations in the coastal areas of Bahia state. Ecuador, though less important than formerly, is still a significant producer, and Ecuadorean cocoa is of high quality. Small amounts are also grown in the Dominican Republic, Trinidad, and Venezuela.

TABLE VIII

*Cocoa: Chief Producers (thousand metric tons)*

| | *1960* | *1968* |
|---|---|---|
| Ghana | 307·7 | 339 |
| Nigeria | 195·0 | 186 |
| Brazil | 180·0 | 156 |
| Ivory Coast | 90·0 | 145 |
| Cameroon | 71·5 | 103 |
| Ecuador | 44·0 | 65 |
| World total | 1085·0 | 1249 |

Exports, both from Africa and South America, go chiefly to the United States and to Western Europe, especially to the United Kingdom, Holland, and West Germany.

**YERBA MATÉ**

Yerba maté or, as the Brazilians call it, *herva mattee* is a tree which grows widely over the southern part of the Brazilian Plateau in groves or thickets. When cultivated on plantations, the tree is kept to shrub size. Harvesting is carried out during the winter season, between April and August. At this time of the year many of the farmers and farm workers of the eastern coastal areas move westwards into the forests and plantations to undertake seasonal work as pickers. In southern Brazil the town of Curitiba is the chief processing and packing centre. The total annual production is in the region of 100,000 tons, of which about half is exported, most of it to Argentina and Uruguay, but some to Paraguay (even though Paraguay produces some of her own) and Chile. These countries drink maté much as the English drink tea. The drink is traditionally imbibed through a tube called a bombilla, inserted in a silver flask or gourd known as the maté.

## *SUGAR*

Until about four hundred years ago honey was widely used in Europe for sweetening purposes since sugar was not known. Nowadays sugar is

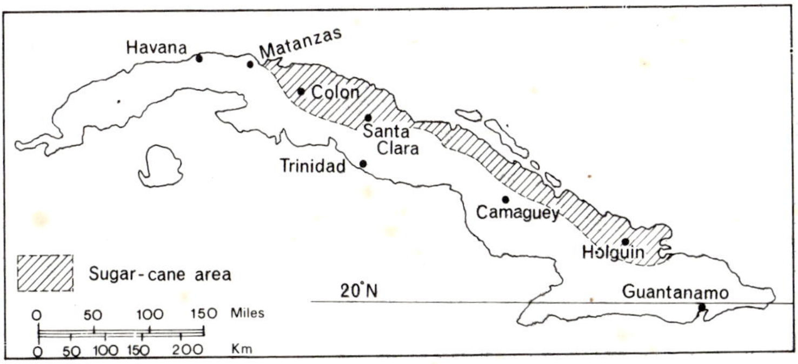


FIG. 52.—THE CUBAN SUGAR BOWL

Cuba, in both the past and the present, has been dominated by sugar. Probably no other country has been dominated for so long by a single crop. The entire economy has been largely geared to, and been dependent upon, the growing and export of sugar. Cane production not only controlled the economy: it may be said to have dominated the whole life of the country in the sense that human activity and prosperity pulsated to the rhythm of the growing and harvesting of the cane. For about half the year a large labour force was fully occupied harvesting and transporting and processing it in some 160 mills. For the remainder of the year the workers were unemployed and money became scarce. This reacted upon other industries which, through lack of consumer demand, had to cut down production. Accordingly, widespread insecurity followed from this seasonal instability.

Until 1959, when she was surpassed by the Soviet Union, Cuba (along with India) had been the world's largest sugar producer. She accounted for about 12% of total world production of some 46 million tons a year. She has long been, and remains, the world's largest exporter. Until the Cuban Revolution (1959), a large proportion of the sugar output was sent to the United States. United States corporations at that time controlled about 36% of the Cuban output.

The Cuban sugar industry has suffered from two major drawbacks: (i) world demand for sugar fluctuates (with wars, stock-piling, trade depressions, etc.) and so, therefore, does the price of sugar; and (ii) the Cubans have neglected sugar production: yields per acre are low, irrigation is rarely used, and there is little research work undertaken. As a result, the sugar industry is in a state of decline, notwithstanding the bumper harvests which were being reaped in the years immediately prior to the revolution. Current production is around 5 million tons.

Sugar-cane production in Cuba is essentially a large-scale industry with mass production as its keynote. Although the great companies have their own plantations, about three-quarters of the cane is produced by small-holders or *colonos* upon *colonias* (cane farms). The small farmers are usually under contract to the great sugar corporations and send their cane to the nearest *central* (cane processing plant). The large plantation, the *colonias*, and the *centrals* are all integrated into a single producing unit. An aerial photograph of such a unit would show the mill, where the cane is processed, a village or township, where the workers live, and the surrounding fields where the cane is grown.

consumed in large quantities in all countries possessing a high standard of living, for not only is it used for sweetening food and drink but it is required in the preparation of jams, canned fruits, and confectionery; elsewhere sugar is often quite beyond the means of the poorer peoples. "The great expansion in sugar consumption was made possible by the development of low cost production in sugar-cane growing countries, which provide more than 95% of total exportable supplies."* Desirable as sugar is for adding taste to drink and comestibles, it is also an important foodstuff in itself, being an energy-providing substance.

Sugar may be obtained from the juices of many plants but there are three main sources: (*a*) from a variety of trees, *e.g.* the sugar-maple, the datepalm, and several other palms; (*b*) from the sugar-cane, a tall, tropical, thick-stemmed perennial grass; and (*c*) from the sugar-beet, a root which is a plant of the cool temperate lands. Although some sugar is obtained from the sugar-maple, which grows in eastern Canada and in New England, and in Eastern countries from a variety of palms, the vast bulk of the commercially used raw sugar is derived from cane, which supplies some three-fifths of the total world production, and from beet, which supplies about two-fifths. The proportions supplied from these two sources have varied greatly during the present century.

## SUGAR-CANE

Eastern Asia appears to be the original home of the sugar-cane. It was made known to Europe by the Arabs who brought it first to Egypt and then introduced it into Sicily and southern Spain (this being the only area in Europe where it continues to flourish). From Spain it was taken to the Atlantic islands of the Madeiras and Canaries, whence it was transplanted to Brazil and the West Indies. From these two areas it has spread widely throughout the inter-tropical belt of the Americas and even just beyond it, for it is successfully grown in Louisiana in the United States. During more recent times sugar-cane has been introduced into South Africa, where it is important in Natal, and into Australia, where it grows along the coastal plains of Queensland and northern New South Wales.

Conditions necessary for the cultivation of sugar-cane are: (*a*) a hot climate, with temperatures between 21°–27° C (70°–80° F) nearly all the year round; (*b*) abundant rainfall with a minimum of about 50 in. (1270 mm) unless cultivated by irrigation; (*c*) a relatively dry, sunny season for ripening (to ensure maximal sugar accumulation) and harvesting; (*d*) deep, rich, well-drained but water-retentive soils; a high level of fertility must be maintained; (*e*) an abundant supply of cheap labour is desirable since cane growing makes heavy demands upon labour.

* *Oxford Economic Atlas*, 2nd edition, 1959, p. 24.

Sugar-cane is mainly grown by monocultural methods on large plantations often owned by foreign (usually European or United States) companies. As we have just noted, cultivation makes heavy demands upon labour (where labour supply is plentiful and cheap) and on most of the sugar estates of tropical America large numbers of coloured people,

[*Courtesy: Studio Cole, for Tate and Lyle Refineries, Ltd.*

FIG. 53.—SUGAR-CANE PLANTATION

Cutting sugar-cane in the West Indies. The cane is cut by hand, the workers using machetes. The cane is taken immediately by either ox-cart or light railway to the sugar-mill. Here the sugar-juice is squeezed out of the canes by crushing machines. Lime is then added to the juice which is then boiled in large, closed pans. Sugar crystals form and these produce the crude sugar which is exported, the actual refining usually being done in the importing countries.

especially Negroes, are employed. It is interesting to compare the situation here with that in Australia where there is no coloured labour available; in Queensland and New South Wales all the arduous field tasks have to be done by white men.

Sugar-cane is a perennial and the cane can be cut for a period of up to five or six years (known as the ratoon period); then the plant is uprooted, new canes being planted, usually by taking cuttings. After about the second season the quantity and quality of the juice of the cane quickly deteriorates, hence the desirability of having new rootstock every few years. When the cane is cut, it is taken to the factory, which is attached to the plantation. The cane is cut into pieces, crushed between rollers, boiled with lime, and then allowed to crystallise out. This forms the crude raw sugar. Sugar is exported in this state, the refining taking place in the importing country. Molasses and rum are by-products of raw sugar production.

TABLE IX

*Raw Sugar Production (thousand metric tons)*

| | *1962–3* | *1970* |
|---|---|---|
| Cuba | 3820 | 7559 |
| Brazil | 3304 | 5019 |
| U.S.A. | — | 5327 |
| Australia | 1875 | 2507 |
| China | — | 2900 |
| India | 2384 | 4634 |
| Mexico | 1732 | 2402 |
| S. Africa | — | 1649 |
| Philippines | 1555 | 1980 |
| Taiwan | — | 814 |
| Dominican Rep. | — | 1014 |
| World total 1968 | | 72,951 |

India, it will be noted, is a very large producer, but seldom has any surplus available for export. The sugar is usually consumed within the country, mostly in its unrefined form known as *gur*. The main exporter is Cuba, followed by the other West Indian islands and countries, notably the Dominican Republic, Puerto Rico, Jamaica, and Barbados.

## SUGAR-BEET

A German, in the middle of the eighteenth century, discovered how to extract sugar from a variety of the beet family, a large white or yellow root now familarly known as the sugar-beet. The development of the sugar-beet industry has been closely bound up with strategic considerations. Its greatest expansion occurred during the early part of the present century due to Germany's desire to be self-sufficing in sugar supplies in the event of war. Beet sugar production in Europe has been greatly stimulated by government subsidies and bounties, *e.g.* in Britain. A crop of cool temperate latitudes, sugar-beet is practically confined to Europe, the western part of the Soviet Union, the United States and Canada.

Beet sugar demands: (*a*) a long growing season of five to six months; with temperatures between 16° and 23° C (60° and 73° F) for three months of summer; (*b*) a regular supply of water during the growing period; this means about 25 in. (635 mm) of rainfall unless the crop is irrigated; (*c*) abundant sunshine and preferably a cool autumn to produce a high sugar content; (*d*) well-drained, fertile, loamy and stone-free soils, treated with lime and containing salt.

Sugar-beet is a very exhausting crop, hence the soil must be well fertilised before and after cropping. Naturally rich soils such as the loess and limon soils of central Europe are ideal for its cultivation as are the alluvial soils of the English Fenland. Furthermore, these deep, stone-free

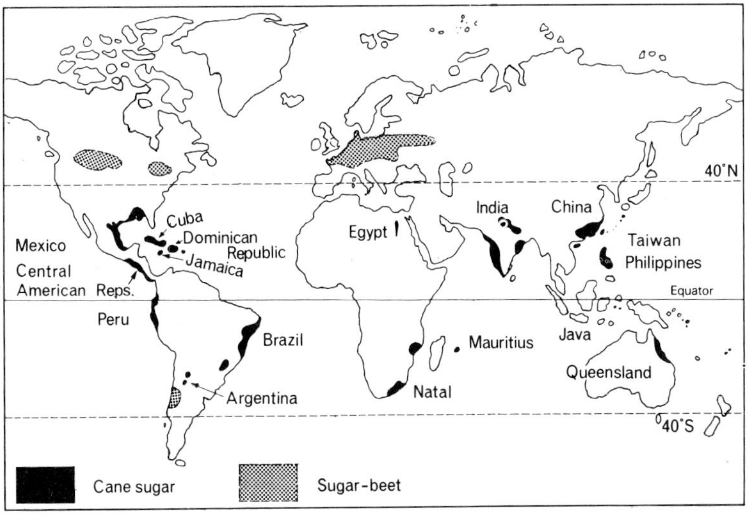


FIG. 54.—CANE- AND BEET-SUGAR AREAS

soils are valuable since they allow proper root development. The cultivation of beet, which is usually done in a rotation system, has certain advantages for, in addition to being a valuable cash crop in itself, the "tops" provide a useful feedstuff, the beet pulp forms a nutritious cattle feed, the factory waste can be used as a fertiliser, and the cultivation of beet helps to "clean" the soil.

Nearly all the world's supply of beet comes from either the large sugar-consuming countries which share the more fertile soils of the Great North European Plain, especially Russia, France, West Germany, Poland, and Czechoslovakia, or from the United States, particularly California, Colorado and Utah where it is grown under irrigation.

Beet sugar enters into international trade scarcely at all.

Cane and beet sugar are complementary rather than competitive in their distribution. They both undergo the same refining processes, and there is no difference in the final sugar-product. Cane usually yields a greater quantity of sugar per acre, but, weight for weight, beet is more profitable.

TABLE X

*Sugar-beet Production (thousand metric tons)**

| | *1960* | *1970* |
|---|---|---|
| U.S.S.R. | 57,728 | 78,324 |
| U.S.A. | 14,897 | 23,766 |
| France | 18,000 | 17,435 |
| Poland | 10,262 | 12,742 |
| West Germany | 12,860 | 13,458 |
| Italy | | 9518 |
| Czechoslovakia | 7853 | 6644 |
| United Kingdom | | 6412 |
| East Germany | | 6135 |
| World total 1968 | | 251,244 |

## *SPICES, ESSENCES, ETC.*

Spices may be defined as flavouring stuffs, usually aromatic substances of tropical or sub-tropical vegetable origin, which come from the leaves, flowers, berries, seeds, bark, and roots of plants. They are widely used in cooking to flavour and season, and make more piquant and palatable dull and uninteresting foods. Occasionally they are used in confectionery, *e.g.* aniseed, ginger, and sometimes in the making of liqueurs, *e.g.* Pernod, Kümmel.

The word spice comes from the French *épice*, which, in turn, derives from the Latin *species*, meaning "sort" or "kind." A French grocer still calls his shop an *épicerie*. Even the English word grocer can be traced to the spice trade, for it comes from the medieval merchants who bought and sold in "gross," *i.e.* in large quantities, as distinct from those who engaged in retailing. The Pepperers formed one of the earliest medieval English guilds, for they were in existence as early as A.D. 1180. In 1345 they were incorporated into a group which became the Grocers' Company.

Spices, along with silks, gems, and perfumes, were the most important of the early articles of trade. The Romans imported considerable quantities of spices and even in the centuries after the collapse of the Roman Empire spices continued to find a market in Europe and throughout medieval times they formed a major (and from the point of view of value the most important) article of trade. In Europe spices were almost necessary for health, and certainly desirable for seasoning the salted food of winter and the salted fish of Lent. In the days before canning and refrigeration, food, if it was to be kept for any length of time, had to be pickled or preserved in some way to keep it eatable.

These days, apart from pepper and mustard, which are widely used as condiments, spices are by no means as important, relatively speaking, as they were four or five hundred years ago. In Europe and in English-

* Note these figures refer to production of beet, and not to sugar production from beet.

speaking countries, spices are mostly used in the meat-packing industries and in the manufacture of pickles and sauces. Vanilla, which is a flavouring essence, rather than a spice, is much used in the confectionery business. Some peoples, particularly in the tropical regions, use much spice to give some variety to the monotony of their diet; for example, the peoples of India make great use of curry powder and the Latin Americans use chilli powder.

## TROPICAL AND SUB-TROPICAL SPICES

Some of the better-known spices are:

*Allspice.* The dried, unripe berry of the pimento tree, a member of the myrtle family, which is a native of Central America and the West Indies; most of it comes from Jamaica and it is frequently called Jamaican pepper.

*Cassia.* The dried bark of an evergreen tree belonging to the laurel family.

*Cinnamon.* The dried inner bark of a small tree which belongs, like cassia, to the laurel family and is a native of Ceylon.

*Cloves.* The dried flower-buds of the clove tree; the flower has a highly aromatic, pungent smell and a hot taste; most of the world's supply comes from the East African islands of Zanzibar and Pemba.

*Ginger.* The dried, tuberous roots of the ginger plant which originally came from southern China; the ground-up roots produce "brown ginger"; "preserved ginger" is made by pickling the roots in syrup.

*Mace.* The dried, ground-up outer red membrane covering the kernel of the fruit of the nutmeg tree; mace is similar but slightly stronger than nutmeg; only $\frac{1}{4}$ oz (7 grammes) is obtained from 1 lb (0·4 kg) of nutmeg, hence it is expensive.

*Nutmeg.* The kernel or seed of the fruit of the nutmeg tree, a tropical evergreen originating from the Banda Islands in the East Indies; the world's supply comes mostly from Indonesia though a little is produced in Grenada in the West Indies.

*Pepper.* The most important of all the spices, pepper comes from the berries of a creeping vine native to southern India; it is now mostly cultivated in Cambodia, Indonesia, and Sarawak; there are white, black, and red peppers, the last being the most pungent of all spices.

*Cardamom.* The dried ripe fruit of several plants related to the ginger plant; much grown in India, it is one of the prime ingredients of curry powders.

*Capsicum.* The dried pods of a dwarf plant found in Central and South America and also grown in India; the pods are known as chillies or red peppers; ground, they produce the so-called Cayenne or "red" pepper.

*Turmeric.* The dried ground stem or root of an East Indian plant belonging to the ginger family; similar in flavour to ginger, but more discreet.

*Vanilla.* This is a flavouring essence rather than a spice; it comes from the dried pods of a climbing orchid which yield a flavouring material much used these days in the making of chocolate and ice-cream; much of it comes from Mexico but it is also produced in Madagascar and Java.

## TEMPERATE SPICES

Although the majority of spices comes from tropical and sub-tropical regions, a number are products of the more temperate lands. The most noteworthy are:

*Aniseed.* A liquorice-flavoured spice which comes from the Middle East, much used in sweet-making; aniseed is also distilled to make the liqueurs *Anisette* and Pernod.

*Caraway.* The seeds of a dried, ripe fruit which grows wild in Asia; it is used in the making of the liqueur Kümmel and in confectionery.

*Coriander.* The dried fruits of a plant native to Southern Europe and Asia Minor; it is mainly used as an ingredient in sauces.

*Mustard.* The ground seeds of several species of annual plants, mainly the white and black mustard; first introduced into France by the Romans, it is now widely grown throughout northern Europe and in North America, too.

*Cumin seed.* Similar to, but stronger than, caraway seed, this forms an important ingredient of curry powders; cumin seed is probably native to Western Asia.

### EXERCISES

1. With reference to *either* cacao *or* bananas discuss the characteristics of plantation agriculture in the tropics in terms of physical and economic conditions.
2. Describe the conditions in which tea, coffee, and cacao are grown. For *one* of these commodities show how it enters into world trade.
3. Compare and contrast the world production and world trade in cane sugar and beet sugar.
4. Discuss the part played by sugar in the economic geography of the West Indies and show how it has influenced developments in Cuba.
5. With reference to two widely separated areas, examine the varying conditions under which *either* bananas *or* citrus fruits are produced commercially.
6. Give a brief description of the cultivation of *two* of the following: (*a*) tea in Assam; (*b*) cacao in Ghana; (*c*) coffee in Colombia; (*d*) yerba maté in South America.
7. Give an account of coffee production and coffee exports in south-eastern Brazil. Illustrate your answer with a sketch-map.
8. Write an essay on the cultivation of the vine and the production of wine in *one* of the following countries: Spain, France, West Germany.

# Chapter XII

# ANIMAL FOODSTUFFS

## *THE DOMESTICATION OF ANIMALS*

EXACTLY how, when, and where animals came to be domesticated is not known with any certainty. The circumstances leading to the domestication of animals is a matter of speculation. The dog appears to have been the first of the wild animals to be tamed and there are suggestions that it was domesticated by Palaeolithic man perhaps 10,000 years, or even more, ago. The ancestor of the dog is the wolf and many people believe it became tamed through its association with early human settlements as a scavenger on the food discarded by primitive man. In this way the wolf, being fed, would become less fierce and would also become familiar with man. In due course, especially when cubs were born and brought up in association with man, the animals would become tame and, ultimately, domesticated. The domestication of cattle and sheep may well have been due to man and beast being compelled to use the same water-holes when the phase of drier conditions set in over the present arid belt of the world. But precisely how animals came to be domesticated is problematical and it is unlikely we shall ever find the answer.

Exactly when domestication of cattle and sheep took place is also unknown but there is some evidence to suggest that it occurred around 5000 B.C., approximately at the same time as the discovery of the art of growing crops. These two important developments helped to usher in the Neolithic Revolution which saw the development of civilised life. Wall carvings in Ancient Egypt indicate quite clearly that dogs, cats, cattle, and sheep had all been domesticated very early on in history. The horse and the camel seem to have been domesticated later; certainly they do not make their appearance in Europe until about the beginning of the Christian era.

Opinion inclines to the belief that domestication, at least of cattle, sheep, goats, and horses, and probably the camel, too, first took place in the Near or Middle East or perhaps on the grasslands of southern Russia or Turkestan. Cattle have their ancestor in the urus, or wild ox, which continued to live in Europe until medieval times. The progenitor of the horse was the small steppe pony which roamed the grasslands of Eurasia many thousands of years ago and was, at first, hunted by Palaeolithic man. Its use as a riding animal may date from the Bronze Age. Some creatures, *e.g.* the pig, water buffalo, elephant, and fowls, were not domesticated in the semi-arid lands; rather do they appear to have come originally from South-east Asia.

## USE OF DOMESTIC ANIMALS

One of the interesting facts about animal life is that relatively few animals have been domesticated and that no addition to the list of domesticated animals has been made for two or three thousand years (although man has tried to domesticate certain other animals, *e.g.* the caribou and the zebra). It would seem that some animals are amenable to domestication; others are not. Yet another interesting fact about animals is that only twenty-four species are of much economic significance and "six of these are cattle; three belong to the camel family, and two each to the horse, goat, hen, and duck families. The deer, pig, dog, cat, and pigeon families have one representative each. The other two species are insects, the bee and the silkworm."* Elephants should perhaps be added to this list (see Fig. 80).

Animals, chiefly domestic animals, are used for a variety of purposes: (*a*) for carriage or draught; (*b*) for food, chiefly meat and milk; (*c*) for wool and hair and silk; (*d*) for hides, skins and other raw materials. These are their principal uses but they have minor uses, *e.g.* for hunting, for protection, for scavenging, for fertiliser, for ritual practices.

Animals contribute much to the world's economy; in fact, the annual value of their contribution is greater than that provided by all the minerals. Although some animals are declining in importance, *e.g.* the horse, animal life as a whole is contributing more each year to the world's economy; this is partly due to human needs, partly to human efforts. "Biological science has enabled the development of new breeds and provided for numbers and for quality. The control and improvement of the animal population by man has been as revolutionary in its effects as was the revolution in mechanical appliances in the nineteenth century. As sources of meat and milk, or as draught animals and as providers of hair, wool and hides, furs and skins, the animals, wild and domesticated alike, are extremely important in the modern world. It is the western industrial states which have utilised them most, and it may be noted that two-thirds of the world's human population largely abstains from meat or any extensive use of animal fibres."†

In this chapter we are concerned only with animals as producers of foodstuffs, that is with the production of meat, milk, and dairy produce, and commodities of minor importance such as honey.

* SHAW, E. B. *World Economic Geography*, John Wiley & Sons, Ltd., 1955, p. 170.
† GAULD, W. A. *Man, Nature, and Time*, Bell, 1946, p. 105.

## *CATTLE*

Cattle are reared for three different reasons: (*a*) as draught animals; (*b*) for beef production; and (*c*) for dairy products. It is unusual for cattle to be reared for a dual purpose. Cattle for draught purposes are reared mainly in south-eastern Asia and such beasts yield poor milk and meat. Elsewhere in the world, cattle are mostly reared for beef and milk.

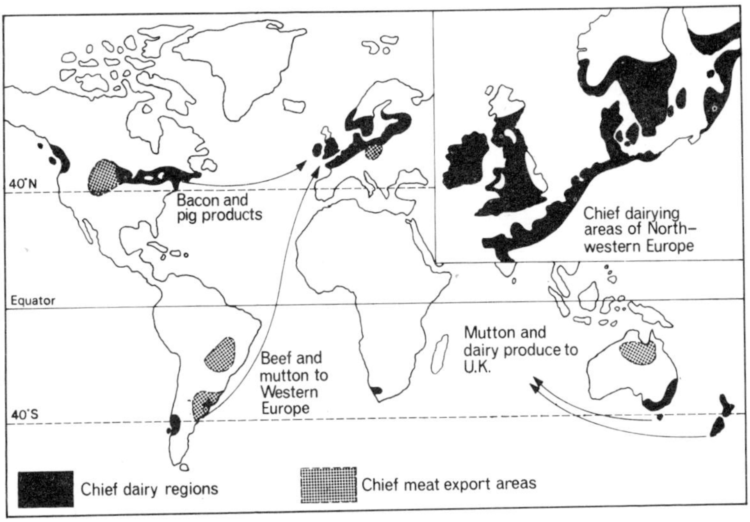


FIG. 55.—DAIRYING AND MEAT PRODUCTION

Cattle are, in fact, the world's chief providers of meat and dairy produce. The demands of beef cattle and dairy cattle in respect of feedstuffs, stalling, attention, and labour differ greatly and commonly there is specialisation in one or the other. But in regions having a mixed farming economy it is not unusual for both beef and dairy cattle to be raised on the same farm.

Cattle are widely spread throughout the world. Large numbers are reared in both temperate and tropical grassland areas, but, in general, more and better beef and dairy produce are obtained from animals raised on the former, whose grasses are more nutritious. Most cattle per acre, however, are found not on the world's natural grasslands but in the cool moist temperate lands of Western Europe and the north-eastern part of the United States, areas which were originally under forest, but which are now under "artificial pasture," *i.e.* man-made grasslands. In these regions plentiful rainfall evenly distributed throughout the year provides superior pasture to that found on the natural temperate grasslands. Moreover, in these regions man grows fodder crops, *e.g.* oats, clover, roots, which can be used as extra feed. Animal rearing here tends to be part of an integrated mixed farming system unlike that which is typical of the natural grassland regions. Again, animal farming is stimulated in these regions because the peoples enjoy high standards of living and large urban markets are present. Some countries or parts of countries, *e.g.* Denmark, Holland, Normandy, Eire, have tended to specialise in the production of meat and dairy produce.

One country, India, has more cattle than any other country in the world: some 188 million. India is not a country of either natural or artificial grasslands; why, then, has it such a large cattle population? Why, especially when the peoples of India eat little meat or dairy produce? The reason is that cattle have a ritualistic status; they are sacred and cannot be killed. The result is that India is plagued with large numbers of under-nourished, scrawny beasts producing very little milk and inefficient as draught animals, yet consuming large quantities of food that can ill be spared.

On the tropical grasslands of West and East Africa large numbers of cattle are kept by native herders. In Africa, however, cattle are used as a yardstick of wealth; a native measures his wealth in terms of the numbers of cattle he possesses. Note, however, that it is quantity, not quality of cattle, that is important. These animals are seldom slaughtered and native African cattle-raising lies outside the field of commercial stock-rearing.

Above we mentioned there were six types of cattle, but of the 1000 million cattle in the world two types predominate: (*a*) European cattle, of which there are many breeds, which are raised almost exclusively for the production of milk and meat; and (*b*) Zebu or humped cattle, which are native to, and adapted to, tropical conditions and which are used principally as draught animals.

**BEEF**

Although the finest beef cattle producing the best beef are raised in the cool temperate lands of Europe and North America, these areas are incapable of supplying the great demand for beef; hence the meat deficiency must be imported from elsewhere. In the case of North-western Europe beef is mainly supplied by the southern hemisphere beef producers of Argentina, Uruguay, and Australia. Large beef herds are reared on the *pampas* or natural grasslands of Argentina, Uruguay, and southern Brazil, where both alfalfa and maize are grown and used as supplementary feeding stuffs. The beef is prepared in *frigorificos* (slaughter-houses and freezing plants) and *saladeros* (canning factories) and the chilled, frozen,

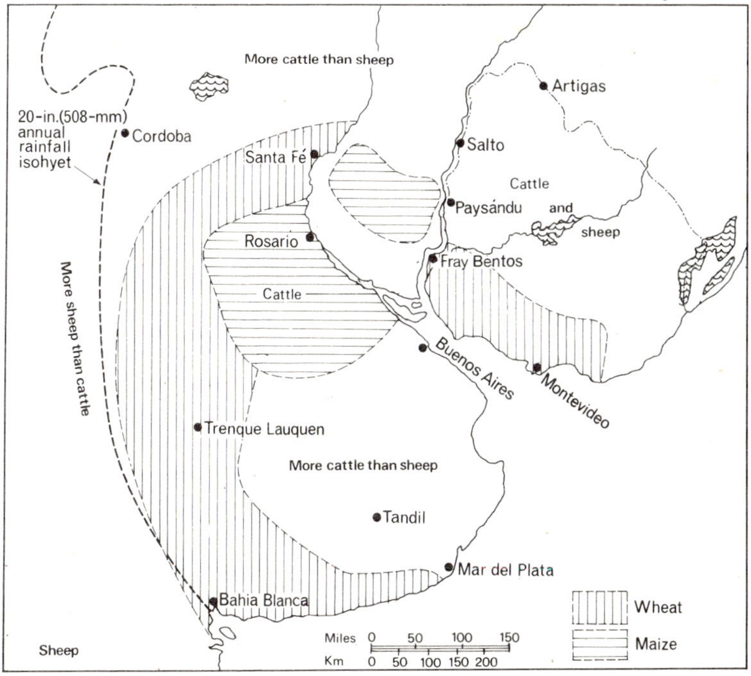


FIG. 56.—CATTLE IN ARGENTINA AND URUGUAY

The warm temperate grasslands or *pampas* of Argentina and Uruguay form one of the world's chief grazing areas. Cattle, first introduced by the Spaniards, multiplied rapidly and by the beginning of the nineteenth century millions of half-wild beasts roamed the *pampas* tended by cowboys called *gauchos* who lived a semi-nomadic life. At first the cattle had no value apart from their hides. Cattle-ranching for beef production was stimulated by a series of inventions and developments; the enclosing of the ranges by barbed-wire fences; the introduction of alfalfa, a nutritious plant of the clover family; the invention of beef extracts (*e.g.* Bovril, Oxo) and of canning; and the development of refrigeration which made possible the export of frozen and chilled meat.

The large cattle farms are known as *estancias*; the smaller estates are called *chacras*. The *estancias* are divided into twenty or thirty large fields each of which is served by a wind pump. Cattle, mostly of the British Shorthorn breed, have been improved by importing prize bulls. Periodically the cattle are rounded up and despatched to the city *saladeros* (processing and canning plants) or *frigorificos* (the slaughtering and refrigeration plants). The names of such towns as Fray Bentos and Paysandu are very familiar but Buenos Aires is the great centre of the meat-packing industry. Argentina has some 50 million cattle and 48 million sheep; Uruguay 7 million cattle and 22 million sheep. Argentina produces more than 2 million metric tons of beef a year. Some 45% of the United Kingdom's beef imports come from Argentina.

and tinned meat is exported to European markets. The tropical grasslands or *campos* of Brazil and the *llanos* of Venezuela hold out great possibilities for beef production and, although improvements and developments are gradually taking place, these areas are still of only relatively minor importance as beef producers because of inferior breeds, poor feed, troublesome insect pests, such as the cattle tick, and inadequate communications and packing plant. The Australian and South African grasslands are largely given over to sheep, but substantial numbers of beef cattle are herded on large ranches in the tropical parts of Australia and frozen beef from the Northern Territory and Queensland is shipped to Britain.

In the United States, which is the largest beef producer in the world, beef cattle are concentrated in the Corn Belt but are by no means confined to it. Large herds are grazed on the prairies and then moved into the Corn Belt for fattening prior to slaughter. Most of the beef cattle are reared some distance from the great consuming centres, in areas where it is more economic to use the land for maize-growing or where the climate is unfavourable for crop production. American beef is good quality beef and a large number of slaughter-houses and meat-packing plants turn out prime beef, tinned meats, and meat products.

The principle of cold storage, which was largely developed between 1860–80, completely revolutionised world trade in beef. Prior to this development, beef could be exported only as dried, salted, or canned beef. The invention of cold-storage plant, the refrigerator car, and the refrigerator ship made it possible for far-distant meat producers, such as Argentina, Australia, and New Zealand, to export carcasses to the markets of north-western Europe and the eastern United States.

TABLE XI—*Cattle and Beef Production*—1970

| | *Cattle: numbers (in millions)* | *Beef production (thousand metric tons)* |
|---|---|---|
| India | 176 | ? |
| U.S.A. | 114 | 10,008 |
| U.S.S.R. | 99 | 5400 |
| Brazil | 97 | 1900 |
| China | 63 | 2200 |
| Argentina | 49 | 2647 |
| France | 21 | 1624 |
| Australia | 24 | 966 |
| West Germany | 14 | 1322 |
| United Kingdom | 12 | 900 |

## DAIRY PRODUCE

The production of dairy produce entails a much more intensive type of farming than does beef, and the areas devoted to dairying are very

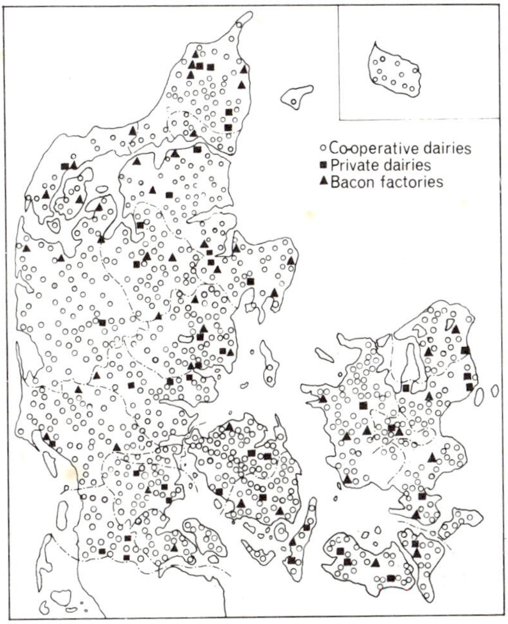


FIG. 57.—DAIRYING IN DENMARK

Denmark's pre-eminence as a producer and exporter of "breakfast table products" began about 1870. Up to that time she was a grain-producing country but competition from the vast new wheatlands of the New World threatened ruin and Denmark was compelled to change the basis of her agriculture. She turned to livestock-rearing with special emphasis upon dairying. For this Denmark had two natural advantages: cool, moist conditions desirable for the rearing of milch cows and a geographical position near to the densely-peopled, highly-industrialised countries of Europe which provided good markets. From the earliest days dairy farming was greatly helped by the introduction of the co-operative system which undertakes the preparation and marketing of the dairy produce and the bulk buying of animal foodstuffs, fertilisers, etc. Today, Denmark has about 1250 co-operative dairies, which deal with 90% of the milk produced and which produce some 154,000 tons of butter and 124,000 tons of cheese annually, and more than fifty co-operative bacon factories. There is relatively little land under grass; mostly the soil is intensively cultivated to produce grain, roots, and other animal feeding stuffs. Denmark's limited area necessitated that the most intensive and economical methods of farming should be employed and the farmer contrives to get the maximum return from his land and his animals. Great attention is paid to animal breeding, feeding, milking and to the quality of the products produced. Butter, cheese, eggs, bacon, and canned milk and meat products are the major items of the dairying industry.

This is an example of a simple dot distribution map: each dot represents a correctly located dairy, each triangle a bacon factory. For further details of this type of map see the caption to Fig. 76.

restricted. Milk, in its liquid state, has an economic limit of distribution of about two hundred miles, but in its processed form of butter and cheese it can be carried much farther (particularly if the aid of refrigeration is invoked), and in its condensed, tinned, or powdered form it can be transported over very long distances. The demand for fresh milk necessitates its production on the spot or at least in close proximity to the consuming areas. If markets are not close at hand, there is a distinct tendency for the milk to be turned into butter or cheese. Dairying has thus tended to become very specialised, each dairying area concentrating upon either milk or butter or cheese.

[*Courtesy: New Zealand House.*

FIG. 58.—DAIRY FARM, NEW ZEALAND

Most dairy herds are found in cool, moist regions enjoying equable climates where rich pastures are available and fodder crops can be grown, and where large numbers of people with high living standards are grouped together. Such areas include north-western Europe, north-eastern United States, the south-east coastlands of Australia, and the well-watered plains of New Zealand. In all these areas there is plenty of labour to look after milch cattle, which demand much more attention than beef herds.

Dairying is very highly developed in many parts of north-western Europe, especially in Denmark, Holland, and Britain, and to a lesser extent in Eire. The large industrial urban populations of north-western Europe (especially those of the United Kingdom and West Germany) provide big, nearby markets. It is interesting to note that Denmark tends to specialise in butter and bacon, whereas Holland and Switzerland concentrate upon cheese. The United Kingdom, though a large producer of milk, cannot satisfy all her needs of butter and cheese, and so must import large quantities. Considerable amounts of butter are imported from New Zealand which has a large surplus.

In North America, cattle grazed on the cool, moist northern Atlantic coastlands produce large quantities of milk for sale in New York, Boston, Philadelphia, and other big cities. States farther west, such as Wisconsin, Minnesota, and Michigan, have a smaller milk sale and so tend to concentrate more upon the production of butter and cheese. In the United States and in the St Lawrence lowlands of Canada, too (another dairying area), dairying is hampered somewhat by the cold winters which necessitate winter shedding and stall feeding with hay and silage.

During recent years many other countries, becoming educated to the value of milk as an item of the diet, are urging their peoples to "drinka pinta milka day." Argentina, India, Israel, and Japan are but a few of the countries which have begun to develop dairying industries, although the output of milk and milk products is still small. Valuable and interesting as these developments are, we should note that few of these countries possess natural conditions favourable to large-scale dairying.

TABLE XII

*Butter and Cheese Production—1970*

| *Butter (thousand metric tons)* | | *Cheese (thousand metric tons)* | |
|---|---|---|---|
| U.S.S.R. | 1067 | U.S.A. | 1329 |
| France | 506 | France | 792 |
| West Germany | 510 | U.S.S.R. | 530 |
| U.S.A. | 518 | West Germany | 500 |
| New Zealand | 237 | Italy | 460 |
| East Germany | 216 | Netherlands | 278 |
| Australia | 224 | Poland | 245 |
| Denmark | 131 | Argentina | 167 |
| Canada | 153 | U.K. | 130 |
| Poland | 201 | Denmark | 111 |

## *MUTTON*

Sheep are seldom reared for a dual purpose; normally they are raised either for their wool or their meat. Generally speaking, sheep reared for their flesh require better pasture and are, therefore, concentrated in the wetter areas. Mutton and lamb (this is a butchers' distinction, "lamb" being the meat produced by sheep killed before they are eight months old) are much less important than beef. Apart from the people of Britain, Australians and New Zealanders and the peoples of the arid zone of the Old World, mutton and lamb are of little significance as a foodstuff. The Continentals and the Americans, for instance, eat very little.

Most of the world's mutton and lamb is obtained from the southern hemisphere and New Zealand dominates the world market. The cool, moist lowlands of New Zealand are ideally suited to the rearing of sheep for their meat and in this respect New Zealand contrasts with Australia which is predominantly concerned with the production of wool. New Zealand does produce some wool but this is largely a by-product of her meat industry. Other producers of mutton and lamb are Argentina, Uruguay, Chile, and Australia.

Britain is the leading world importer of mutton and lamb, taking some 90% of the world export.

## *PIGS AND PIG PRODUCTS*

The pig, as we have already noted, was first domesticated in South-east Asia, probably by the Chinese, and it is interesting to find that approximately a third of all the pigs in the world are still to be found in China. Pigs are essentially animals of the forest, not grassland; unlike cattle and sheep they do not ruminate and, though they will eat grass, they cannot live purely on a grass diet. The pig is by nature a scavenger, eating acorns,

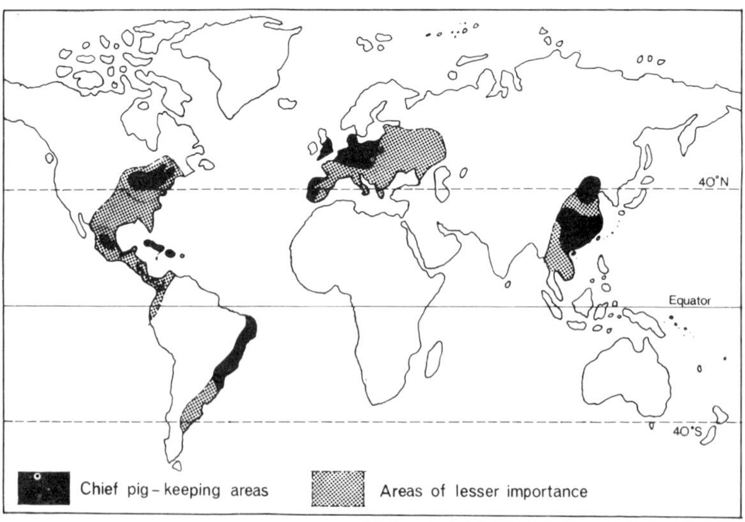


FIG. 59.—WORLD DISTRIBUTION OF PIGS

beech mast, roots, etc., and in this respect is a most useful animal as it will feed successfully on domestic garbage and turn it into good food. Although "back-yard" and "hen-run" pigs, fed on "pig-swill" are still reared, most pigs these days are raised on pig farms and fed on a diet of cereals, potatoes, and skimmed milk. For this reason, pigs are frequently associated with cattle on dairy farms.

The chief pig-rearing regions of the world are the United States, where they are mostly found in the Corn Belt, sometimes called the Hog Belt; north-western and central Europe, where they are chiefly reared on potatoes and skimmed milk; Argentina and Brazil where they are fed on maize; and China where they forage on whatever domestic or field left-overs are available. The Moslems, Jews, and Hindus regard the pig as an unclean animal and will not eat pig meat, hence pig-keeping is largely absent through much of Africa and Asia.

In addition to the production of pork, *i.e.* fresh pig meat, bacon, ham, and lard are obtained from the pig. Bacon and ham are cured forms of pig meat. The bulk of the pig meat entering into international trade is in the form of bacon and ham. Important exporters are Denmark, Eire, and Canada.

## *POULTRY*

The term "poultry" embraces hens, ducks, geese, and turkeys. Of these various birds, the hen is the most important. It is widely spread because of its tolerance of a wide variety of climatic conditions, because it can thrive on all sorts of food, because of its high egg productivity, and because eggs cannot easily be transported. Denmark and Holland are the chief exporters of shell eggs. The Chinese have long had a great egg export trade but this has mostly been in liquid or dried egg form much used in bakery. Poultry meat, especially chicken, is very popular, and the increasing demand for "oven-ready" and "spit-roasted" birds has led to the rearing of chickens under the "battery" system.

Waterfowl, such as ducks and geese, are much less important than hens except in one or two areas. In Europe, they are more particularly important in the Low Countries, Germany, and Hungary. They are especially significant in the Far East and are to be found wherever there is a trickle or patch of water.

## *HONEY*

The common honey-bee is found in nearly every part of the world. The keeping of bees for their honey is an age-old practice and goes back at least 5000 years, for it was carried on in Ancient Egypt. Honey production is a local activity of importance on the heathlands of the North European Plain. In the United States, California, Texas, and Iowa are the most important commercial beekeeping states. "The bee industry in California," writes E. B. Shaw, "has developed some interesting customs and problems. In this state it is quite customary for the beekeeper to move from the orange district into one with an abundance of sage, then from the sage into the bean fields, or into localities where alfalfa is being grown."* The chief exporters of honey are Australia and Canada.

## EXERCISES

1. Show how geographical conditions have brought about differences in the nature and purpose of dairy farming between Switzerland and Denmark.
2. Describe the geographical and economic conditions which are necessary for successful dairy farming, giving suitable examples from the British Isles. (*East Midland Educational Union Examinations.*)
3. Outline the pattern of world production and trade in meat. What geographical conditions favour production and how far have transport problems influenced the development in some areas?
4. Examine the economic and physical factors which have influenced the development of commercial beef production in North and South America.
5. It is often said that the savanna lands may well be the future great stock-raising areas of the world. Describe the difficulties which would need to be overcome before this could become true.
6. India has more cattle than any other country, yet is not an important producer of either beef or dairy produce. Explain this seeming paradox.
7. Give a reasoned account of the cattle rearing and meat production industries of Argentina.

* *World Economic Geography*, p. 199.

# Chapter XIII

# THE FISHING INDUSTRY

THE practice of fishing is one of man's oldest activities. It is still a widespread activity undertaken by primitive and advanced societies alike. Among some peoples, such as the Nootka Indians who lived on the coasts of British Columbia, the Yaghans of Tierra del Fuego, many of the Eskimo tribes, and the South Sea Islanders of the present day, fishing was the basis of their mode of life and their chief food. Fish still forms an important part of the diet of many peoples, *e.g.* the British, the Norwegians, and the Japanese. Among many Oriental peoples it is the most cheaply produced form of protein food and on this account is of particular importance since animal foodstuffs are scarce. Fishing is also a major industry in some countries and contributes substantially to the national exchequer; this is especially so in the cases of Iceland and Norway.

The modern fishing industry is a skilful and highly organised industry. Important as fishing is, it must be recognised that it supplies only a small fraction of the world's total food supply and occupies but a very minor proportion of the world's people. Although some areas have been overfished with the consequent depletion of catches, it remains true that the seas could provide greatly increased quantities of fish. The research and scientific development which have long been applied to agriculture are only now beginning to be applied to fishing but already, as a result of the pioneer work being done, there are indications that the fishing industry is going to assume a role of much greater importance in the future.

Let us look, in turn, at the nature of marine life, at the methods of fishing, the major fishing countries, and at the chief marine products other than fish.

## *MARINE LIFE*

It has been said that where food exists for fish there will be fish for food. Just as people on the earth's surface tend to congregate in certain areas, usually where the living conditions are best, so also marine life tends to live where the living conditions are best. These conditions are best on the continental shelves in temperate latitudes. The sun-lit, aerated waters of these shallow shelf areas are favourable for plankton (minute forms of animal and plant life) upon which fish feed. Sometimes the sea is so rich in plankton as to have the consistency of thin soup.

Plankton is especially plentiful in certain parts of the ocean: (*a*) in the shallow seas of the continental shelves; (*b*) in cold polar waters; (*c*) where warm and cold currents meet; and (*d*) where cold bottom water wells up to the surface. Consequently, in these areas, which are rich in organic life providing abundant supplies of food for fish, fish will thrive and such areas form great fishing grounds. Such grounds have not automatically given rise to important fishing industries: much has depended upon their nearness to densely peopled countries needing to augment their food supplies.

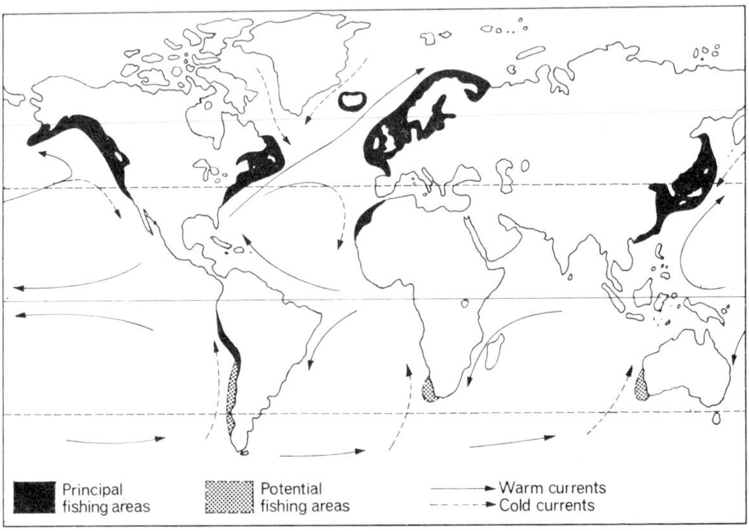


FIG. 60.—WORLD FISHING GROUNDS

Fish may be grouped in various ways. A common method is to divide them into pelagic fish such as herring, sardine, and mackerel, which live at or near the surface and feed on plankton, and demersal fish such as cod, haddock, plaice, and sole which live and feed on or near the sea-bed—the so-called "white" fish. Another method of grouping is to divide them into those which live and move in shoals, such as herring, and those, the majority of fish types, which do not associate together. A third way is to divide them into fresh-water and salt-water types although some species, notably the salmon, are anadromous, that is they live in the sea for the greater part of their lives but swim up the rivers to spawn in fresh water. These differences are important for upon them fishermen have had to base their methods of catch: the use of drift-nets to catch the surface types and of trawl-nets for the sea-bed types. The fishing season is also often determined by the seasonal migrations of certain fish types which have distinctive migratory habits. The antics of the herring and the capers of the cod, for example, impart a seasonal rhythm to the fishing industry.

Finally, though not fishes in the biological sense, we should note that marine life includes crustaceans, such as crabs, lobsters, crawfish; molluscs, such as oysters, cockles, mussels, and clams; creatures of the sea, such as dolphins, seals, turtles, and whales; the seaweed, sponges, and corals. Altogether, these various marine fauna and flora account for some 10% of the world fishery products.

## *METHODS OF FISHING*

Man has used numerous devices for catching fish, among them spearing, trapping, and netting. Native peoples in tropical regions still use spears and bows and arrows while ingenious traps are used by many simple fisherfolk. In the Orient a wide variety of nets is used. The Chinaman's ingenuity has not only led him to invent elaborate fishing-nets but even to enlist the aid of animals. The fish-eating cormorant, for example, has been cleverly used from the earliest times to catch fish; the birds have been trained, like hawks, to catch fish and bring them back to the fisherman and to ensure that they do bring them back and do not swallow the fish, rings are slipped round their necks. Viscount Kelburn recounts seeing one old fisherman at Ichang in the Yangtse valley who used an otter to scare fish into his net. Truly, fishing techniques may take many surprising forms.

The methods by which fish are captured in commercial fishing fall into three groups: line-fishing, netting, and trapping. Traps, usually in the form of wicker-work baskets, are used mainly for catching shellfish. Fishing by line consists of shooting overboard a line from which hang baited hooks. Long-lines may be several miles in length and carry two or three thousand hooks. In areas where the sea-bed is rough or is fouled by numerous wrecks, making the use of trawl nets impossible, long-lining is often resorted to. Cod off the Newfoundland coast are usually caught by line. Small boats called dories are sent out from the "mother" vessel to do the actual lining. The Scottish haddock industry also uses lines.

Netting is done in three main ways (Fig. 61). First, by drift-nets these are nets which hang curtain-like in the water and are kept vertical by weighting the lower end and buoying the upper end with floats; the nets drift with the tide and the fish are caught by their gills becoming entangled in the mesh of the net. Shoal fish are captured in this way. A modern herring drifter can lay down several miles of netting. The position of the suspended net can be adjusted by manipulating the weights and the

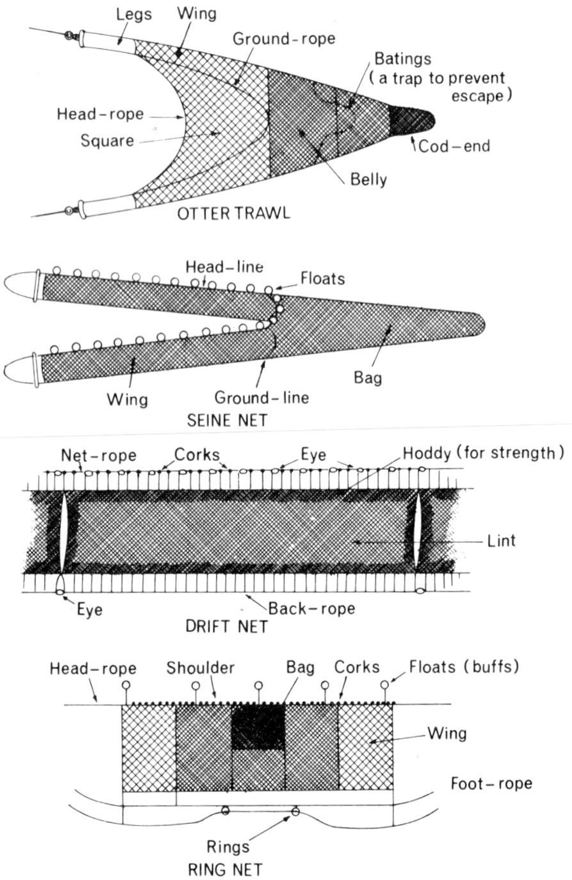


FIG. 61.—TYPES OF NET

*Otter trawl:* a sisal bag dragged along the sea-bed. The fish are trapped in the fine-meshed "cod-end." The otter trawl is used on long-distance trawlers but for short-distance work has largely been replaced by the more efficient and simpler Seine net.

*Seine net:* made of cotton twine, it is much cheaper than the trawl. It is not dragged on the sea-bed so is less liable to damage from rocks and wrecks. It floats just below the surface and does not hold as much as the trawl, so it has to be hauled in very frequently. On the other hand it can be used by smaller vessels, requiring fewer men.

*Drift net:* used for herring fishing. They hang down from the surface like a tennis net in front of the herring shoal, which swims into them. Many drift nets belonging to different vessels are connected like an immense curtain, sometimes two miles long.

*Ring net:* also used for herring, but by smaller boats and in more sheltered waters. The ring net is connected to a vessel at each end and the boats keep the net floating in an arc shape. When the catch is completed the boats come towards each other, forcing the fish into the strong part of the centre panel, known as the "bag."

buoys. Usually it is kept close to the surface but if fishing is carried out during daytime the net is lowered since the herring swim deeper by day.

Demersal fish, which live on or close to the sea-bed and which rarely wander far from their feeding grounds, are caught by trawl net. The trawl net is in effect a large conical-shaped bag, the mouth of which is kept open by a wooden beam or by otter boards. Such nets may be 80 feet (27 m) long and the mouth of the net 50 feet (17 m) wide. These nets are dragged along the sea-bed usually at a speed of between 2 and 5 miles (3–8 km) an hour. This method of fishing is restricted to shallow water areas and to fishing grounds with an even bed. Obviously submerged rocks and wrecks would tear the trawl to pieces. "A major factor in the decline of the trawling industry at one small West of England port has been the great number of wrecks, dating from the First World War, which littered its trawling ground."*

The third type of net, usually used close to the shore, is the seine net. Herring are sometimes caught by seining. The technique in seining is to encircle a school of fish. The net is spread between two vessels which work against the tide and gradually draw closer causing the fish to become trapped in the enclosing net. The purse-seine, which is a variation of the seine net, is used in Britain for catching mackerel in deep water. The purse-seine is also much used in the United States.

## *THE MAJOR FISHING GROUNDS*

As we have already indicated, the major fishing grounds are found in temperate latitudes. The chief reasons for this are: (*a*) fish food is more abundant in cooler water than in warmer water and, accordingly, fish are more plentiful; (*b*) the continental shelves are better developed in mid-latitudes, especially in the northern hemisphere, than elsewhere; (*c*) the coastal regions in higher latitudes are either infertile or overcrowded, hence the people have turned to the sea to augment their food supplies; (*d*) it is easier to keep fish fresh and therefore to trade in fish in cooler climates than in warm regions.

Within a broad belt in temperate latitudes, approximately between 30 and 66½ degrees N., bordering the continental areas of the northern hemisphere, are to be found the world's major fishing regions; they are four in number: (*a*) the marginal seas off north-western Europe with which may be included the Icelandic area; (*b*) the shelves and banks of north-eastern North America between Cape Cod and Newfoundland; (*c*) the north-western Pacific coast region of North America between Oregon

* POUNDS, N. J. G., *An Introduction to Economic Geography*, p. 84.

and the Bering Strait; (*d*) the marginal seas of eastern Asia around Japan, Korea, and off the coast of China.

In the southern hemisphere neither the fishing grounds nor the fishing industries are so well developed although considerable potentialities exist off northern Chile and Peru (now being rapidly developed), off the western coast of South Africa, and off the western coast of Western Australia.

## *THE MAJOR FISHING COUNTRIES*

The accompanying table gives the catch of the chief fishing countries for the years 1960 and 1968.

TABLE XIII

*Fish Catch (thousand metric tons)*

| | *1960*★ | *1970*† |
|---|---|---|
| Peru | 3531 | 10,520 |
| Japan | 6193 | 9309 |
| U.S.S.R. | 2340 | 7252 |
| China | 5800 | 5800 |
| Norway | 1599 | 2980 |
| U.S.A. | 2797 | 2714 |
| Spain | 935 | 1497 |
| Canada | 927 | 1378 |
| Iceland | 593 | 850 |
| United Kingdom | 924 | 1099 |
| Denmark | 581 | 1227 |
| West Germany | 674 | 680 |
| France | 596 | 650 |

(Source: ★ *World Fisheries Yearbook*; † F.A.O.)

From the above figures it is clear that the East Asian region is of first importance as regards the weight of the annual catch. The Japanese fisheries have long been the most important in the world. Since the end of the Second World War the Chinese fishing industry has increased by leaps and bounds and China now runs Japan a fairly close second. The Japanese are the greatest fishing nation on earth; the reasons are: (*a*) the meeting of warm and cold currents in the shallow Japanese seas favours abundant supplies of fish; (*b*) the broken, indented coasts provide numerous bays and coves for fishing villages; (*c*) the large population which is mainly concentrated on the coastal lowlands has caused the people to turn to the sea to supplement their livelihood; and (*d*) the limited amount of cultivable land has stimulated the industry. It is estimated that some 5 million people are engaged, directly or indirectly, in the fishing industry. All types of fish are caught and eaten, for the Japanese are much less fastidious than the British about the kind of fish they eat. Ancient and modern methods of fishing are strangely mixed; much offshore fishing is done from small boats with only crude equipment; on the other hand, Japan has modern motor vessels which fish far afield in the Pacific. Nagasaki is the principal trawler base. The Japanese eat more fish per head of the population than any other people in the world.

The Chinese, like the Japanese, love fish, but because of transport difficulties fish is common only in the coastal regions or in the vicinity of the great inland lakes. Fish-farming, that is the rearing of fish, mainly

[*Courtesy: Eric Kay.*

FIG. 62.—NETTING FISH

carp, in ponds and even in the flooded rice-fields, is common in parts of China. The Chousan archipelago is the chief fishing centre, but Tientsin, Hai-chow, Amoy, and Foochow are important fishing ports.

The fisheries of North-west Europe, which extend from Portugal to the White Sea, form the world's second greatest fishing region. The well-mixed waters of the broad continental shelf, together with the deeply indented coasts, and the populous countries of the region, which provide a large market, are the chief reasons for the development of the industry, although in the cases of Iceland, Norway, Scotland, and Holland the infertility of the soils also greatly stimulated the fisheries in earlier times. Norway, with an annual production of about $2\frac{3}{4}$ million tons of fish, is the leading fishing nation in Europe. Some 60,000 people live by fishing. The Lofoten Islands are the great centre of the cod fishery; farther south are the herring and sprat fisheries which focus on Bergen. Bergen, Trondhjem, Tromsö, and Hammerfest are the leading fishing ports.

In Britain, once the second most important fishing nation in North-west Europe, the industry is relatively much less important than formerly although the value of the catch is still very great. The leading varieties of fish caught are cod, herring, haddock, hake, plaice, and sole, with

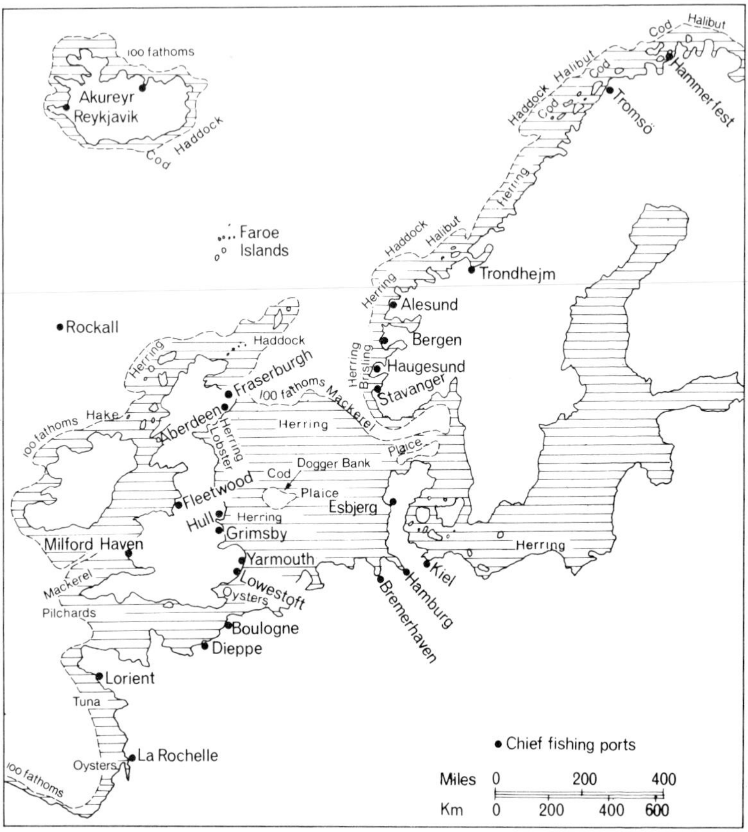


FIG. 63.—FISHERIES OF NORTH-WESTERN EUROPE

some mackerel and pilchard in the warmer waters of the English Channel and the Bristol Channel. Longshore fishermen raise oysters (especially in the tidal estuaries of the Thames and the Essex rivers), trap crabs and lobsters (particularly along the north Yorkshire and south Cornish coasts), net prawns and shrimps (especially in Morecambe Bay), and gather mussels, cockles, whelks, and winkles. Hull and Grimsby are the chief trawler ports. Aberdeen and Fraserburgh are centres of a declining herring fishery. On the west coast of Britain, Fleetwood and Milford Haven are the chief ports.

France, Holland, Western Germany, and Denmark share in the North Sea herring fishery and France, Spain, and Portugal share in the cod fisheries of Iceland and Newfoundland. Sardine fisheries are important in the Bay of Biscay and off the Spanish and Portuguese coasts.

Both the Atlantic and Pacific coasts of North America are important for fishing. Especially noteworthy is the "banks" region between Cape Cod and Newfoundland where warm and cold currents meet in shallow waters. Here is an historically famous and almost fabulously rich cod fishery. Fishing villages abound in Newfoundland, the Maritime Provinces, and the New England States although the numbers engaged in fishing have declined to less than half what they used to be—largely because of improved fishing techniques and the introduction of power-driven vessels. Much shellfish is also caught in this region. Large quantities of menhaden and shrimps are caught in the Gulf of Mexico.

The Pacific fishery is mainly concerned with salmon but halibut and sardines are also important. The trapping of salmon is carried on on a big scale on the Fraser, Skeena, Columbia, Sacramento, and other rivers of the north-west coast. The deep red sock-eye salmon is especially important and greatly prized for the canning industry. Almost every creek along this indented coast has its cannery. Halibut, not so very long ago almost fished to extinction, ranks next to salmon in value. Prince Rupert, which is the leading fishing centre on this coast, is also the chief centre of the halibut fishery. San Diego, in California, is a major sardine fishery focus.

In the southern hemisphere the fisheries have not been developed to any appreciable extent except in Peru. Until very recent years Peru's fishing industry was extremely small, but it grew with gigantic strides from 84,000 tons in 1950 to 3,531,000 tons in 1960, 8,789,000 tons in 1967 to over 12,000,000 tons in 1971. More than 90% of the catch, however, is processed into fish-meal for sale as fertiliser. Can you explain why the waters off the Peruvian coast should be so rich in fish? The fishing industry is growing in South Africa, which now nets over 1 million tons annually.

## *INDUSTRIES BASED ON FISHING*

A number of industries have grown out of the fishing industry. This means, very often, that as many people are indirectly employed as are directly employed in the fishing industry.

Broadly speaking, these industries fall into two groups: (*a*) those concerned with the marine products themselves such as the drying, curing, and canning of fish and the processing of fish to make fish-oil, fish-glue, and fish-manure. Such industries obviously grow up in the fishing ports. Bergen in Norway, for instance, exports its fish in various forms: as frozen, salted, dried, pickled, and canned fish; but because of the enormous herring catch and the impossibility of processing all of it for human consumption, some two-thirds of it is reduced to oil or converted into fish-meal for fertiliser. (*b*) The ancillary industries form a wide group and include such diverse activities as boat-building, the manufacture of nets and other fishing tackle, the making of barrels, boxes, and containers (tin cans), salt refining, and the manufacture of ice. Many, too, are engaged in the trade and transport of fish and a few find jobs in fish culture (*e.g.* the great plaice hatchery in Liim Fjord in Denmark) and fish research. Bergen, for example, to quote it again, has several interesting and important functions in connection with the fishery: it is the centre of the Norwegian fisheries administration and the centre for oceanographical research.

## *WHALING*

Of the various marine mammals which are commercially hunted the whale is by far the most important. There are two main types: the baleen or whalebone whale and the toothed whale. The former, which feed on krill (small shrimp-like creatures), though found in all the oceans, are now mostly hunted in the Southern Ocean. The baleen whales include the Rorquals, Humpback, and Right Whales. The so-called Blue Whale, which is the largest creature of the animal kingdom and which may reach nearly 100 feet (30 m) in length, is a member of the Rorqual group. It is now scarce. Of the toothed whales, the two most important are the Sperm Whale and the Killer Whale. Whales are mostly found in the colder waters of high latitudes but in winter they migrate into the temperate zone and it is then that they mate.

The chief whale fisheries are now in Antarctic waters, especially in the region between the southern tip of South America and New Zealand. Formerly, whales were plentiful in Arctic waters but they proved to be such profitable prey that by the end of the nineteenth century they had been fished almost to extinction. Modern, easy methods of catching are rapidly leading to the same state of affairs in the Southern Ocean in spite of international agreements which have placed certain restrictions on whaling. For example, in the 1946–47 season, over 9000 Blue Whales were shot in Antarctic waters while in the 1961–62 season the number was only 1500.

Norway and the United Kingdom were the traditional whaling nations but in recent years the U.S.S.R. and Japan have dominated the industry. During the whaling season of 1962–63 (the season extends from 15th December to 7th April), 22 factory ships ("floating factory" ships were first developed in 1923), comprising 9 from Norway, 7 from Japan, 3 from the Soviet Union, 2 from the United Kingdom, and 1 from Holland, with a total of 253 catcher-boats, were in Antarctic seas hunting whales. Since this time the U.K., Norway and the Netherlands have all withdrawn from the whaling industry.

FIG. 64.—WHALING FACTORY SHIP

This whaling factory ship, the *Southern Harvester*, was originally a British ship but is now part of the Norwegian fleet. The tug is a converted catcher-ship. Largely because of competition from Japan and the Soviet Union and the decline in the whale population, Britain has withdrawn from the whaling industry and sold her last whaling ship. Note the aperture at the stern through which the captured whales are hauled.

Whaling is a hard, tiring, and unpleasant job but if catches are good it can be very rewarding. A good harpooner can earn an average of about £7,000 a season. The star harpooner of the Norwegian Christensen fleet earned £27,270 during the 1961–62 season! Approximately 50,000 whales are caught each year. They yield about 400,000 tons of oil, obtained from the blubber, worth some £35 million. In addition, they give some 60,000 tons of by-products, the most important of which are wax from the Sperm Whale (used as the basis of cosmetics), whale meat (mainly eaten by the Japanese although a few years ago an attempt was made to market it in Britain), whale liver oil (a valuable source of vitamin A), whalebone (now of little value but sixty years ago in great demand for corsetry), ambergris (a valuable base for the making of perfume), and fertiliser (made from the ground-up bones and flesh).

## *MISCELLANEOUS MARINE PRODUCTS*

A number of miscellaneous marine products may be noted.

### PEARLS

Natural pearls, obtained from the pearl oyster which lives in shallow waters in tropical seas, are the most valuable of these marine products. The pearl beds are found chiefly in the Persian Gulf, off the northern coast of Ceylon, and in the seas off northern Australia. There is little pearling carried on nowadays. Pearl shell, used for making buttons, etc., is collected. A notable industry is the artificial or cultured pearl industry of Japan. Cultured pearls are produced by inserting a grain of mother-of-pearl into three-year-old oysters; the grain acts as an irritant and the oyster secretes a substance which forms a pearl. The first "cultivated" pearl was produced by Kokichi Mikimoto, a Japanese, in 1913. As a result, a new industry was born and when Mikimoto died in 1954 more than 100,000 people in Japan were employed in pearl culture.

### SPONGES

Sponges are marine animals or rather the tissues of such. Many varieties of sponges occur but only a few are suitable for domestic use. The sponge fishing industry has greatly declined, partly due to the production of cheap synthetic "sponges". At one time sponge fishing was the staple industry of the Bahama Islands; now it is of little, if any, importance. Sponges continue to be collected in the eastern basin of the Mediterranean Sea, chiefly off the Turkish coast, around Greece, and in the Gulf of Sidra off the Libyan coast, but even here the industry is now of only minor significance.

### SEAWEED

There are numerous kinds of seaweed, many of which are of value. The Japanese put to good use the various seaweeds which grow around the shores of their homeland. Both edible and industrial seaweeds are collected. Seaweed is used as a relish for soups and rice. From it is also produced agar-agar, a gum resembling gelatine, which is used as a medium for bacteria culture and for making isinglass. Formerly, seaweed was used as a source of iodine for seaweeds extract iodine from the sea-water. Seaweed is also used frequently as a fertiliser. Not only the Japanese but the Bretons of Brittany have long collected seaweed and spread it on their fields as a manure.

## *FISH CONSERVATION*

In Chapter VII we referred to the possibility of increasing the quantity of fish taken from the oceans to help increase the world's food supply. It may seem strange, therefore, to discuss here the need for conservation measures with respect to fisheries. The apparent paradox is explained by the fact that over-fishing is confined to certain areas and to certain species of marine life. Catches from the North Sea have been declining for several decades and the fisheries have been severely depleted through over-fishing. Now, however, the same threat faces the North Atlantic fisheries, and the situation could become critical unless some international agreement is reached. Ironically, greater quantities of fish are being caught in the North Atlantic than ever before, for more and more countries are despatching their fleets into that ocean. Today eighteen countries (Great Britain, Norway, Iceland, Denmark, West Germany, East Germany, Poland, Sweden, the Soviet Union, Belgium, Holland, France, Spain, Portugal, Eire, the United States, Canada, and Japan) are engaged in the North Atlantic fishery. But, notwithstanding the greater total catch, the individual catches are declining.

Fishery experts have gone so far as to say that the greatest problem facing the world's fishing fleets now and in the near future is not fishery limits or increased markets but conservation. Fishery scientists draw attention to a number of facts which are having important repercussions upon the North Atlantic fishery: (i) year by year the total catch has increased until in 1968 approximately 12 million tons of fish were caught; (ii) there is no major stock of fish anywhere in the North Atlantic which is not being fished to near exhaustion-point; (iii) both the Soviet Union and Poland plan substantial increases in the size of their fishing fleets which, without doubt, are to fish in the North Atlantic; and (iv) the Japanese have now arrived to join in the fishery although it is 14,000 miles (22,500 km) from home.

The threat of exhaustion arises not merely from the greatly increased quantities being caught but also from the methods of fishing that are now being used. Many countries, including the Soviet Union, East Germany, Poland and Japan, use what has been called "vacuum cleaner" techniques of fishing; such methods draw all fish, including the immature ones, into their nets. This means that natural reproduction and replacement are prevented. Countries using these methods are interested mainly in securing bulk supplies of protein food; they are not much interested in the quality of the fish. Neither are these countries particularly conservation conscious, hence it is difficult to persuade them to participate in international control. But if the fishery is to be preserved, some international control aimed at (i) limiting national catches, (ii) preserving breeding grounds, (iii) regulating net mesh-size (to allow immature fish to escape) must be agreed upon.

## EXERCISES

1. Write a short essay on the fishing industry along the east coast of Britain, giving special attention to the principal fishing grounds and markets. (*Union of Educational Institutions.*)
2. Give an account of the fishing industry in Great Britain and the problems facing the industry today.
3. Mark on an outline map of the British Isles five principal fishing ports and indicate reasons for their growth.
4. Japan is the greatest fishing nation in the world. Explain why this is so.
5. Explain why there is relatively little development of the fisheries in (*a*) tropical latitudes, (*b*) the southern hemisphere.
6. Why are the world's major fishing grounds found in temperate latitudes? Locate the chief fishing grounds.
7. In view of the world's food shortage how might the world's fisheries be more effectively utilised?
8. Write an essay on the whaling industry indicating the important whaling countries, the products of whaling, and the problems facing the industry.

PART THREE

# INDUSTRIAL MATERIALS AND MANUFACTURE

## *Chapter XIV*

## INDUSTRIAL CROPS

MANY commodities are collected or grown for industrial uses. Modern industry requires large quantities of numerous vegetable products and such vegetable products, which formerly were frequently collected in their wild state (*e.g.* rubber, palm oil), must now be systematically cultivated in order to meet the demand. In this chapter we are concerned only with the vegetable raw materials consumed by industry. Broadly these materials may be grouped as follows: (*a*) fibres used in textiles; (*b*) vegetable oils; (*c*) rubber and gums; (*d*) tobacco. Textile manufacturing in its various forms is one of the leading, as well as one of the oldest, industries in the world. In many of the less-industrialised countries it forms the largest single source of industrial employment. A variety of fibres of animal and vegetable origin, augmented in more recent times by artificial man-made fibres, are used in the making of textiles. Of all the fibres (including animal and artificial) cotton is the most important. Vegetable fibres of much lesser importance are flax, jute, and hemp. A number of fibres, chiefly abaca, sisal, and henequen, are known as hard fibres; they are chiefly used for making string, cordage, etc.

Vegetable oils, used in ever-increasing amounts these days, serve two purposes: they are used as edible oils, chiefly for the production of margarine and cooking fats, and as industrial oils used in the preparation of soap, paints, varnishes, lubricants, etc. "As prosperity increases, the consumption of vegetable oils rises: more fats are added to the diet, more technical uses are discovered for oilseed products as industry advances, and more soap is needed as cleanliness changes from being a privileged eccentricity to a universal necessity."*

A number of vegetable products, which are mostly so-called forest products, are of great significance in modern industry. Ignoring timber, rubber is the most important of all these vegetable products. Of lesser importance are resins, gums, waxes, and vegetable dyestuffs. Some of these are dealt with in the chapter on forest resources. (*See* Chapter XVI.)

* *Oxford Economic Atlas*, 2nd edition, 1959, p. 32.

Finally, tobacco, widely grown and widely consumed, has given rise to an important farming and industrial activity.

## *COTTON*

Cotton is the most important of all the industrial vegetable fibres. Though long known—it was familiar to the Ancient Egyptians—its large-scale use as a textile fibre is recent. Prior to the mid-eighteenth century, cotton fabrics were considered a luxury in Europe. However, as a result of the invention, in 1793, of a cotton ginning machine and of the introduction of mechanised spinning and weaving, cotton cloth quickly came into its own and soon became the leading textile.

There are seven commonly recognised species and numerous varieties but cotton is usually grouped into two main types—American and Asiatic—which differ in various respects but which are primarily distinguished according to their staple (*i.e.* fibre length). American types are long-stapled, Asiatic types short-stapled; but plant-breeding and selection have done much to improve both the staple and quality of cotton. American Sea Island cotton, which is exceptionally long-stapled, fine, and silky, is the mostly highly esteemed of all cottons, but is cultivated in only relatively small quantities. Egypt produces the finest of the long-stapled cottons which are cultivated on a large scale.

### CONDITIONS FOR PRODUCTION

Ideal conditions for good cotton production are: (*a*) equable, warm conditions during the growing period with temperatures reaching 25° C (77° F) in summer; (*b*) a minimum growing period of at least 200 days free from frost since the cotton plant is very susceptible to frost; (*c*) approximately 25–40 in. (635–1016 mm) of rain during the growing season or the equivalent supply of water by irrigation; (*d*) dry, sunny conditions during the maturing period since rain at this time is apt to spoil the bolls; (*e*) a deep, rich, well-drained soil but one with a high moisture-holding capacity; cotton is an exhausting crop; (*f*) a considerable supply of labour to meet the heavy demands of ploughing, sowing, weeding, and picking.

### WORLD PRODUCTION

World production and export, notwithstanding a large home consumption, is dominated by the United States. About a third of the total world production and a third of the cotton entering into international trade comes from the United States. Although the acreage devoted to cotton has declined in the States, the output has been maintained by growing higher-yielding varieties. Details of production are given in

Fig. 65. China, the Soviet Union, India are the next most important producers, followed by Brazil, Egypt, Mexico, and Pakistan.

Recently China has greatly expanded her output and now closely rivals the United States. The bulk of the cotton grown in China is cultivated on the Great Plain and in the lower Yangtse valley; little, however, is exported. The Soviet Union has fostered the growing of irrigated cotton in Soviet Central Asia (Turkestan). Indian production is concentrated in the north-western part of the Deccan where the rich water-retentive

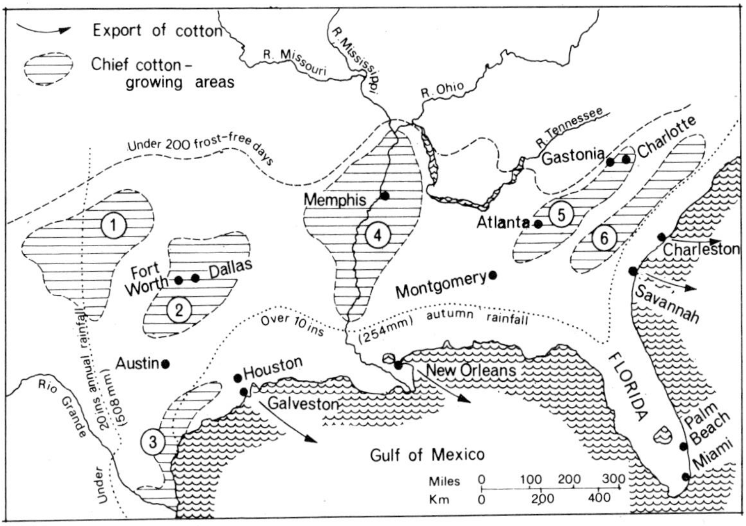


FIG. 65.—THE UNITED STATES COTTON BELT

Note the climatic controls: the 200 frost-free days line, which also corresponds closely with the summer isotherm of 25° C (77° F), marking the northern limit; the 20-in. (508-mm) isohyet which marks the western limit; and the occurrence of heavy autumn rains in the south which have prevented the extension of the Cotton Belt to the Gulf coast. The chief producing areas are: 1. the High Plains of north-west Texas; 2. the Black Waxy Prairies; 3. the coastal plain of southern Texas; 4. the Mississippi flood-plain; 5. the middle Tennessee valley; and 6. the Atlantic coastal plain and Piedmont Plateau. Cotton is also grown under irrigation in California.

Black Cotton soils, derived from the weathered basaltic rocks, favour cotton cultivation. Most of the cotton is consumed by the home looms but a little is exported to Japan. Pakistan grows long-stapled American varieties on the irrigated alluvial soils of the Indus valley and is a considerable exporter. In Egypt, a major world exporter, production has been linked with the development of perennial irrigation, for all crops must be grown by irrigation in that country. Egyptian cotton is of very good quality and is much in demand; moreover, yields are high; but increased production is handicapped by the need to grow subsistence crops for Egypt's rapidly expanding population. Though not a major world producer we should note the importance of Sudan's Gezira cotton area. Here, on the flat plain between the confluence of the Blue and White Niles,

[*Courtesy: U.S. Information Service.*

FIG. 66.—MECHANICAL COTTON-PICKER

Much cotton is still hand-picked, especially in the "Old South" of the United States, where Negroes do the work. Mechanical cotton-picking is gradually replacing manual methods: this is especially true in Texas one of the "newer" growing areas in the United States. Most of the machines are operated by white men, and use suction devices for stripping the bolls.

some 5 million acres (2 million ha) irrigated by water stored by the Sennar Dam are growing cotton. Cotton is patchily cultivated throughout much of East, Central, and West Africa. In South America, Brazil, Argentina, and Peru are the only important cotton growers. Cotton has long been grown in Brazil. Tree cotton comes from the "shoulder" of Brazil; bush cotton from the São Paulo region. Since the Second World War there has been considerable expansion in the latter area. In recent years Argentina has developed cotton plantations in the Chaco region. High quality cotton is grown in the Peruvian oases. Mexico is also a producer of note.

TABLE XIV

*Cotton Production (thousand metric tons)*

| | *1960* | *1970* |
|---|---|---|
| U.S.A. | 3107 | 2213 |
| U.S.S.R. | 1482 | 2340 |
| China | 2410 | 1518 |
| India | 959 | 954 |
| Brazil | 437 | 672 |
| Mexico | 483 | 304 |
| Pakistan | 478 | 531 |
| Egypt | 304 | 509 |

Rather less than half of the world's cotton crop enters into international trade. The chief exporters, roughly in order of importance, are the United States, Egypt, Mexico, Brazil, and Pakistan. The bulk of the exported cotton goes to the important cotton textile manufacturing countries of Europe—to the United Kingdom, France, West Germany, and Italy—and to Japan.

TABLE XV

*Cotton: Chief Importers (percentage of world total)*

| | |
|---|---|
| Japan | 20 |
| United Kingdom | 14 |
| France | 12 |
| West Germany | 11 |
| Italy | 6 |

## *FLAX*

The flax plant, a member of the nettle family, has been used for the making of linen cloth from very early times and may well be the oldest of all textile fibres. Linen has been found in the tombs of the early Egyptians and fragments recovered from the pre-historic lake-dwellings of Switzerland. Linen is made from the fibrous material of the flax plant, a plant which is also grown for its seed (linseed). Flax is a plant of moist cool temperate lands but if it is grown for its seed it is usually cultivated in warmer latitudes. Linen is much less important nowadays as a textile than formerly although fine linen cloth has much to commend it.

### CONDITIONS FOR PRODUCTION

Here we are concerned only with the growing of flax for its fibre and the geographical conditions required for its cultivation in this respect are: (*a*) cool temperate conditions with summer temperatures of between 16°–18° C (60°–65° F); fairly uniform temperatures are desirable; (*b*) moderate rainfall of between 20 and 30 in. (580–762 mm) and a fairly high atmospheric humidity; (*c*) level land with a firm, moist, but well-drained soil which has been well fertilised; (*d*) a cheap and abundant labour force for much work is involved in the preparation of the fibre. It is partly for the last reason that flax cultivation has almost disappeared from Northern Ireland and why it continues to be grown in the Low Countries, Poland, and the Soviet Union. Flax is a very exhausting crop—so much so that it is seldom grown on the same land more than once in about eight years. Uniform temperatures during the growing season appear to be the crucial factor; a hot and dryish summer will result in the flax plant producing a coarse fibre.

## FLAX PROCESSING

The flax plant is usually harvested by hand; it is hand-pulled to ensure a maximum length of fibre. It is piled up to dry and "rippled" or combed to rid the stalks of leaves and seeds. Next, the flax is immersed in water and left to rot, a process known as "retting." In former times bundles of flax were left to ret in streams, ponds, or field-side ditches (and one could always recognise the presence of the flax by the nauseating stench which the decaying vegetable matter gave off). Nowadays the flax is commonly retted in tanks of water to which chemicals are added to help speed up the decomposition of the vegetable matter. After a long period of soaking, the green, fleshy, resinous matter breaks down and the fibrous material is then scraped or "scutched" to clean it of the soft vegetable matter. The fibre is then dried and the longer fibres (or "line") separated from the shorter (or "tow"), for flax fibres may be anything from 10 to 15 in. (254–381 mm) in length. Only the longer fibres are used in the manufacture of linen, the tow being used in rope-making. Flax fibre, though very strong and durable, has the serious drawback, from the point of view of manufacturing, of lacking uniformity.

## WORLD PRODUCTION

The cultivation of flax for its fibre is virtually confined to the North European Plain extending from north-eastern France and the Low Countries to Russia. Of the total world output some 725,000 metric tons, the Soviet Union alone accounts for approximately two-thirds. In spite of the preponderance of production in the Soviet Union, most of the flax entering international trade comes from Belgium and Holland which produce fibre of very high quality. Formerly flax was grown in Northern Ireland to support the local linen industry but cultivation has died out and the industry depends almost entirely upon imported supplies. It is interesting to note that during the Second World War, as a result of the inability to import supplies, the cultivation of flax was resuscitated and the acreage greatly expanded; but this proved to be only a temporary matter and as soon as flax imports became available again the growing of the fibre rapidly declined.

TABLE XVI

*Flax Fibre Production (thousand metric tons)*

| | *1960* | *1970* |
|---|---|---|
| U.S.S.R. | 425·0 | 456 |
| France | 37·8 | 38 |
| Poland | 43·0 | 52 |
| Belgium | 35·1 | 10 |
| Netherlands | 31·3 | 10 |
| Czechoslovakia | 20·1 | 14 |
| World total | 592·3 | 619 |

## *JUTE*

Jute is the coarse fibre obtained from the stem of a tall, reed-like annual plant which grows in tropical swamplands. It requires hot, moist conditions; indeed, it is cultivated under flood conditions. It requires soils

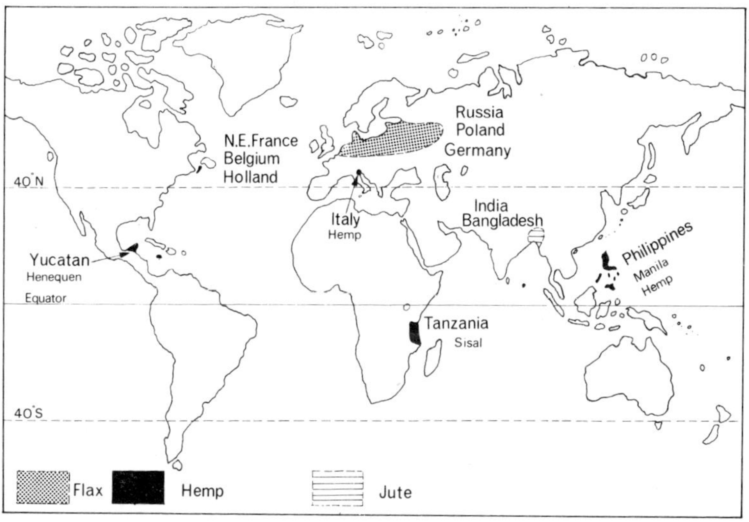


FIG. 67.—DISTRIBUTION OF JUTE, HEMP, AND FLAX

of high fertility and thrives on river alluvium. Planted in early spring, it grows rapidly, up to heights of 12 feet (4 m), and within three or four months it is in flower. The jute plant is harvested when it is in bloom. Jute-growing demands much hand labour and at harvest time the labourers are often compelled to work in water which may be up to their waists. After the jute is cut, it must be processed in much the same way as flax.

A notable feature of jute production is that some four-fifths of all the jute grown in the world is cultivated in the Ganges–Brahmaputra delta (in some periods during recent decades this area has produced more than 95% of world output). In this region is to be found the perfect combination of tropical climate, flooded land, alluvial soil, and abundant cheap labour for the large-scale production of jute and, accordingly, it has become a very important crop.

A large proportion of the jute is woven (in the mills around Calcutta) into "gunny cloth" used for the making of sacks and bags. It finds other uses in the manufacture of carpets, linoleum, tarpaulin, upholstery, and twine.

TABLE XVII

*Jute and kenaf production*
*(thousand metric tons)*

| | |
|---|---|
| India | 1104 |
| Bangladesh | 1261 |
| China | 500 |
| Thailand | 366 |
| Brazil | 53 |
| U.S.S.R. | 50 |
| Nepal | 35 |
| Burma | 28 |
| World total | 3502 |

The bulk of the raw jute is processed in the Indian sub-continent but some is exported to Britain (the British import goes mainly to Dundee and Barnsley where it is manufactured into hessian), West Germany, France, and Belgium. The United States also imports considerable quantities of jute.

## *HEMP*

True hemp, sometimes called "soft" hemp to distinguish it from the so-called Manila hemp—a tropical "hard" fibre which is not really a hemp at all—comprises two species, one grown in Europe, the other (known as sunn-hemp) chiefly in India.

The true hemp which is cultivated in Europe is closely related to, and grows under much the same conditions as, flax. Furthermore, the fibre is separated from the plant stems by processes very similar to those used in the preparation of flax. Italy has a tradition of producing the best quality hemp fibre, but it is also grown in Poland, Hungary, Rumania, Yugoslavia, and Turkey. It is chiefly used in the manufacture of cordage and canvas.

The Indian variety, sunn-hemp, is used as a substitute for jute. India produces about one-fifth of the total world production of hemp.

## *THE HARD FIBRES*

Abaca (Manila hemp), henequen, and sisal are known as hard fibres; they are mainly used in the manufacture of cordage and matting. The hard fibres differ from hemp, jute, and flax in that the fibrous material is obtained from the leaves instead of the stems of the plants.

*Abaca*, a plant related to the banana family, requires warm, humid conditions and fertile, well-drained soils for its successful growth. It is cultivated almost exclusively in the Philippine Islands, whence it gets its alternative name—Manila being the capital of the Republic. Abaca yields a tough, strong, sea-water resistant fibre, hence it is especially suited for the making of ships' hawsers.

*Henequen* and *sisal* are really the same thing; they are species of agave native to Central America. Henequen is grown in the Yucatan peninsula of Mexico. During the present century agave was introduced to East Africa where, as sisal, it is grown as a perennial crop in dry, tropical climates. It is an important crop in Tanzania and is grown in Kenya and Angola. Appreciable quantities are also cultivated in the dry north-eastern "shoulder" of Brazil. It is used in the making of coarse twine and matting.

## *OTHER VEGETABLE FIBRES*

There are a number of other vegetable fibres of lesser importance which may be mentioned.

*Ramie*, or *China grass*, is a rough strong fibre which may be woven into a textile. It is grown throughout central China, where it is much used.

*Raffia*, or *Bass*, comes from Madagascar; it is obtained from the leaves of the raffia palm.

*Phormium*, or *New Zealand flax*, is a fibre derived from the narrow, sword-like leaves of a plant which thrives on poor, boggy soils. It can be used for the same purposes as hemp fibre but very little of it is used.

*Kapok*, or vegetable down, is the floss obtained from the pods of the so-called silk cotton trees, species of which grow in tropical America, India, and Indonesia. Kapok, on account of its buoyance is much used in life-saving equipment. It is also much used as a stuffing material in bedding and upholstery.

*Coir*, the coarse, brittle, fibrous material obtained from the coconut, is much used in the manufacture of brushes, matting, and ropes. Most of it comes from Ceylon and India.

*Esparto grass*, also known as *alfa* in North Africa, is used in the manufacture of high quality paper. Most of it comes from Algeria, Tunisia, and southern Spain.

## *SYNTHETIC FIBRES*

The production of man-made or synthetic fibres is now an important industry in many parts of the world. Synthetic fibre production may be said to lie half-way between the making of textiles and chemicals; in fact, it has been amusingly but rather aptly described as a "kind of chemical sheep-farming." Man-made fibres may be grouped into two main kinds:

1. Those derived from so-called "natural" materials, for example, wood pulp or cellulose, groundnut protein, and casein (a waste product of milk) which, respectively, are used in the making of the fibres known as rayon, Ardil, and Fibrolane.

2. Those based on "true" synthetics which are made from hydrocarbons, *i.e.* coal and petroleum, especially the latter, and which give us such materials as Terylene, Tricel, Acrilan, and Orlon, and, above all, Nylon, now one of the most widely used of all man-made fibres.

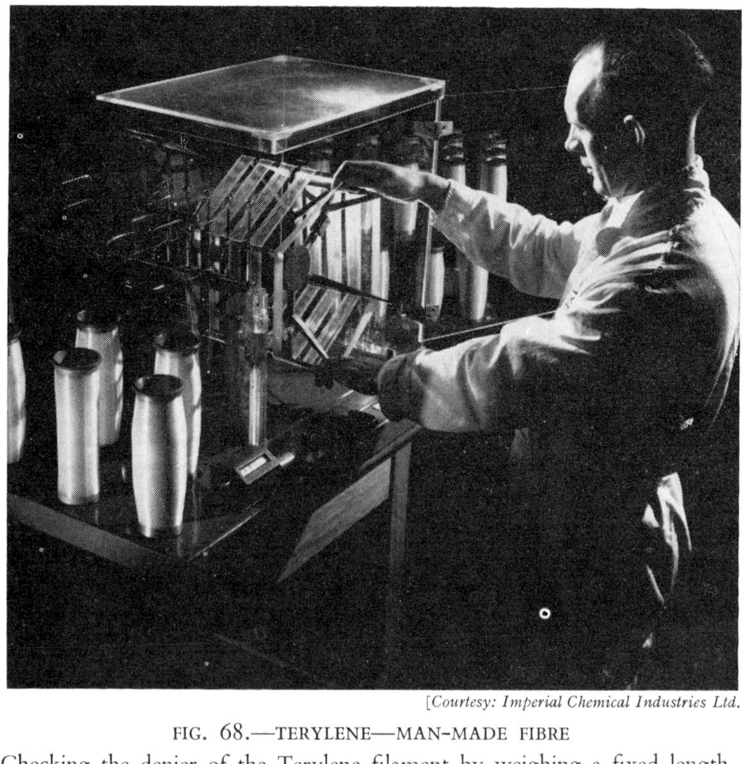

[*Courtesy: Imperial Chemical Industries Ltd.*

FIG. 68.—TERYLENE—MAN-MADE FIBRE

Checking the denier of the Terylene filament by weighing a fixed length. Wilton, in north Yorkshire, is the great Terylene-making centre of I.C.I.

Rayon, or artificial silk as it was once popularly called, is the oldest of the man-made fibres. Chance played a large part in its development as a textile. About eighty years ago, Sir Joseph Swann, a pioneer in electricity, was experimenting with rayon thread as a possible filament of light bulbs. Spare threads were used by his daughters for their crochet work and this gave Swann the idea of using the rayon thread as a textile material. Although artificial silk was made as early as 1884, it was not until the turn of the century that its commercial production got under way and the biggest and most rapid developments have occurred since 1920.

Rayon is made by treating wood-pulp, cotton, and other vegetable materials with chemicals to produce cellulose. The process of making rayon is the same in principle as that of the silkworm making a cocoon; the vegetable matter is reduced to a pulp by machine and chemical and then forced through a fine nozzle, the cellulose solidifying upon coming into contact with the air and forming a thread. The world production of rayon continues to increase rapidly, largely because of the relative cheapness of the fibre. Its production surpassed that of real silk in 1925 and of wool just before the Second World War. Total world production is now about 3½ million tons annually—almost twice that of wool. The chief producers are the United States, Japan, the United Kingdom, West Germany, Italy, East Germany, the Soviet Union, and France.

The true synthetic or non-cellulosic man-made fibres are largely a development of the last twenty-five years. In 1938 came an historic event for women: the first pair of nylon stockings was displayed at the World Fair in New York! Since that date, advances have been rapid. Production of purely synthetic fibres, of which nylon is easily the most important, has increased from nearly 100,000 tons in 1950, to 195,000 tons in 1954, to about 500,000 tons in 1958, to over 1½ million tons in 1964, and to over 2 million tons at the present day.

While the past in this new world of man-made fibres may be said to belong to the products of wood-pulp, cotton-waste, groundnuts, milk by-products, and other "natural" materials, there seems to be no doubt that the present and the near future belong to the products of petroleum. The petroleum chemist has wrought miracles during recent years and added a profusion of new fibres to the textile industry. Thus, within the space of about half a century, man has enlarged enormously the limited choice of fabrics imposed by the four basic natural fibres: wool, cotton, silk, and linen.

## *VEGETABLE OILS*

One of the most important developments in economic production and trade during the twentieth century has been related to the increasing use of vegetable oils. This increased use results mainly from three developments: (*a*) the production of animal fats lagging behind the increasing world demand for fats; (*b*) the gradually extending uses to which vegetables oils can be put; and (*c*) technological improvements in the extraction and processing of vegetable oils.

The introduction of margarine as a substitute for butter and the growth of the soap-making industry were the two main factors stimulating the development of vegetable oil production. But during the past half century vegetable oils have found an increasingly wide range of uses, *e.g.* in paints, varnishes, lubricants, plastics. Between 1850 and 1950 the consumption of vegetable oils increased fourfold, between 1939 and 1956 world production of vegetable oil rose from 14 to 20 million tons, while in the early nineteen-sixties it reached over 30 million tons. This increased demand for vegetable oils led to a great expansion of tropical plantation agriculture.

It is of interest to note the changes which have taken place in the types of vegetable oil used and in the pattern of world trade which resulted from the Second World War. The war, for instance, cut off supplies of Chinese soya-beans and Malaysian copra; this meant that consumers turned to alternative sources such as groundnut, sunflower, and linseed oil. The United States turned to cotton seed oil, greatly expanded its acreage of soya-beans, and introduced tung trees to supply tung oil. Wartime shortages led to the substitution of one vegetable oil for another, especially as scientists improved processing and refining. A further development has been the growth of oilseed crushing and oil expressing in the producing countries in contrast to the pre-war practice of doing this mostly in the consuming countries.

Vegetable oil may be extracted from the fruits, nuts, seeds, or roots of certain plants. Let us now briefly review the different vegetable oils that enter into international trade.

## GROUNDNUT OIL

This is obtained from the groundnut plant. The groundnuts, commonly known as peanuts or "monkey nuts," are really the seeds of a plant belonging to the pea family. The flowering stalks bend over after fertilisation and the "nuts" form in the soil around the base of the plant. The nuts are a palatable foodstuff and native peoples eat them and also grind them up to make a kind of flour. Peanuts and peanut "butter" are common both in Britain and in the United States. The groundnuts when crushed yield a valuable edible oil which is much used in the making of margarine. It is also used in the manufacture of soap and plastics. The residue forms an excellent cattle feedstuff. Groundnuts are widely grown in the tropical and warm temperate regions; they are specially important in West Africa (in Nigeria and Senegal), where they thrive in the drier parts of the savanna lands, in India, China, and Central America. Considerable quantities are now grown in the United States.

## PALM OIL

The oil palm is a native of West Africa, especially Nigeria, whence comes three-quarters of the total world production of palm oil. The palm grows wild and the industry is more in the nature of large-scale commercial collecting rather than planting, although there are now some palm plantations in West Africa. In the East Indian region, *i.e.* Malaysia and Indonesia, the oil palm is mostly raised on commercially-owned and scientifically-managed plantations. The oil palm produces great clusters of pulpy plum-like red fruits containing hard kernels. Oil is expressed from the fruit by crushing and boiling; this, the ordinary palm oil, is used in the manufacture of soap and candles, and as a lubricant, and also has a special use in the manufacture of tin-plate. The kernels within the fruit also contain an oil and these, when crushed, yield palm-kernel oil which is a valuable ingredient in margarine manufacture.

## COCONUT OIL

Coconut oil is expressed from the dried flesh or copra of the fruits of the coconut palm, which is characteristic of the sandy coastlands of tropical

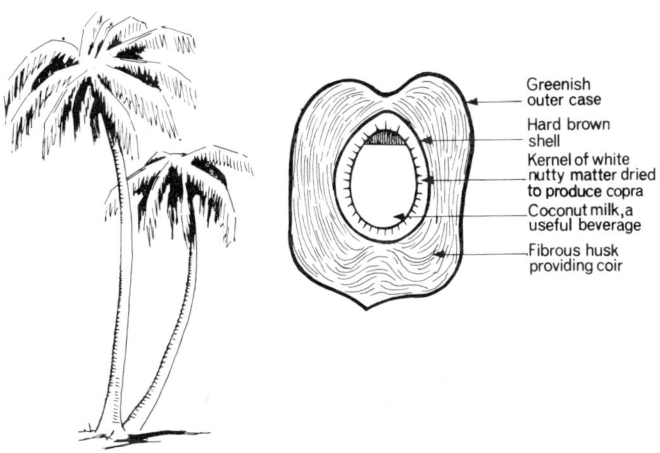


FIG. 69.—DIAGRAM OF A COCONUT

The coconut palm and its fruit. Coir is used for making brooms and coconut matting, the shell provides a substitute for native crockery, the flesh is the source of desiccated coconut, and the dried flesh (copra) is pressed to yield its oil.

regions, especially in and around the Indian and Pacific Oceans. Coconut oil is highly valued and usually forms the basis of high-grade margarine and quality toilet soaps. The dried coconut flesh is usually exported as copra to the consuming countries where the oil is then squeezed out and refined. The residual pulp of copra forms a nutritious animal feedstuff. Although copra is collected by traders from the scattered islands of the Indian and Pacific Oceans, the chief commercial sources are the coconut palm plantations of Southern India, Ceylon, Malaya, Indonesia, and the Philippines. The coconut palm also yields coir, the matted fibre which encloses the nut and which is used commercially in the manufacture of matting, brooms, etc.

### BABASSU OIL

Another vegetable oil, little known as yet, but one which we shall probably hear much of in the future, is that from the babassu (or babacu) palm. This palm, which yields a high-grade oil, grows in great groves in the north-eastern part of Brazil. Thousands of millions of these wild palms provide an enormous crop potential. Present production is small. The nuts are collected as they fall to the ground. They are broken by hand with sledge hammers, each worker producing about 18 lb (8 kg) of kernels a day. One of the major problems of the industry lies in the hardness of the nuts: they are extremely difficult to crack. However, experiments have been undertaken to devise a nut-cracking machine which will break the nut shell without damaging the kernel. If cheap processing could be introduced the whole pattern of the vegetable oil industry might be changed, and Brazil might find its economy revolutionised by the development of an oil industry larger than its coffee industry. Babassu oil has many applications: it can be used as an edible fat, as an ingredient in soap, as a lubricant, and provides an excellent substitute fuel for diesel engines.

### OLIVE OIL

Olive oil is derived from the fruit of the olive tree which flourishes in, and is restricted to, regions having a "Mediterranean" type of climate. In the lands around the Mediterranean Sea olive oil has long been used as a foodstuff, taking the place of animal fats in human consumption. The olive tree thrives under dry conditions, indeed, it is intolerant of summer rain, and flourishes in thin, stony, limestone soils. Once an olive tree has become established, it will continue to yield several hundreds of pounds of fruit each year for as long as a century. Although some olives are used for food, the bulk of the crop is pressed for oil. Production is very largely confined to the lands surrounding the Mediterranean Sea: Italy, Spain

Tunisia, Greece, Portugal, and Turkey, roughly in that order, are the chief producers. Some olive oil is used in soap-making.

## SOYA-BEAN OIL

The soya-bean is a plant of warm temperate areas and native to the Far East. It has long been an important crop in the East where the bean has been used as a vegetable, made into flour and bean-curd, and its oil content used for cooking purposes. At the beginning of the present century methods were found to rid the oil of its rancid flavour and as a result it came to be increasingly used in the preparation of margarine.

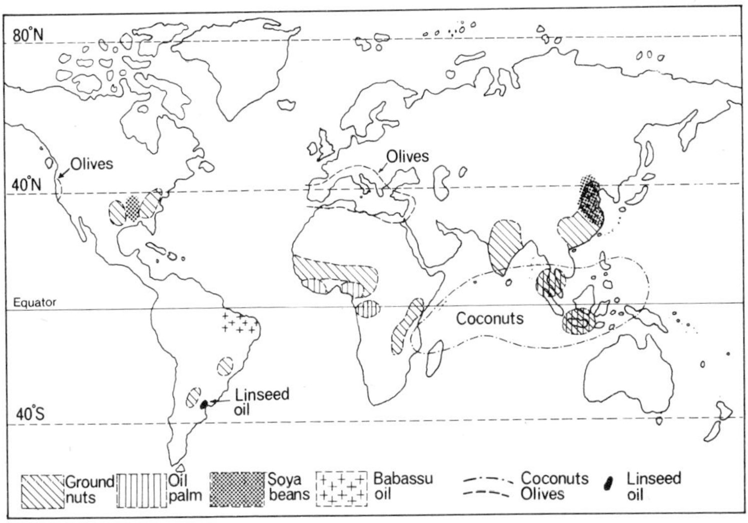


FIG. 70.—DISTRIBUTION OF VEGETABLE OILS

Soya-beans are an important crop in Manchuria, China proper, Korea, and Japan. During the inter-war period a big trade developed in the export of soya-bean oil, especially from Manchuria which contributed over a half of the total. During recent decades the United States has become a major grower and exporter; she now accounts for more than half of world total production. Soya-bean oil is used not only in the preparation of foodstuffs, chiefly margarine and cooking fats, but is put to many industrial uses, *e.g.* in the making of paint and varnish and printing-ink.

## LINSEED OIL

Linseed oil is obtained from the crushed seed of the flax plant. When flax is grown for oil, it is usually cultivated under much warmer conditions than are desirable for fibre. Argentina and the United States are the leading producers of linseed, but India and Canada are not far behind. Linseed oil is especially valuable as a "drying oil" and for this reason is much used in the preparation of paints and varnishes. It is also used in the manufacture of linoleum, oilcloth, and soap. Linseed is very rich in oil, yielding about a third of its weight in oil. The chief exporters of linseed oil are, roughly in order of importance, Canada, India, Argentina, Uruguay, and the United States.

### COTTON-SEED OIL

The seeds of the cotton plant, which are left behind after the ginning of cotton, contain an oil which is now extracted and used for a variety of purposes. The oil is used as a lubricant, in the manufacture of soap and, in its refined state, as a cooking oil, in the making of margarine, etc. What was once a waste product of cotton cultivation has now become a valuable by-product, much used as a substitute for olive oil which it strongly resembles in flavour. Cotton-seed oil is much used in the United States as might be expected from its importance as a cotton grower. A number of cotton-growing countries, *e.g.* Sudan, Uganda, Argentina, export either the seed or the extracted oil. The port of Hull has become a great centre of cotton-seed oil refining.

Many other plants yield useful oils but here we can do no more than merely mention some of them by name: sunflower seed, rapeseed, sesame seed, poppy seed, castor seed, and perilla seed all yield oils which are used in greater or lesser quantities, while a particularly useful "drying oil," which forms an excellent substitute for linseed oil and dries even more quickly, is derived from the nuts of the tung tree, a tree native to southern China but one which during the past two or three decades has come to be much grown in the hotter parts of the United States and Argentina.

## *RUBBER*

### THE HISTORY OF RUBBER

Rubber is one of the strategic materials of the world today. It has a wide variety of uses—from the making of motor tyres to hot-water bottles—and has come to be looked upon as an essential raw material in numerous manufactures. Rubber has several invaluable properties, *e.g.* elasticity, resistance to water, and electrical non-conductivity, which have made it virtually indispensable in modern life. It is true that modern chemistry has provided us with a substitute to natural rubber in synthetic rubber and that plastics have replaced rubber for certain purposes; even so, natural rubber still has innumerable essential uses.

Known to the native peoples of America as *caoutchouc*, rubber was first introduced to Europe by the French explorer, la Condamine, in 1735. Apart from its property of erasing pencil marks, it received little interest or attention until Charles Mackintosh began to experiment with it as a waterproofing agent. In 1823 Mackintosh discovered a method of waterproofing cloth by using rubber, but unfortunately this did not prove to be very successful, for the rubber became sticky in hot weather and hard in cold. This difficulty was not overcome until 1844 when Goodyear in the United States and Hancock in Britain discovered the method of vulcanisation, the heating and mixing together of rubber and sulphur which eliminated the former disadvantages.

Until about 1890 the quantity of rubber used was small and most of it came from the Amazon Basin where the rubber tree grew wild. Two developments in the last decade of the nineteenth century led to a greatly increased demand for rubber: first, the invention of the pneumatic tyre by Dunlop followed by the invention of the motor car and, secondly, the development of electrical power which led to the use of rubber for wire and cable insulation. The increased demand for rubber led to a boom in Amazonia and the rubber barons made big fortunes. The boom lasted until the early years of the present century when, quite suddenly, it collapsed due to the appearance of plantation-grown rubber. Amazonia suffered an economic setback from which it has never recovered.

The history of rubber is intimately linked with several names, with la Condamine, Mackintosh, Goodyear, and Dunlop, already mentioned, and with Sir Henry Wickham. Wickham, so the story goes (though this has recently been discredited), was responsible for smuggling some rubber seeds out of Brazil in 1876. These were sent to London and propagated in the hot houses of Kew Gardens. Later plants were taken to Ceylon, and subsequently to Malaya, where it was known that the climate and soil conditions were ideal for rubber cultivation. Within a very short time large, efficiently planned estates had been developed which were able to produce rubber in large quantities and of superior quality to that produced in Amazonia by haphazard collection and primitive processing. Soon other plantations were set up in the then Dutch East Indies and, later, in Sarawak, Indo-China, West Africa, and the Congo.

## CONDITIONS OF GROWTH

The principal requirements for the production of rubber are those characteristic of equatorial lowland areas: (*a*) high temperatures averaging 27° C (80° F) and not falling below 21° C (70° F); (*b*) heavy rainfall, at least 60 in. (1524 mm), evenly distributed throughout the year; (*c*) undulating land or gentle hill slopes allowing good drainage. These are the three physical pre-requisites. A rich, deep soil is very desirable but rubber will grow on a variety of soil types as, in fact, it does in Malaya. Much labour is entailed in rubber production—land has to be cleared and prepared for cultivation, the rubber saplings have to be reared in nurseries and transplanted, the land must be kept clear of weeds and undergrowth, and systematic collection of the rubber undertaken. All this calls for much cheap and reliable labour.

[*Courtesy: Malaya House.*

FIG. 71.—RUBBER TAPPING

The rubber-tapper uses a specially designed knife to make an incision into the inner bark of the tree. The latex or juice flows out of the cut and is collected into a receptacle fastened to the trunk. Cutting is usually carried out early in the morning. The cups are collected as soon as possible after the latex has stopped flowing, emptied into buckets, and the latex is then taken to the plantation factory for treatment. Note where the outer bark has been stripped for previous incisions.

## COLLECTION AND PROCESSING

Rubber is the milky juice or latex exuded by a number of equatorial forest trees, the most important species of which is the *Hevea brasiliensis* which is a native of Amazonia. This is the tree mostly cultivated. Saplings, reared in nurseries, are transplanted out in rows about 12–20 ft (4–7 m) apart in newly cleared forest land. After about six years they are ready to yield. The latex is extracted from the tree by "tapping." This involves making an incision in the outer bark, care being taken not to cut the inner layer which would damage the tree. Small cups, aluminium receptacles or, sometimes, coconut shells, are attached to the trunk into which the sap trickles. Tapping takes place daily, commonly early in the morning. Several hours later the latex is collected and taken to the plantation factory where it is emptied into tanks containing formic acid. The acid causes the latex to coagulate. Water is then squeezed out of the coagulated rubber by passing it through rollers. Finally, it is dried to make crepe rubber, some of it also being smoked to produce sheet rubber.

## EXPORTERS AND IMPORTERS

Today, Malaya and Indonesia produce most of the world's natural rubber, a large proportion of it being shipped from Singapore which is

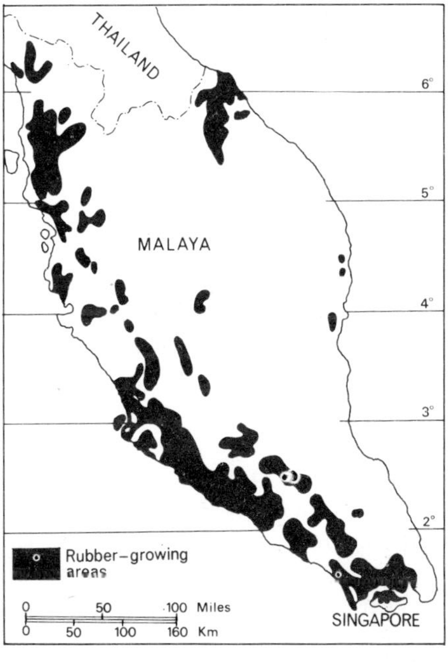


FIG. 72.—RUBBER IN MALAYA

The map shows the location of the rubber-growing areas in Malaya. These are mostly along the western coastlands of the peninsula between sea-level and the 750-feet (225-m) contour. Some $3\frac{1}{2}$ million acres (1·4 million ha) are planted to rubber and Malaya produces about one-third of the world's natural rubber. The rubber tree grows on a variety of soils in Malaya: in coastal clays and alluvium, and in quartzite and shale soils. Peaty soils are avoided. More recently newer plantations have begun to be developed on the eastern side of the peninsula but the only really important area so far is in the north-east in Kelantan.

the great collecting centre. Important contributions to the world's rubber supply are also made by other South-east Asian countries, notably Thailand, South Vietnam, Sarawak, Sabah, and Ceylon. Very little rubber now comes from South America, where there are scarcely any plantations. In the nineteen-thirties Henry Ford established some plantations in Amazonia but these did not prove to be very successful and the project was discontinued. In Africa, the Congo Basin is of little significance and Nigeria and Liberia (where an American company has set up plantations) are the only important producers.

The following countries, which are the major world producers, are the chief exporters.

TABLE XVIII

*Rubber: Chief Producers (thousand metric tons)*

| | *1960* | *1970* |
|---|---|---|
| Malaya | 719·7 | 1223 |
| Indonesia | 639·8 | 811 |
| Thailand | 171·0 | 287 |
| Sri Lanka (Ceylon) | 98·8 | 158 |
| Liberia | 42·4 | 76 |
| Nigeria | 59·5 | 59 |
| Khmer Republic (Cambodia) | — | 46 |
| S. Vietnam* | 76·6 | 33 |

The leading importers are the United States, accounting for approximately 40% of the total world export, the United Kingdom about 13%, West Germany about 8%, and France about 7%.

### SYNTHETIC RUBBER

The cutting-off of Germany and the United States from their normal rubber supplies during the Second World War greatly stimulated the production in those two countries of synthetic rubber. Both countries had already experimented with synthetic substitutes and during the war years they both produced synthetic rubber in large quantities; indeed, by 1945 their combined output was equal to the total world production of natural rubber in the immediate pre-war years. The United States continues to produce synthetic rubber on a large scale as does the Soviet Union and several other industrial countries. Germany is no longer a major producer.

Synthetic rubber, made chiefly by the chemical treatment of petroleum, coal, and limestone, is being produced in increasing quantities and is becoming more and more important. This is especially true of the United States. So rapid has been the post-war development in synthetic rubber production that its output now substantially exceeds that of natural rubber.

## *TOBACCO*

The tobacco plants—there are several species—are all native to the Americas, and when Europeans first visited the New World they found

* The war in South-east Asia has adversely affected production in South Vietnam and also in Cambodia.

the practice of smoking was widespread among the native Indian peoples. Sir John Hawkins is credited with the introduction of tobacco into England but it was Sir Walter Raleigh who popularised the habit of smoking. The popularity of tobacco-smoking waxed and waned during the seventeenth and eighteenth centuries but the nineteenth century brought renewed popularity and since then the habit has continued to grow. Nowadays, the habit is almost world-wide, though some religions, *e.g.* Islam, proscribe such indulgence. The nicotine present in tobacco acts as a mild stimulant and narcotic and, whatever view may be held regarding its ill-effects, the habit of tobacco-smoking shows little sign of any declining popularity. The growing and processing of tobacco have given rise to a major industry, while many governments derive a substantial revenue from the tax they impose upon the sale of tobacco.

## CONDITIONS OF GROWTH

Tobacco thrives best under sub-tropical conditions but will grow under a fairly wide range of climates from equatorial to cool temperate. The most favourable conditions for tobacco plants are fairly constant temperatures, absence of frost, regular water supply, a humid atmosphere, and well-drained soils which are rich in plant foods and not too acid.

The tobacco plant requires considerable attention during its growth period. It is raised from seed in seed-beds which must be protected from both frost and heavy rain. Transplanting into the fields is carried out when all risk of frost is passed. Thereafter, constant weeding and hoeing are required.

The flavour and quality of tobacco depends upon the plant type, the climatic conditions under which it is grown, the soil, and the processing of the leaf. Tobaccos vary widely according to whether they are cigar, pipe or cigarette tobaccos.

## PRODUCTION AND CONSUMPTION

Over $4\frac{3}{4}$ million metric tons of tobacco are grown annually. The United States is at once the largest single producer and exporter of tobacco. As Table XIX indicates, the United States accounts for more than one-quarter of all the tobacco grown and this comes chiefly from the states of Virginia, Kentucky, North and South Carolina, and Georgia. Approximately one-third of the United States output is exported and the United Kingdom takes about half of this. None of the other major producers, however, play any important part in the international tobacco trade; most of their output is consumed internally. Countries supplying considerable quantities of tobacco for export include Cuba and Jamaica, both of which specialise in the production of cigars, Canada (Lake Peninsula area) which exports tobacco to Britain, and Rhodesia, Malawi, and Zambia which also export to the United Kingdom. The Netherlands and Germany are also major importers of tobacco. Two small points of interest are that "Turkish" cigarettes are made from a tobacco with a special flavour of its own, although more "Turkish" tobacco is grown these days in Greece than in Turkey, and that the so-called "Egyptian" cigarettes are, in fact, made from imported tobaccos.

TABLE XIX

*Tobacco Production (thousand metric tons)*

| | *1961* | *1970* |
|---|---|---|
| China | 422·0 | 785 |
| U.S.A. (1959) | 881·8 | 865 |
| India | 285·5 | 338 |
| Brazil | 162·0 | 244 |
| U.S.S.R. | 197·4 | 270 |
| Japan | 121·0 | 150 |
| Pakistan | ? | 164 |
| Turkey | ? | 138 |
| Bulgaria | ? | 122 |
| Canada | 97·1 | 100 |
| Greece | ? | 90 |
| Rhodesia | ? | 62 |
| World total | 3680 | 4674 |

EXERCISES

1. Locate the world's chief sources of *two* of the following: jute, palm oil, rubber. Describe the climatic conditions and methods of production for each of the two commodities you select.

2. Review the different vegetable oils that enter into international trade.

3. Give an account of the conditions of production, processing, and world trade in *either* flax *or* jute.

4. What are the geographical conditions necessary for cotton production? Indicate the chief producers and exporters of raw cotton.

5. What is meant by the term "hard fibres"? Give an account of them, noting their origins and uses.

# *Chapter XV*

# HAIR, HIDES AND SKINS

Clothing, like food, is one of man's basic needs. Clothes and shelter are very necessary under certain conditions of climate for human survival. As early man moved into environments which were colder and wetter than those to which he was accustomed, he felt the need for some bodily protection and doubtlessly took to wrapping himself in animal skins. The Eskimoes provide us with the prime example of a people making the best possible use of animal skins. The making of cloth, through the spinning and weaving of fibres, whether animal or vegetable, was one of man's earliest arts and the use of cotton, flax, silk, and wool go back for many millenia. The making of textiles has been one of man's major industries throughout historical times, although the relative importance of the materials used in the manufacture of cloth has varied considerably throughout the historical period. Changes in the availability of raw materials, in their uses, in fashion, and in production cost have led to changes in textile production. We dealt with the vegetable fibres in the last chapter; let us now turn to the animal fibres and deal first with wool which until the eighteenth century was the predominant textile.

## *WOOL*

Sheep, as we have already noted (p. 177), are kept principally for their meat or wool, although in the lands around the Mediterranean Sea they are sometimes raised for their milk for the production of cheese, while in some countries they are reared chiefly for their skins. Sheep are usually reared for either their mutton or their wool, though they may be reared for both. The best meat producers, however, are not the best wool producers; and vice versa. So, in general, there is an emphasis upon either one commodity or the other. The conditions under which sheep are reared have much to do with this specialisation. Broadly speaking, if the climatic conditions are damp and cool, sheep will be reared mainly for mutton; if dry and warm, primarily for wool. Mutton sheep, if they are to produce good meat, require fairly rich pastures which, of course, are only to be had in the wetter areas. On the other hand, wool sheep, to produce a good fleece, need fairly dry conditions; and since they can graze on relatively poor and scanty pastures, they are to be found in many of the drier regions of the world.

Native, it would seem, to central Asia and probably domesticated there, sheep now have a virtually world-wide distribution. Europeans have taken them to every continent, all of which now have substantial numbers. But, although sheep can be reared successfully under fairly widely differing conditions, they are ill-suited to, and commonly absent from, the hot, moist tropical regions, the true desert areas, and high latitudes where there

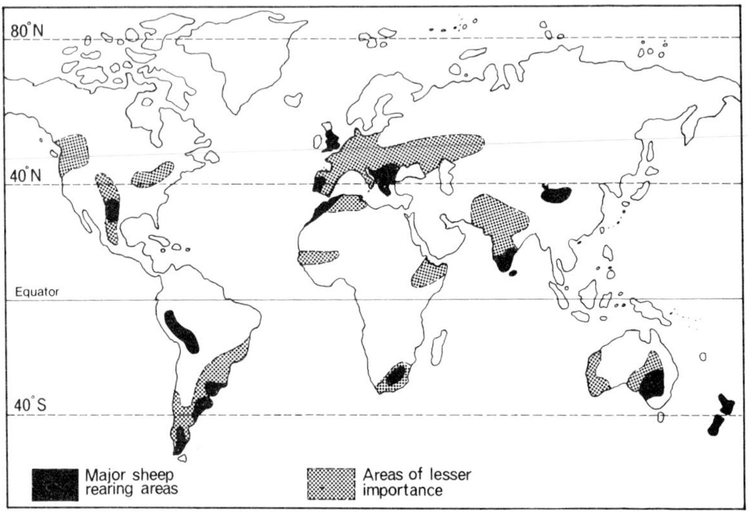


FIG. 73.—WORLD DISTRIBUTION OF SHEEP

is insufficient food supply. Mostly, therefore, they are creatures of the cool temperate and sub-tropical regions, especially areas of grassland or where there is an herbaceous growth. Largely because of economic considerations, sheep tend to be reared in those areas where it is too dry for cereal cultivation or for cattle raising, although it should be noted that sometimes sheep are reared in cereal-growing areas where they form part of a mixed economy, *e.g.* the barley-sheep economy of the Yorkshire Wolds district and the wheet-sheep economy of the Murray–Darling Basin in Australia.

The wool yielded by sheep is usually classified into three broad categories:

1. *Merino wool*, the finest quality wool, which is derived from the merino strain of sheep, originally of Spanish origin. Merino wool comes mainly from areas of low rainfall, though the animals are

sensitive to drought and unusually dry conditions affect the yield. This explains why the production of the finer quality wools often varies from year to year. About one-third of the total world wool production is of the merino type. Merino wool is mostly used in the production of knitting wools and in high quality clothing such as worsted suitings.

2. *Crossbred wool*, of medium quality, is produced usually in association with meat production. Crossbreds, which are reared in areas of higher rainfall and therefore on the richer pasturelands, normally account for between 40% and 45% of the total world output of raw wool. Most of the wool exported by New Zealand and Argentina is of the crossbred variety. Australia, which produces the finest crossbred wool, is increasing her numbers of crossbreds.

3. *Carpet wool*, so-called because it is mainly used in the manufacture of carpets, is a coarser type of wool of lower grade. Much of it comes from the more backward countries where sheep-rearing is an incidental rather than a specialised activity. Countries such as India, Iran and Ethiopia produce carpet wools; Argentina and the Soviet Union are also notable producers. Carpet wool accounts for between 10% and 15% of the total raw wool output.

## WORLD PRODUCTION AND TRADE

Table XX shows that in 1968 there were more than 1000 million sheep in the world. They are the most numerous of all the world's animals, out-numbering cattle by some 100 million. Australia and the Soviet Union possess the largest numbers, having between two and three times as many sheep as the other main sheep-rearing countries.

TABLE XX

*Sheep Population (thousands)*

| | *1950* | *1970* |
|---|---|---|
| Australia | 112,900 | 178,287 |
| U.S.S.R. | 80,000 | 137,940 |
| China | ? | 71,000 |
| New Zealand | 33,900 | 58,913 |
| Argentina | 47,000 | 43,800 |
| South Africa | 31,900 | 30,671 |
| Turkey | ? | 36,470 |
| Iran | ? | 35,000 |
| U.K. | 19,500 | 25,998 |
| U.S.A. | 30,700 | 19,560 |
| Uruguay | 23,000 | 19,700 |
| World total | 800,000 | 1,074,667 |

The numbers of sheep both in Europe and North America have shown a decline since pre-war days; elsewhere, however, there have been increases, notably in Australia and South America.

The world export trade in raw wool is dominated by four countries: Australia which accounts for about 50% of the total trade, New Zealand about 15%, Argentina about 10% and South Africa about 5%.

The United Kingdom is the leading importer of wool, taking nearly one-third of the wool entering into international trade. British imports come chiefly from Australia, New Zealand, South Africa, and Argentina. Japan and France rank next, accounting for about 20% apiece of the international trade. Belgium and Italy each account for about 10% of total world imports. It will be noted that most of the wool goes to Western

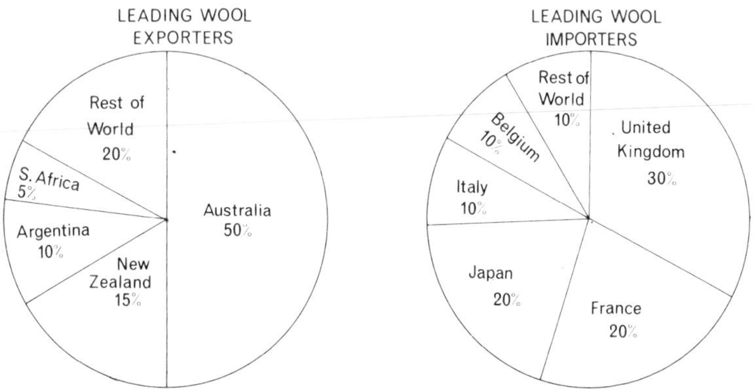


FIG. 74.—WOOL EXPORTS AND IMPORTS

Europe where are located the major woollen textile industries. A notable development in recent years has been the substantial import by Japan of Australian wool. The United States is a fairly substantial importer, but the imports consist mainly of carpet wool from Argentina and India.

## OTHER WOOLS

Several other species of animals yield hair or wool which is used in the textile industry. The most important are:

1. *Mohair*, long, silky fibres obtained from the Angora goat, a native of Asia Minor. The Angora goat was successfully introduced into South Africa which is now the chief centre of production of mohair.

2. *Cashmere*, the soft downy wool from the Kashmir goat which lives in the Himalayan regions. It is used in the making of expensive Cashmere shawls and other costly garments.

3. *Camel hair*, used in combination with other fibres, is employed in the manufacture of coats, blankets, rugs, and carpets. Much of it comes from China.

[*Courtesy: Australia House.*

FIG. 75.—SHEEP MUSTERING

Merinos being mustered for shearing on Benangaroo, a 40,000-acre (16,000-ha) property, on the Southern Tablelands of New South Wales, overlooking the Murrumbidgee River. This station is near Coolac, close to the Hume Highway which runs between Sydney (N.S.W.) and Melbourne, the capital of Victoria. The land hereabouts consists of rolling hill country with many scattered eucalyptus trees giving it a parkland appearance. The rounders, mounted on horseback, are assisted by dogs. Few of the flocks on Australian sheep farms are under 5000 head and most of them number between 10,000 and 50,000.

4. *Alpaca and vicuña wool* comes from the Andean plateaus of South America where these animals, relatives of the llama, live. The wool from these animals is very fine, soft, and resilient, but is in limited supply.

## *SILK*

Silk is the "aristocrat" of all textiles, incomparably the richest and most beautiful. It has been highly esteemed for several thousand years and in ancient times was sold for more than its weight in gold. Although the use of silk as a textile fibre has declined, partly because of its costliness and partly because of its replacement by man-made fibres, it still possesses properties not shared by other natural or synthetic products. All the

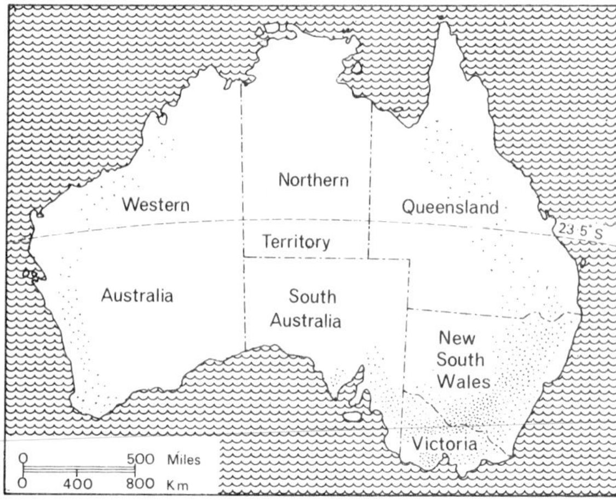


FIG. 76.—SHEEP-REARING IN AUSTRALIA

Australia is the greatest sheep-rearing country in the world. She has about 166 million head, over 70% of them being merinos. She produces about one-third of the world's total wool production, nearly half of the total world's wool exports, and about 10% of the total exports of mutton and lamb.

By their nature sheep are unsuited to tropical heat, hence most of them are found south of the Tropic of Capricorn and more than half of them are in New South Wales and Victoria. Most of the wool sheep (chiefly of the merino breed) are grazed on the drier pastures of the interior, although there are few in those areas where the rainfall drops below 10 in. (254 mm). The sheep which are reared for mutton are raised as part of the general farming system in the moister areas of the south-east. The size of the sheep stations varies: in the moister eastern parts of the Murray-Darling Plains they may be only a few square miles in extent; in the drier interior they may be several hundred square miles in area.

When shearing time approaches the flocks are mustered and driven to the station headquarters. Here they are mechanically shorn by parties of travelling shearers who are expert at the job. Every year more than half a million tons of wool are produced. The wool is auctioned at twelve centres in Australia. To these centres come buyers from all over the world. Between a quarter and a third of the total exports of wool goes to the United Kingdom; France and Japan each buy about 20%, and Italy and Belgium each about 10%.

Like Fig. 57 this is a dot distribution map. In constructing a dot map two requirements are needed: (*a*) statistics of the item to be represented, *e.g.* animals, people, etc. and (*b*) an areal unit against which the distribution figures are plotted, *e.g.* parish, county. When drawing a map a suitable dot value must be determined (in the above map each dot represents 500,000 sheep), a suitable dot size must be decided upon and the statistical unit or area chosen. A map with the appropriate areas marked in must be drawn and then the dots inserted in the appropriate areal units. The dots should be located as accurately as possible (note, for example, the distribution of sheep in Queensland; there is an absence of dots in the wet coastal areas where sheep are not found).

ascertainable facts confirm that China was the earliest centre of the silk industry—indeed, the ancient name in the West for China was Serica, "the land of silk."

Tradition has it that the art of spinning and weaving silk was discovered by the wife of the Chinese Emperor W'Hang sometime about 2500 B.C. The secret of sericulture, the rearing of silkworms for silk, was carefully guarded to preserve the monopoly of silk production by China. Foreigners enquiring about the origin of silk were fobbed off with the tale that it was obtained from sheep which, on being sprinkled with water in the sunshine at certain seasons of the year, grew these fine glossy hairs. It appears that the secret was successfully kept until about A.D. 300. Somewhere about this time silk was introduced into Japan and tradition attributes the spread of sericulture to India at the end of the third or early in the fourth century A.D. Silk was introduced to the West during the reign of the Emperor Justinian in the sixth century. The story goes that he persuaded two Persian monks who had lived in China to return thither and procure some of the precious silkworm eggs. This they did, managing to smuggle them out of China in the hollows of their bamboo staffs.

## SERICULTURE

There are several species of silk spinning moths but the *Bombyx mori* is the most important. It is popularly known as the Mulberry Spinner because it feeds on the leaves of the mulberry. In early summer the moth, after leaving its cocoon, lays its eggs, several hundred of them, from which the silkworms hatch out. These eggs are so small that 30–40,000 weigh only 1 oz (28 grammes). The silkworm grub, when hatched out, is very small but during its lifetime of only four to five weeks it grows to a size of about 3 in. (76 mm), eating voraciously all the time and changing its skin four times. After about a month, the silkworm spins itself a cocoon in which it seals itself during the transformation to chrysalis and moth. In about three weeks the transformation is completed and the moth then eats its way out. To prevent this happening, the cocoons are heated in ovens thus killing the moth.

The cultivation of the silkworm is possible wherever the mulberry tree will grow, but in practice sericulture is confined to sub-tropical and "Mediterranean" regions. The caterpillar itself cannot stand temperatures lower than 16° C (60° F); if, therefore, rearing is carried on in cooler areas, it must be done in slightly heated premises. The so-called "wild silk" or Tussore silk is produced by wild moths (as distinct from those reared "in captivity") which feed on oak leaves. The unwinding of the gossamer-like silk fibre from the cocoon is a delicate and tricky business requiring an abundant supply of highly-expert and cheap labour. It is this, rather than the appropriate climatic conditions, which limits the production of silk.

Each cocoon is built up of fine filaments of silk, coated with a gummy substance to help them stick together, which may be up to two miles in length. To unwind this silk the cocoon is dropped in hot water. Lengths of silk, varying from 300 to 550 yards (275–500 m) in length, are "reeled off" in continuous filaments. Of these continuous threads several are wound together on to a creel to produce nett silk raws. This process may be done by hand or in factories known as filatures. The raws, two or three or more according to the use to which the silk is to be put, are then twisted together, a process called "throwing." The silk thus produced is known as nett silk. The shorter, weaker silk threads that are left from the cocoon are washed, degummed, combed, and then spun in much the same way as cotton or wool to produce a yarn called spun silk.

## PRODUCTION AND TRADE

The total world output of raw silk is small, a mere 30,000 metric tons. Table XXI gives the chief producers.

TABLE XXI

*Raw Silk Production (metric tons)*

| | *1960* | *1970* |
|---|---|---|
| Japan | 18,048 | 20,515 |
| China | (1959) 6225 | 10,200 |
| U.S.S.R. | 2358 | 3024 |
| S. Korea | 500 | 2846 |
| India | 1154 | 2250 |
| Italy | 892 | 850 |
| World total | 30,000 | 40,268 |

It will be seen that silk production comes mainly from two areas:

1. Monsoon Asia, which supplies about 90% of the total output, where Japan is the dominant producer.
2. The Middle East and the Mediterranean countries, where Italy is the only significant producer.

World production of silk is less than a third of what it was in pre-war days and output has declined everywhere except in Italy and the Soviet Union.

Japan, China, and Italy are the chief exporters, with 62%, 16%, and 8% respectively of the total export trade. It is interesting to note that until 1933 silk was Japan's most important export item accounting for one-fifth, by value, of the total export trade. But by 1938 it had dropped to 13%, now it is less than 1%.

The United States is the largest importer, absorbing about 45% of the silk entering into international trade; she takes about two-thirds of the Japanese export and about half of the Italian export. France is the second largest importer accounting for about 16% of the import trade. France, once a producer of raw silk herself, is now dependent upon imported supplies. The Swiss silk industry is also dependent upon imported raw silk.

## *HIDES AND SKINS*

Hides and skins are obtained from many different animals and used for a variety of purposes. Many hides and skins are used locally and figures of total production are impossible to come by. The amount entering into international trade is much greater than is generally realised. Hides and skins are mostly used for the production of leather which is used in the manufacture of footwear, bags, cases, gloves, clothing, belts, and book-bindings. Of the hides and skins used in leather-making, the most important are the hides from cattle, buffalo, and horses, and the skins from sheep, goats, and pigs. Minor varieties of animal skin, which are used for special purposes, include deerskin, and the skins of crocodiles and other reptiles.

Hides are cured or preserved for and during transport by either drying them or pickling them in brine when they are termed respectively as "dry" or "wet" hides. Hides are converted into leather by tanning, a process discovered very early on in human history. Tanning involves the treatment of hides and skins with tanning materials which fall into two groups: (*a*) those of mineral origin such as alum, sodium hyposulphite, and various compounds of chromium; and (*b*) those of vegetable origin such as extracts (tannins) of the quebracho, oak, hemlock, sumach, wattle, and mangrove plants. Tannin is an acid procured from the barks of most trees.

All countries which have a pastoral industry will have supplies of hides and skins, but the chief producers and exporters clearly will be those countries having large numbers of animals and highly developed pastoral industries. The United States, the Soviet Union, India, Australia, Argentina, and South Africa are the most important of these countries. Cattle hides are exported chiefly from Brazil, Argentina, India, Australia, and South Africa. Sheepskins entering world trade come principally from Australia, New Zealand, Argentina, Uruguay, and South Africa. Goatskins come from India, East Africa, and the Middle East.

The United States is the most important leather producing country in the world with a home production of over 300,000 tons; even so, she imports annually something of the order of 100,000 tons of hides and skins. The United Kingdom has a substantial home production of cattle-hides and sheepskins, which amount to some 80,000 tons, but imports usually exceed this figure. Most of the countries of Western Europe have well-developed leather industries and, though they have home supplies, they are often fairly substantial importers of hides and skins.

[*Courtesy: National Film Board of Canada.*

FIG. 77.—FUR-TRAPPING IN THE NORTHERN FORESTS

Like lumbering, fur-trapping is mainly a winter occupation, as it is in the coldest season that the animals' coats attain their best condition. Particularly valuable to the furriers are the pelts of mink, musk-rat (for musquash fur) beaver, marten, squirrel, ermine, and fox. In the photograph the trappers are hanging up musk-rat pelts to dry on stretchers.

**FURS**

Numerous animal species are fur-bearing but the creatures providing the finest furs come from the colder and more inaccessible parts of the world, especially the northern coniferous forest lands and the tundra and arctic regions. Among the mammals yielding the most valuable furs are the sable, stoat (ermine), mink, squirrel, and silver fox. The high quality of their fur together with their scarcity make their skins very valuable. Certain marine creatures, such as the sea-otter (which was hunted almost to extinction) and the fur seal, also yield high quality fur. But the animals supplying the largest number of skins to the fur trade are the more common creatures such as the rabbit, hare, musk-rat, coypu (nutria), and seal.

Fur garments are much worn in cold climates, as in central Europe and the Soviet Union, but elsewhere furs are more in the nature of a luxury commodity used for decoration and adornment. Because of this, the fur trade is very susceptible to the whims of fashion. Some thirty years ago silver fox fur was all the rage. The demand for it was so great that foxes came to be reared in captivity on "fur farms." Canada led the way in this practice. Fashion changed, however, and the demand for silver fox declined. Nowadays the big demand is for mink and chinchilla, and both these animals are reared on fur farms. There are, in fact, many such farms in Britain.

London is the world's greatest fur market, receiving furs from all over the world. Leipzig is another major European market. In North America, New York (which, like London, is one of the world's greatest clothing centres) is the major fur market.

## *HORN, IVORY, ETC.*

From time immemorial man has used animal horn for making weapons. trinkets, etc. More recently, horn and ivory have been used in the manufacture of a great variety of objects. It is sometimes said that the introduction of plastics has done away with the need for animal horn but the truth is that the world demand, at least for high quality deer, cattle, buffalo and rhinoceros horn, is so great that it can no longer be supplied.

Britain, for example, imports horn from every continent. Black buffalo or rhinoceros horn is used for walking stick or umbrella handles, cattle horn for combs, whistles, and pipe mouthpieces, and the beautiful grey South African ox horn for such special jobs as pocket-knife inlays. The toughness and resilience of rams' horn makes it a valuable material in high class golf clubs. Horn is often used for the handles of tools because its yielding texture is kind to the hand and prevents blisters. For the same reason horn pipe mouthpieces are popular among musicians.

The finest horn to be had comes from the ox—bulls' horn is coarse, and cows' of poor quality because the goodness goes into the milk. The horn of the great South African cattle, the nearest approach we have to the original wild ox, fetches up to £130 a ton. This is exceeded only by rhinoceros horn, which is becoming very scarce. Horn will never be cheap, because the high cost of the raw material combined with the amount of workmanship that goes into a good finish is bound to put it into the luxury class.

Ivory comes from the tusks of such creatures as the elephant, the hippopotamus, the walrus, and the narwhal. Hippopotamus ivory is the finest, for it is harder, of superior quality, and of a more enduring whiteness than that obtained from other animals. But most of the ivory used comes from the African elephant. As a result of the reduction in the big game of Africa and the measures taken to try and preserve it, the amount of ivory entering into international commerce has greatly declined. Because of its value, the native peoples of Africa do a considerable amount of poaching. The Chinese have always made much use of ivory and it is interesting to note that for many centuries a proportion of the ivory was supplied by the remains of the Siberian mammoth.

Of other animal products, bones are much used and there is a considerable trade in them. Bone-ash is used in the manufacture of pottery (bone chinaware); bones are used in the making of glue; and, since they contain much phosphate of lime, they make valuable fertiliser (bone-meal). Argentina and India are important sources of bone and most of Britain's imports come from these two countries.

## EXERCISES

1. Discuss the geographical background of the fact that over 90% of world exports of wool come from the southern hemisphere.
2. Discuss the reasons for the difference in the development of the sheep-rearing industry in Australia and New Zealand.
3. Give an account of the production and trade in raw silk.
4. "Australia possesses special advantages for wool production and these have given her a leading position among the sheep-raising countries of the world." Discuss.
5. Write an essay on fur-trapping, fur-farming, and the fur trade.
6. Discuss carefully the relation between the sheep industry of Australia and temperature and rainfall, and explain by what routes the wool reaches Australia's chief customers.

# *Chapter XVI*

# FOREST PRODUCTS

## *THE WORLD'S FORESTS*

Over a quarter of the earth's land surface is covered with forest. Formerly the forested area was much larger but the need of more land for crop-growing, especially when the world's population began seriously to grow, and the many valuable uses of timber, *e.g.* for house building, shipbuilding and other constructional purposes, for making charcoal for smelting, as well as for fuel, led to widespread clearing. Lands long settled, such as the Mediterranean Lands and Western Europe and in the Far East, China, have very little natural forest left.

Although man has greatly modified and altered the distribution and nature of the forest in many areas, we must recognise the importance of climate as a basic factor influencing the distribution of forests and forest types. Moisture, rather than temperature, is the dominating factor in tree growth: the quantity of the rainfall and the length of the wet season are the paramount influences. Trees are very susceptible to drought, especially when young. Temperature is more important in influencing the type of forest that grows, although it should be noted that an average temperature of 10° C (50° F) for the warmest month is the "low temperature" for tree growth. The regions of the world in which trees really flourish and form great forests are, generally speaking, those having a moderately long warm season with fairly abundant moisture.

It is interesting to note that in some parts of the world, other than those long and closely settled, forests are absent even though the climatic conditions are apparently suited to tree growth. Can any reasons be put forward to explain this? Dry, porous soils certainly might explain the absence of forest locally but where extensive treeless areas exist it seems very likely that fires and animals have played a major role. Man, both at present and in the past, has often fired the forest quite deliberately to clear ground for tillage or to extend and improve pasture for his grazing animals. What has been called the "slash and burn" technique is widely followed by the primitive cultivators of the forest lands of South America. Many parts of the African savanna lands were once forested but were burned by the native pastoralists to provide more pasture for their cattle.

In spite of progressive and widespread forest destruction over many parts of the earth's surface for at least two thousand years there remains a substantial forest cover. Table XXII gives the estimated distribution by continents of the world's forests.

TABLE XXII

*Estimated Distribution of World Forests (million hectares)*

| Region | Total forested area | Accessibility | | Composition | | Utilisation | | Forest as % of land area |
|---|---|---|---|---|---|---|---|---|
| | | *Inaccessible forests* | *Accessible forests* | *Conifers* | *Non-conifers* | *Forests in use* | *Non-exploited* | |
| Europe | 136 | 3 | 133 | 79 | 57 | 130 | 6 | 28·4 |
| U.S.S.R. | 743 | 318 | 425 | 583 | 160 | 350 | 393 | 33·9 |
| N. America | 656 | 344 | 312 | 463 | 193 | 220 | 436 | 36·1 |
| Latin America | 927 | 584 | 343 | 30 | 897 | 90 | 837 | 41·1 |
| Asia | 567 | 308 | 259 | 120 | 447 | 205 | 362 | 21·0 |
| Africa | 801 | 519 | 282 | 5 | 796 | 115 | 686 | 27·0 |
| Pacific area | 85 | 65 | 20 | 8 | 77 | 17 | 68 | 9·9 |
| World total | 3915 | 2141 | 1774 | 1288 | 2627 | 1127 | 2788 | 29·5 |

(*Source:* A Geographer's Reference Book, 1955.)

## HARDWOODS AND SOFTWOODS

Trees are of two main types: hardwoods and softwoods. First, we should make it clear that these terms "hardwoods" and "softwoods" have nothing to do with the actual hardness of the wood, although it is true that many of the better known hardwood trees also produce hard timber and many of the softwoods soft timber.

Generally speaking, the hardwoods have broad leaves and they are deciduous, that is they shed their leaves during seasons of cold or drought. Oak, ash, beech, elm, poplar, maple, magnolia, and hickory, for example, are well-known hardwoods and most of them are common to Britain. But not every broad-leaved tree is a hardwood; the American gingko, for instance, though it has a broad leaf, is not a hardwood. Hardwood trees are found in both temperate and tropical latitudes; hence we commonly distinguish between temperate hardwoods, such as the species mentioned above, and the tropical hardwoods which include such well-known trees as teak, ebony, greenheart, logwood, mahogany, and ironwood.

Most softwoods are conifers, have needle-shaped leaves, bear cones, and are evergreen. Some of the more common coniferous types are pine, spruce, fir, cedar, balsam, and redwood. But, once again, there are difficulties: not all conifers are evergreen, for example, the tamarack, western larch, and southern cypress all shed their leaves annually; again, not all conifers are softwoods—for example, the Paraná pine of southern Brazil is a hardwood.

All this is very confusing but at least it will teach us not to divide trees simply and clearly into two exclusive groups.

## *DISTRIBUTION OF THE FORESTS*

Let us look briefly at the distribution of the main types of forest (Fig. 78).

### CONIFEROUS FOREST

Nearly all the great coniferous forests are situated in the northern hemisphere. A world vegetation map shows that a broad zone of coniferous forest extends roughly between 50 and 70 degrees N. In North

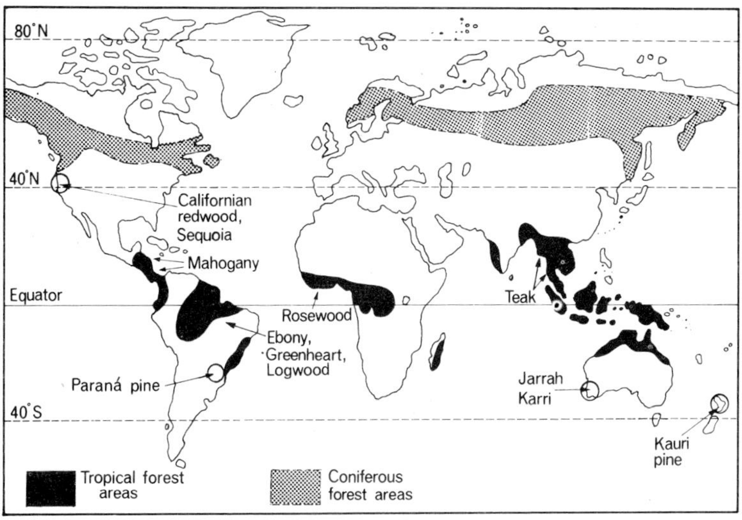


FIG. 78.—THE WORLD'S MAIN FOREST AREAS

America the forest stretches from Alaska and British Columbia, through the northern parts of the Prairie Provinces of Canada, through Ontario and Quebec, to Labrador and Newfoundland. The forest extends southwards in the western part of the United States in the Cordilleran region and near the Pacific coast where the generally mild and moist conditions are especially favourable to tree growth and allow the forest to reach its finest development in the magnificent Douglas firs and giant redwoods. A similar tongue of coniferous forest extends part way down the Appalachian Mountains in the east. We should note, too, the occurrence of conifers on the sandy soils of the south-eastern part of the United States.

In Eurasia the main belt of coniferous forest runs from Scandinavia, through Finland, northern Russia, central and eastern Siberia to the Pacific. In Europe coniferous forest is found in the uplands of central Europe and in the Alps farther south. Conifers are also found at high levels in the Mediterranean region.

In the southern hemisphere the coniferous forest is strictly limited to relatively small areas in southern Chile, southern Brazil, and New Zealand.

## TEMPERATE HARDWOOD FOREST

The temperate hardwood forests are usually found between approximately 30 and 50 degrees N. and S. These forests were formerly of much greater extent but since they occurred in mid-latitudes in those lands which were early occupied by man, large-scale felling to provide space for agricultural, pastoral, and industrial development has taken place and today only small, discontinuous patches exist. This is true of China and the eastern part of the United States as well as of Western and Central Europe.

In the northern hemisphere temperate hardwoods are found chiefly in four areas: the north-eastern part of the United States: in Western Europe extending eastwards in a wedge-shape as far as Moscow; southern Western Siberia; and in Manchuria, Korea, Japan, China, and Kamchatka. In southern Europe the temperate hardwood forests give way to the "Mediterranean" forest, a type classed as broad-leaved evergreen forest consisting of cork-oak, olive, carob, fig trees, etc. In eastern Asia, where a monsoon climate prevails, the temperate hardwood forest merges into a distinctive modified type possessing some tropical characteristics.

In the southern hemisphere temperate hardwood forest is limited to small areas: to central and southern Chile, the south-western corner of Western Australia, Victoria and New South Wales, Tasmania, and New Zealand.

## TROPICAL HARDWOOD FOREST

The tropical forests include the equatorial rain-forest and the tropical monsoon forest. The former occurs in equatorial latitudes where temperatures are constantly high, around 27° C (80° F), and where there is abundant rainfall, around 80 in. (2032 mm), well distributed throughout the year. On tropical trade-wind coasts, abundant precipitation gives rise to a luxuriant forest growth very similar to the equatorial rain-forest. In monsoon regions, where there is a rainfall of 40–80 in. (1016–2032 mm) or more, a deciduous type of forest occurs, the trees shedding their leaves in the dry season.

Three main areas of tropical forest are found: in the Amazon basin, in West Africa and the Congo Basin, and in South-east Asia and the East Indian region. In the Americas, areas of tropical forest also occur in Central America and the West Indies, along the Pacific coastlands of Colombia, and along the eastern coast of Brazil. In Africa a dry type of tropical forest occurs in Northern Rhodesia while the coast regions of East Africa south of the equator and eastern Madagascar have much forest cover. India still has a considerable area under monsoon forest and there are patches of tropical forest along the northern and north-eastern coasts of Australia.

## *TIMBERS AND THEIR USES*

Man's demand for timber appears to be insatiable; year by year increased quantities are required and the cutting of timber takes place on a greater scale than ever before, in spite of the many substitutes that are now being used for wood. The principal uses of timber are for fuel, for constructional purposes, for props and poles, for sleepers, for pulp, and for tannin extracts. Approximately 40% of the total wood production is used as firewood; the remaining 60% is absorbed by industry. Until very recently the quantities of hardwood and softwood felled were about equal, but now the total production of softwood is substantially greater.

Softwoods are obtained almost entirely from the coniferous forests of the northern hemisphere. The softwoods supply considerably more than three-quarters of the world demand for industrial timber and forest products; only a relatively small amount of softwood is used as fuel. Softwoods are commonly light, strong, fairly durable, and easy to work; they are used in constructional work, in the making of cheap furniture, for box-making, for match-making, for pit-props, and for the making of wood-pulp used in the manufacture of newsprint and rayon. The huge demand for softwood and the extravagant exploitation of the coniferous forests in the more accessible regions have greatly reduced world reserves and necessitated the adoption of re-afforestation schemes. In some countries, such as Norway and Sweden, cutting is carefully controlled and re-planting rigorously enforced. Britain, during more recent times, has begun seriously to establish plantations.

About 70% of hardwood production is used for fuel. Some countries which are short of energy resources, *e.g.* Brazil, consume large quantities of timber as fuel wood. Hardwoods not used for fuel have special uses, *e.g.* they are used as "cabinet woods" in the furniture trade, for harbour piles and dock gates because of their great strength and resistance to decay, in boat-building where durable timber is needed, and for the extraction

[*Courtesy: Swedish Travel Bureau.*

FIG. 79.—FLOATING LOGS

Floating timber on the Jösseälv River, Värmland, Sweden. The cut logs are assembled into "rafts" containing thousands of logs. From the cutting site in the forest to the saw-mill or pulp factory the logs may be floated over a distance of several hundred miles, controlled all the way by raftsmen and sorters. In the photograph, note how part of the river is kept free for steamer traffic.

of tannin (*e.g.* quebracho) and occasionally, though much less so than formerly due to the cheap mineral dyes derived from coal-tar, for dye-stuffs. Most of the hardwoods used are tropical hardwoods; supplies of temperate hardwoods are now rather limited.

## *LUMBERING IN THE CONIFEROUS AND TROPICAL FORESTS COMPARED*

Most of the timbers of the coniferous forests are softwoods whereas those of the tropical forests are mostly hardwoods. At present the biggest demand is for softwoods for pulp, paper, and constructional work generally, hence there is a much greater exploitation of the coniferous forests. A great advantage of the softwood forests is that the species tend to grow in large colonies; on the other hand, species in the tropical forests are extravagantly mixed and individual species may be widely separated and scattered at random through the forest, a fact which makes their exploitation difficult. Trees of a particular kind are, therefore, not only difficult to find but, when felled, are difficult to extricate from the thick and tangled vegetational growth.

[*Courtesy: Paul Popper.*

FIG. 80.—ELEPHANTS MOVING TEAK LOGS

Elephants are still used for teak-logging in Burma and Thailand, although they are gradually being displaced by machinery. Tractors are now often used to haul the timbers. Note the huge size of the teak log. Teak provides a wood of exceptional hardness and durability and it is very resistant to water.

In both forest regions much use is made of rivers for transporting the logs (Fig. 79). In the coniferous forests, winters are keen, the ground is snow-covered and frozen; the hard, slippery ground facilitates the overland haulage of logs which may be done by horses or tractors. The logs are taken to the river banks and piled up to await the spring thaw which will set the rivers in motion again and carry the floating logs downstream to the saw-mills or pulp-mills. The often wet ground and tangled undergrowth of the tropical forests militates against easy movement; moreover, many of the hardwoods are extremely heavy and of great size which make handling and carriage difficult. Until recent times the elephant was much used in the teak forests of South-east Asia (Fig. 80) but gradually it is being displaced by the tractor. Many of the hardwoods are so heavy they will not float; hence they have to be piled on rafts and rafted downstream.

Another difference in the lumbering is that while conifers are usually cut without any preliminaries other than an initial marking to indicate that it is ripe for cutting, the hardwoods of the tropical forests, such as the teak trees of South-east Asia, are selected for cutting, ring-barked (*i.e.* a cut is made round and through the bark at the base of the tree which kills it), and then left standing for a year or two to enable the timber to dry out.

Though conditions in the coniferous forests are not particularly attractive, in the tropical forests they are usually unattractive and trying to say the least. The conditions of heat, humidity, rain, swamp, and insect pests make lumbering an arduous and usually expensive business. On this account only the valuable species, *e.g.* mahogany, greenheart, iron-wood, can justify the high costs involved in cutting and transport.

The coniferous forests often have another advantage: the rapid streams which float the logs also supply abundant hydro-electric power for the mills at their outlets, which saw or pulp the timber. Finally, whereas the tropical forests lie far distant from the major timber-consuming countries, the coniferous forests of northern North America and northern Europe lie close to them.

## *WOOD PULP AND PAPER*

Paper can be made from a variety of materials, *e.g.* rags, esparto grass, banana trash, but over 90% of all the paper produced is manufactured essentially from wood pulp. The two principal ingredients of wood are cellulose and lignin, and paper is made from the former. Cellulose, incidentally, is also important in the manufacture of numerous other commodities including rayon, plastics, photographic film, lacquers, and explosives. Broadly speaking, wood pulp is made in two chief ways and produces two types of pulp:

1. *Mechanical pulp*, made by grinding the softwood timber and soaking it in water until it is reduced to a mushy, fibrous mass having the consistency of thick porridge.
2. *Chemical pulp*, made by digesting wood chips with an acid solution of calcium and magnesium bisulphite or with caustic soda which produces a similar pulpy material.

Mechanical pulp is mostly used for the production of newsprint and cheap wrapping paper; chemical pulp, which is more costly to produce but which gives a more durable paper, is used for the production of better quality paper.

Wood, water, and power are three basic raw materials required in paper manufacture. A plentiful and continuous supply of softwood timber is needed. Spruce (which provides the best raw material) along with various species of fir and pine are mostly used. Secondly, an abundant supply of clear, soft water, free from chemical impurities, is required. Thirdly, supplies of cheap power are needed, since the manufacture of paper demands rather heavy supplies of energy. To these three conditions

[*Courtesy: Capital Press, Ottawa.*

FIG. 81.—PAPER-MILLS, CORNWALL, ONTARIO

These large Canadian paper-mills employ about 1000 workers. They turn out 275 tons of pulp and 125 tons of paper each day. Note the piles of wooden chips on the right. Cornwall is close to the huge new hydro-electric power house on the St Lawrence Seaway.

needed for the large-scale production of pulp and paper, a fourth may be added: cheap and efficient transport not only for the supply of the raw materials but also for the distribution of the pulp and the paper. Because of these demands, we find that the necessary conditions for large-scale production of pulp and paper are practically confined to the United States, Canada, Norway, Sweden, Finland, Western Germany, Austria, the Soviet Union, and Japan. The United Kingdom has no home supplies of pulp but the paper-making industry is quite important.

The following table gives the chief world producers of wood-pulp and newsprint. Note the outstanding positions of the United States and Canada as producers of each commodity, and note also the rather unexpected appearance of the United Kingdom as a producer of newsprint.

TABLE XXIII

*Wood-Pulp and Newsprint Production*

| *Wood-Pulp* | % | *Newsprint* | % |
|---|---|---|---|
| U.S.A. | 42 | Canada | 40 |
| Canada | 20 | U.S.A. | 10 |
| Sweden | 9 | U.K. | 4 |
| Finland | 6 | Finland | 6 |
| Japan | 4 | Japan | 6 |
| U.S.S.R. | 4 | Sweden | 5 |
| West Germany | 3 | U.S.S.R. | 5 |
| Others | 12 | Others | 24 |
| | 100 | | 100 |

## *OTHER FOREST PRODUCTS*

In addition to timber, the forests yield a wide variety of other products which are gathered or collected and form the so-called "gathering industries." Among the chief of these are the following:

### GUMS, RESINS, ETC.

Although most of the world's rubber now comes from tropical plantations, small quantities of wild rubber are still collected in Amazonia and the Congo Basin. Balata and gutta-percha are juices or gums obtained from trees by tapping and subsequently hardened; they are used for much the same purposes as rubber. Certain types of pine yield turpentine and resin; the juice from the bark is distilled to give turpentine while the remaining residue forms resin.

### CAMPHOR

Camphor comes mainly from a species of cinnamon which grows in Japan, central and southern China, Formosa, and Malaya; extraction is by distillation from the leaves and wood of the plant. Most of the world's supplies come from Taiwan.

### KAPOK, RAFFIA, ETC.

Kapok is a vegetable down which encloses the seeds of the plant (compare the cotton lint which encloses the cotton seed). This soft, light, elastic fibre, which is also resistant to moisture, is too short for spinning purposes and so cannot be used for the making of textiles but it is much used as a stuffing for upholstery, cushions, life-savers, etc. Raffia is a fibrous material obtained from the leaves of the raffia palm which grows in Madagascar. Toquilla fibre from the toquilla palm of Ecuador is used for making "Panama" hats; the hats are so-called because they are sent chiefly to Panama for sale.

### NUTS

Brazil nuts are widely collected in Amazonia; Manaus and Pará on the Amazon are the great collecting and shipping points. Brazil nuts may be processed to yield a vegetable oil. Ivory nuts, which come from Ecuador, are the fruit of the tagua palm; these hard nuts provide what is known as "vegetable ivory," once much used for making buttons, studs, etc., although the development of plastics has much reduced the trade in ivory nuts.

### YERBA MATÉ

Yerba maté, or the Paraguayan "tea-plant," grows wild in the subtropical forests of South America. The dried leaves of the plant are infused to produce "tea." Though still collected as a forest product, yerba maté is now also a plantation crop (*see* p. 160).

### MEDICINAL PLANTS

A number of plants having medicinal properties are collected; the most noteworthy are: cinchona-bark for making quinine, coca leaves for making cocaine, the castor oil plant, ipecacuanha, and sarsaparilla.

### TANNING MATERIALS

Quebracho is an extract from the wood of the quebracho ("axe-breaking") tree which is a native of the forests of the Paraná—Paraguay basin in South America. The most valuable source of tannin, quebracho, is exported from Paraguay in considerable quantities. Wattle-bark and extract, coming chiefly from Natal and Kenya, is another source of tannin. Mangrove-bark, oak-bark, and hemlock-bark are also sources of tanning agents.

### CORK

Cork comes from the thick bark of the cork-oak which grows only in the Mediterranean lands. The countries around the western basin of the Mediterranean Sea have a virtual monopoly of cork production. Spain and Portugal are the chief producers and between them they provide the bulk of the supplies entering international trade.

### EXERCISES

1. Explain the advantages and disadvantages of the world's main coniferous areas in regard to their commercial development.

2. Write an essay on the world's timber industries comparing tropical and temperate production.

3. Give an account of the utilisation of the timber resources of either (*a*) Scandinavia and Finland, or (*b*) Newfoundland and eastern Canada.

4. With reference to Table XXII, which gives the estimated distribution of the world's forests, account for the large areas of non-exploited forests in North America, Latin America, and Africa.

5. Explain (*a*) why all forested areas in temperate latitudes are not important lumbering areas; (*b*) why the forests of the equatorial regions are relatively little exploited.

6. Write explanatory notes on *three* of the following forest products: ivory nuts, quebracho extract, camphor, yerba maté, kapok, cinchona-bark.

7. Compare the coniferous and tropical forests from the points of view of (*a*) types of trees, (*b*) methods of lumbering, and (*c*) economic uses of the timber.

# *Chapter XVII*

# METALS AND MINERALS

## *THE IMPORTANCE OF MINERALS*

THE world's resources are of animal, vegetable or mineral origin. But both animal and vegetable matter are dependent basically upon mineral matter, so we might say that minerals are of crucial importance in our world. Certainly the crude mineral materials, *e.g.* water, rock, soil, metals, coal, oil, which constitute the greater part of the earth on which we live, enter into our lives in manifold ways. This point need not be elaborated for a little thought will readily bring to mind the numerous things that are made from minerals and the many ways in which they influence our health, well-being, and comfort.

For many thousands of years mineral materials have been of direct significance to man: so much so that he has described his cultural development in pre-history and subsequent historical times in terms of the mineral resources at his command—"the Stone Age," "the Bronze Age," "the Irone Age," "the Steel Age," "the Oil Age," etc. Such terms attest, quite clearly, the tremendous direct importance of mineral resources in man's cultural evolution. They also indicate that man has gone through a progressive development associated with the discovery, development, and utilisation of different kinds of mineral resources.

Let us look at this progressive evolution for a moment. In the very earliest times primitive man used a stone—any stone—as a weapon or implement but eventually he began to choose his stone, usually preferring flint because it was especially hard and could be chipped or flaked to produce a sharp, cutting edge. Once he had discovered how to make fire, not only was his attention drawn to copper ores which were reduced or melted in fire but he realised that he could deliberately melt and fashion copper. This led, in turn, to the making of bronze, an alloy of copper and tin which was superior to copper. Later he learned how to smelt iron ores, and artifacts of iron replaced those of bronze because iron was harder, tougher, and took on a keener cutting edge. After two thousand or more years iron was replaced by steel, a superior alloy. It is sometimes said that at the present day we are on the threshold of "the Aluminium–Titanium Age"; the use of aluminium has increased about five-fold within the past 25 years and already the first all-aluminium ship is sailing the seas.

The present-day use of aluminium is only possible because of human advances in technology; this was equally true historically in the cases of iron and copper.

Thus the story of the human use of mineral resources runs roughly as follows. In the beginning man made use of such minerals as were readily at hand, the common rocks; subsequently he came to use the more common metals, such as copper and iron, which sometimes occurred native; later still he came to use the mineral fuels, the more complex metallic minerals, and the rarer minerals. It is a story involving two developments which are mutually related: first, the progressive discovery, development and utilisation of differing mineral resources ranging from the more common and simpler types to the more rare and the more complex kinds; and, secondly, to the development of man's scientific knowledge and technological skill, for upon these depended his ability to recognise, isolate, mine, and process the various minerals.

## *THE CLASSIFICATION, OCCURRENCE AND EXPLOITATION OF MINERALS*

The mineral resources of the world, from the geographical point of view, may be divided into five main groups: (*a*) rocks, used for building stones and brick-making; (*b*) non-metallic minerals, *e.g.* salt, sulphur, nitrates, asbestos; (*c*) metals, *e.g.* gold, iron, copper, molybdenum; (*d*) mineral fuels, *e.g.* coal, petroleum (dealt with in Chapter XVIII); (*e*) water.

The whole of the earth's crust is, of course, composed of mineral matter and no parts of the crust are entirely deficient in mineral matter which is of value at the present time. But it should be remembered that a mineral resource, like any other resource, is only of potential value until man finds a way of extracting it and a use for it. Although mineral substances are widely spread throughout the crust, in general, comparatively few are found in sufficiently concentrated quantities to justify their commercial exploitation. For this reason, the mining of minerals is usually confined to localised areas.

Broadly speaking, minerals may be said to occur in one of three ways depending upon the geological conditions under which they were formed.

### VEINS AND LODES

During times of tectonic disturbance when igneous rocks were squeezed into the crust or metamorphism took place, gases found their way into the cracks and fissures in the rock, eventually cooling and congealing to form veins and lodes, so that local ore bodies came to be formed within the surrounding rocks. Particular minerals solidify at particular temperatures hence it not infrequently happens that a zonation of minerals is developed; for example, in a particular lode, copper may give place in depth to tin. Again, it often happens that certain groups of minerals occur in close association with one another, *e.g.* silver is commonly found with lead and zinc, while lead and zinc are themselves almost inseparable twins. A good example of this close association of minerals is provided by the recently discovered complex ore body at Boliden in northern Sweden where occurs a great variety of minerals, including copper, silver, gold, lead, zinc, nickel, arsenic, and bismuth.

### BEDDED ORES AND DEPOSITS

These commonly occur in horizontal sheets or layers which are said to form beds, although sometimes they may form masses of irregular shape. Coal, common salt, nitrate, potash salts, and some iron ores are found in bedded deposits. Formation of these beds takes place in different ways. Some deposits are due to precipitates being deposited on lake floors; others are the result of the evaporation of water which leaves dissolved salts behind; others again are due to percolating water dissolving mineral matter out of the rocks and then re-depositing it again, at depth, in concentrated form; and yet others, such as coal and oil, result from organic matter becoming trapped and squeezed between layers of rock. Some mineral deposits result from the decomposition of the surface rocks; soluble matter is washed away leaving behind a residual deposit. Bauxite, the ore of aluminium, and kaolin or China Clay occur in this way.

### ALLUVIAL DEPOSITS

Many minerals are to be found as alluvial deposits in the muds, sands, and gravels of alluvial fans at the bases of hills or in valley bottoms. The mineral matter, often occurring originally in lodes and veins in the adjacent hills, has, as a result of the weathering and erosion of the latter, been carried down slope and dumped in the debris which has accumulated at their bases. The alluvial tin deposits of Malaya have originated in this way. It is important to note that only such minerals as are resistant to water corrosion are to be found in alluvial deposits. Gold, tin, and wolfram are the most important of the minerals found in this manner. Many precious and semi-precious stones, *e.g.* diamonds, rubies, and zircons, are also to be found in alluvial deposits.

## *THE EXPLOITATION OF MINERAL RESOURCES*

Before proceeding to describe the principal groups of mineral resources, we should ask the question: what factors must be taken into account when considering their exploitation? There are, in fact, many and these may be enumerated as follows:

1. *The size of the deposit*, which is related to the value of the mineral, for what would be termed a very large deposit in, for instance, the case of a nickel deposit would be very likely a very small deposit in, say, the case of iron. Where a bulky ore, such as bauxite, is required in large quantities then extensive deposits must occur.

2. *The richness of the ore*, that is the abundance of the metal in the ore in relation to the waste material. We speak of "high-grade" and "low-grade" ores; for example, some hematite iron ores contain as much as 65% or more of iron whereas limonite may contain only 25% or less. The richness of the ore body may well determine whether it is worthwhile working it.

3. *The depth at which the ore occurs.* It will be obvious that deep mines are much more costly to establish and maintain than surface or open-cast workings. The low-grade iron ores of the Jurassic belt in the English Midlands are economic only because they can be easily obtained by open-cast methods of working.

4. *The physical accessibility* of the ore body and, linked with this, the existence of adequate transport facilities. There are, for example, very large and rich deposits of iron ore in Bolivia but because of the country's relative isolation, the inaccessibility of the deposits, and the lack of communications, it will probably be a very long time before these deposits are exploited.

5. *The distance of the mineral deposit from the market.* Great distance from a market will involve high transport costs and this may well be a serious handicap to the development of a deposit. Bolivian tin, for instance, has always been at a disadvantage compared with other producers, partly because Bolivia lies so far removed from the markets for tin.

6. *Capital.* These days most mineral exploitation necessitates a very heavy capital investment if it is to be an economic proposition, occurring on a large scale—consider the enormous capital outlay involved in the exploitation of the Alaskan oilfields. Many poor, undeveloped countries have to depend upon Western capital if they wish to develop their mineral resources.

7. *Labour.* Often of great significance is the existence of a satisfactory labour supply, which may be difficult to recruit in isolated, unattractive areas; oil companies who have to operate frequently in hot, wet tropical regions or in arid, burning deserts have considerable problems.

8. *The presence or absence of fuel and power* for working plant and processing the ore, *i.e.* for concentration, smelting and perhaps refining, may be

an important factor in the exploitation of a mineral deposit. A case in point is Katanga copper: once dependent upon the coal supplied from the Wankie coalfield, the expanding industry demanded increased power which was met by the hydro-electric power produced by the Kariba Dam.

9. *World demand.* World scarcity of a particular mineral is likely to raise its price on the world market, hence specific mineral deposits, formerly neglected, may become worth working. On the other hand, over-production of a commodity may lead to a fall in price and so to the closing down of mineral working.

## *BUILDING STONES AND KINDRED MINERALS*

From the dawn of human history man has usually built his more important structures, *e.g.*, temples, palaces, monuments, arenas, aqueducts, and often bridges, in stone, for stone is the most durable of building materials and the one that lends itself best to artistic manipulation. But because stone is heavy and costly to move over distances, locally quarried stone is mostly used in construction. In this sense the cultural environment

[*Courtesy: Eric Kay.*

FIG. 82.—QUARRYING PURBECK LIMESTONE

Purbeck open quarry in the "Isle" of Purbeck, Dorset, England. Blocks of this fine quality building stone are hewn and sent all over the country. It is much used in the construction of important public buildings.

may in some way reflect the physical environment. This explains why the gritstone cottages of the Yorkshire moors seem to grow out of the landscape.

Certain local building stones of special beauty, durability and utility may have a much wider distribution; for example, in England, Portland Stone is much prized for public buildings and so is to be found all over the country. Welsh or Cornish slates, which cleave or split easily, were

[*Courtesy: Eric Kay.*

FIG. 83.—CHINA CLAY MINING

China Clay is a white clay derived from the decomposition of granite. This clay pit is near St Austell in Cornwall where there are large deposits of kaolin. About $1\frac{1}{2}$ million tons are mined annually, some being exported to Canada and the United States. Kaolin has a variety of uses but most is used in the paper industry (50%) and in the making of pottery (25%).

formerly widely used for roofing purposes, but the industry is now almost dead. Marbles, which are metamorphosed limestones, are usually beautifully marked and coloured and are capable of being highly polished and so are extensively used for ornamental stone work. Especially noteworthy are Carrara marble (quarried in Tuscany, in Italy, and much used in statuary), Pentelic marble (from Greece), Connemara marble (from western Ireland), and black marble (quarried at Ashford, in Derbyshire).

Many building materials in use at the present day are manufactured products; they are, so to speak, "synthetic stone." The most important are bricks and concrete which are now almost universally used. During recent years a variety of synthetic stones, made from chippings of other stones and bonded by cement, have been manufactured and are being used increasingly as facing stones and for decorative inlay work.

Sand, clay, and limestone are the three most important mineral materials used in the making of manufactured products. Sand is an important ingredient in the manufacture of concrete and glass, and is used in the making of moulds for iron foundries. Clay finds its greatest use in brick-making and cement manufacture but is much used, also, in the earthenware and pottery industry, *e.g.* for saggars, sewage pipes, tiles, ablution equipment, etc. Limestone has a multitude of uses and is, perhaps, the most widely used of all the minerals: it is used as building stone, supplies lime for cement and concrete manufacture, is used as a flux in iron ore smelting, is used as an antidote for "sour" soils, and finds a role in many industrial processes including sugar-refining and glass manufacture.

Kaolin or China Clay is a soft white residual clay produced by the weathering of granite. It is used in the manufacture of porcelain, tiles, paper, toothpaste, etc. The finest kaolin deposits in the world are found in Devon and Cornwall and large quantities are exported in addition to those sent to the North Staffordshire Potteries. Kaolin is also found in France, Germany, China, Japan, and the United States.

It is not often realised how important are the common mineral products of the earth, nor how important are the movements of such minerals within a country which has a well-developed building trade and a highly developed communications system. It is not easy to procure figures of the quantities of building materials but in Great Britain something like 20 million tons of clay and 20 million tons of gravel are used annually for brick and cement making; 15 million tons of limestone for building, road-making, and cement manufacture; 10 million tons of igneous rocks for building and road-making; 3 million tons of fireclay, mostly used in connection with smelting; and about 1½ million tons of China Clay. Cement production in Britain runs at about 18 million tons annually, of which about 1 million tons is exported.

## *NON-METALLIC MINERALS*

These we can usefully divide into: (*a*) industrial minerals, such as asbestos, mica, graphite, sulphur, and salt; and (*b*) mineral fertilisers, such as nitrates, phosphates, potash, and guano.

## INDUSTRIAL MINERALS

*Asbestos* is a general term applied to certain fibrous minerals but the most common asbestos mineral is chrysotile, a variety of serpentine. Since asbestos will not burn or melt or deteriorate and is also highly resistant to electricity, it forms an ideal fireproof material; accordingly, it is much used for the making of fireproof clothing, safety curtains, and heat-resistant sheeting. Large quantities of asbestos are found in the igneous rocks of Devonian age in eastern Canada, whence comes some three-quarters of the world's supply. The U.S.S.R. is a major producer.

*Mica* resembles asbestos in many of its properties, especially in its resistance to heat and electricity. Mica occurs in crystals which cleave, *i.e.* they can be split into thin sheets. Because of its transparency and heat resistance, mica is used for the windows in the doors of furnaces and stoves. Because of its insulating properties, it has important electrical uses, *e.g.* in the building of electrical plant, in radio equipment, etc. Tiny fragments of mica are used to make "artificial frost," much used in Christmas decorations. The United States and India are the world's largest producers; in the former country there are large deposits in North Carolina, in the latter at Hazaribagh in the Damodar Valley.

*Graphite*, commonly known as plumbago or "black lead" because at one time it was mistaken for lead, is a finely crystalline form of carbon. Silvery-black in colour, soft, and greasy to the touch, graphite, in its powdered form and mixed with fine clay, is used for the "lead" in pencils. Nowadays, graphite has many more, and more important, uses being employed in the making of crucibles for metal smelting, as carbon brushes in electric motors, and as a "moderator" in atomic reactors. Graphite occurs in small quantities in Ceylon and Madagascar and at one time was mined near Keswick in the English Lake District. Supplies of natural graphite are rather limited, however, and nowadays synthetic graphite is made for industrial usages.

*Sulphur* is an essential raw material in the chemical industry. It is obtained from three sources: either native from areas of vulcanism, or from the mineral iron pyrites ($FeSO_2$), or as a by-product from natural gas. Its chief use is in the manufacture of sulphuric acid—a key commodity of the chemical industry—but it is used in the vulcanisation of rubber, in oil-refining, in iron and steel manufacture, in paper- and paint-making, and in the manufacture of innumerable chemical products including insecticides and fungicides. Annual world production is around 10 million tons. Large deposits are found in the U.S.A. (in Texas and Louisiana) and she is the world's largest producer. Other natural deposits are found and worked in Mexico, Japan, and Sicily. During recent years France has become a major producer of sulphur, deriving it from natural gas.

*Salt* (sodium chloride) may occur as a natural brine or as a crystalline solid (rock salt). Salt is a very necessary constituent of the human diet, hence from the very earliest times it has been an important commodity of trade. The evaporation of sea water in salt pans has been a traditional activity in the Mediterranean lands for many centuries. Some 90 million tons of salt are produced annually. Much of it is used for human dietary purposes, but about half is used by the chemical industry in which it is a basic raw material. The U.S.A. is the world's largest producer. There are important deposits in Durham and Cheshire in England, in Lorraine in France, in Lower Saxony in West Germany, around Stassfurt in East Germany, and in Poland and Austria, while the Soviet Union uses the Caspian as a source of salt. China is also a substantial producer.

## MINERAL FERTILISERS

*Nitrates* (sodium nitrate or Chile saltpetre and potassium nitrate or saltpetre) were used originally only in the making of gunpowder but just over a hundred years ago it was discovered that they formed a valuable artificial fertiliser. Natural nitrates are of infrequent and restricted occurrence. The chief deposits are found in the Atacama Desert of northern Chile. Here, in depressions, are found gravelly beds, known as *caliche*, which contain nitrates. For half a century Chile had a virtual monopoly of the nitrate trade. Then, in 1906, a method was discovered of extracting nitrogen from the air which led to the production of synthetic nitrates and nowadays most of the nitrogenous fertilisers are manufactured chemically. Accordingly, the nitrate trade of Chile has greatly declined. The chief producers of nitrates are the United States and those countries of Europe possessing well-developed electro-chemical industries based upon plentiful cheap hydro-electric power, *e.g.* Norway, Sweden, France, and Switzerland.

*Phosphates* are also used as artificial fertilisers. Sources of phosphate include phosphatic rock (a limestone containing phosphorus), the mineral apatite (especially the variety known as phosphorite), and guano (accumulated bird-droppings). Fertiliser known as "super-phosphate"—phosphatic rock treated with sulphuric acid—was originally made from bones. Basic slag, a derivative of the smelting of certain types of iron ores, contains phosphates and is much used nowadays as a fertiliser. Guano, which is naturally rich in phosphate, is found along certain desert coasts, notably on the islands off the coast of Peru and South-west Africa, and on certain ocean islands such as Nauru and Ocean Island in the Pacific. The Peruvian deposits have been over-exploited in the past and controlled exploitation has had to be imposed to conserve the valuable fertiliser. The total annual world output of phosphate rock is around 98 million tons. The chief producers are the United States (Florida and New Mexico) and the Soviet Union; the Atlas countries, especially Morocco and Tunisia, are also important producers.

*Potash*, like rock salt, occurs in beds which were laid down when former inland seas dried up. Natural potash deposits were first found around 1860 when they were discovered in association with the rock salt deposits at Stassfurt in East Germany. Prior to this, wood ash was the only source of potassium salts. Valuable as a fertiliser, potash is also used in the manufacture of explosives, glass, paper, soap, and medicines and finds uses in bleaching, tanning, and photographic processes. The best-known and richest deposits occur at Stassfurt. There are similar, but less extensive deposits in Alsace in France and at Suria in Spain. Potash is also obtained from the Saldura Marsh (Utah) and the Searles Marsh (California) in the United States and from the Solikamsk region of the Soviet Union. Considerable quantities are obtained from the Dead Sea where both Israel and Jordan have set up evaporating plants. World total production of potash runs at just over 15½ million tons annually.

[*Courtesy: Israeli Government.*

FIG. 84.—POTASH WORKS IN ISRAEL

Potash works at Sdom (Sedom) on the Dead Sea. The evaporating pans shown are used to test the rate of evaporation. The commercial pans are much larger. Bromine is produced along with potash.

## *METALS*

Metals are commonly distinctive materials possessing distinctive physical properties such as hardness, heaviness, lustre, etc. Comparatively few metals, however, occur in a "native" or "free" state (gold, silver, and

[*Courtesy: Malm, Stockholm.*

FIG. 85.—IRON ORE MINING, KIRUNA

In Swedish Lappland are vast deposits of iron ore. The photograph shows the open workings in the iron-ore mountain of Kirunavaara. The ores have a high content—60 to 70%—of pure iron, but they are phosphoric. In the long, dark winter, work continues by artificial lighting provided by hydro-electric power developed at Porjus about 100 miles (160 km) to the south.

copper are the most common); more usually they exist as compounds, that is, they are chemically combined with some other mineral as, for example, oxygen, sulphur, silicon. Seldom do such chemical compounds resemble metals. This helps to explain why man was so long in discovering metals. Man, moreover, could not progress from the Stone Age into the Metal Age until he had learned how to extract metals from their compounds. His discovery of how this could be done marks one of the most important steps in his cultural development. It is very probable that the first metal he learned how to use was copper, since it is often found pure, *i.e.* in its native state, and, if not, is easily smelted from its ore. Later, it was found that copper when combined with tin produced bronze,

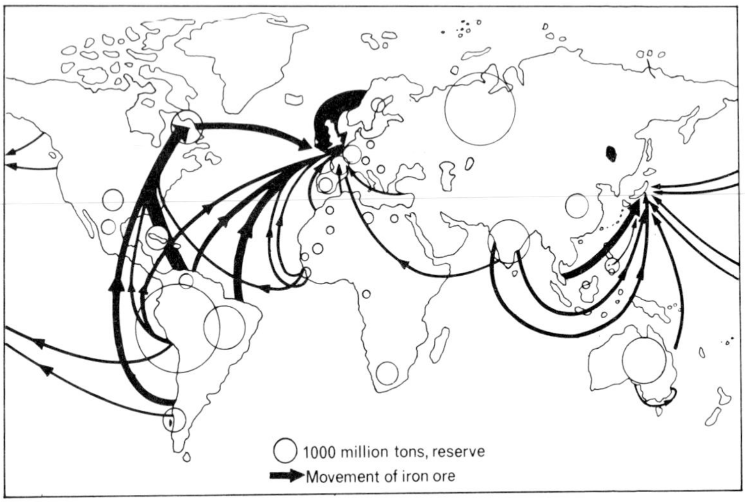


FIG. 86.—WORLD IRON-ORE RESERVES

an alloy metal which was much harder than pure copper. Iron, though far more abundant than copper, is much more difficult to smelt hence there was a long interval before man entered the Iron Age.

For several thousand years now iron, abundant and widely distributed, hard and versatile, has been dominant and its usefulness has steadily increased. As man's scientific and technical knowledge developed, iron accordingly was processed to produce, in turn, cast iron, wrought iron, steel, and, finally, a wide range of alloy steels; also, as time went on, man found ever-increasing uses to which iron and steel could be put. As a result, iron is at the present time by far the most important of all the metals and more valuable than all the other metals combined. Apart from iron, the chief metals, in terms of quantity produced, are lead, zinc, copper, and aluminium. Let us look first at iron, then at these four important non-ferrous metals, and finally at gold and silver, the two most important precious metals.

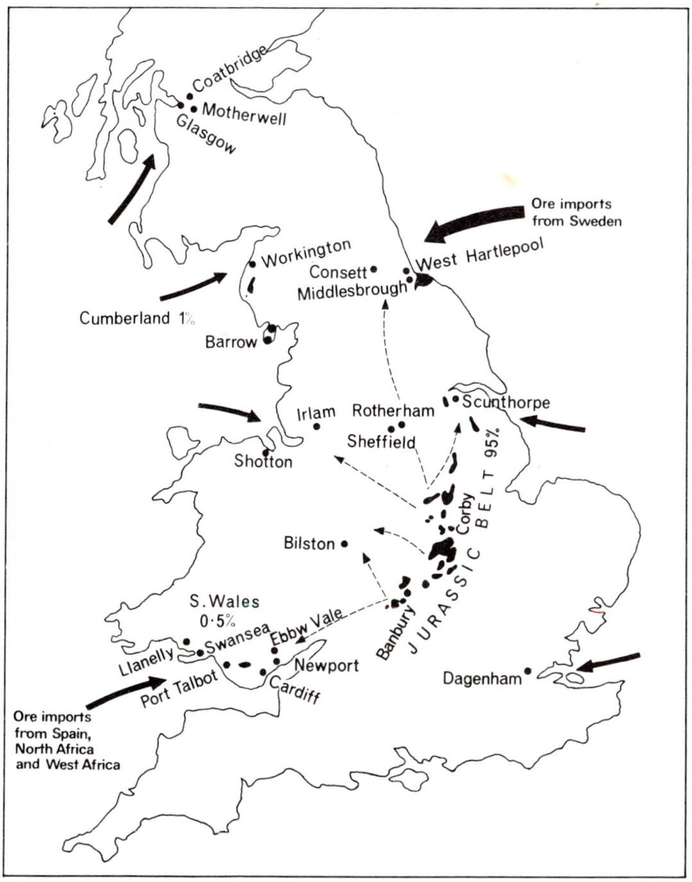


FIG. 87.—IRON ORE, BRITAIN

The chief deposits are shown in black and the figures indicate the proportions mined in recent years. The Cleveland ores are largely exhausted and production ceased several years ago. The low-grade ores of the Jurassic belt alone are now of any significance. As the pecked arrows show, ores are moved from the Midlands area to other smelting areas. Over the years, as the indigenous supplies have become worked out, Britain's dependence upon imported foreign ores has steadily increased and now something like half of the ore used in British blast furnaces comes from abroad. Sweden, Algeria, Canada, Liberia, Venezuela, Morocco, and France are the chief suppliers. Total iron-ore consumption is of the order of 30 million tons annually. The chief smelting centres are South Wales and Monmouthshire, the North-east Coast, and north Lincolnshire. Because of the need to import such large amounts of iron ore, there has been a decided tendency for the iron-smelting industry to migrate to coastal locations.

## IRON

Iron occurs very widely in nature, being probably the fourth most abundant element in the earth's crust; it is estimated that iron accounts for about 5% of the crust. The red, brown, and yellowish colours of rocks and soils are usually due to the presence of iron, and at least a trace of iron is to be found in every soil. Iron ores, however, are very unequally distributed both in respect of quantity and quality. Iron does not occur in nature as a pure metal; it occurs in the form of iron compounds or ores. There are many forms of ore (the chief are hematite, magnetite, limonite, and siderite) and the percentage of iron in the ore varies widely both in the different forms and in different places. To be commercially important, an ore deposit must either contain a high proportion of iron, or, if the percentage of the metal content is low, the deposit must be very large and readily accessible. It is reckoned that it does not pay to work ores containing less than 20% iron, and normally exploitation takes place only where the metal forms 30% of the ore. Note that the low-grade Jurassic ores in England, which contain an average of only about 26–28% of iron, are exploitable simply because they occur in large quantities and can be readily mined by open-cast methods. Some ore deposits, on the other hand, are very rich, *e.g.* the hematite ores of Cumberland, now almost exhausted, contain over 50% of iron while those of northern Sweden, mined at Kiruna and Gellivare, average about 60%.

The following table lists the chief world producers of iron ore.

TABLE XXIV

*Iron-Ore Production (million tons iron content)*

| | *1963* | *1970* |
|---|---|---|
| U.S.S.R. | 79,460,000 | 106,058,000 |
| U.S.A. | 41,542,000 | 53,308,000 |
| Canada | 15,036,000 | 29,687,000 |
| China | 20,000,000 (est.) | 22,000,000 |
| Sweden | 14,182,000 | 19,804,000 |
| France | 18,815,000 | 17,759,000 |
| Brazil | 7,330,000 | 20,400,000 |
| India | 12,464,000 | 19,572,000 |
| Australia | 3,614,000 | 32,732,000 |
| Liberia | 4,312,000 | 15,388,000 |
| Venezuela | 7,518,000 | 14,084,000 |
| U.K. | 4,091,000 | 3,365,000 |

Total world production 1970 418,800,000

Generally speaking, only the richer ores, *i.e.* those with 50% or more iron content, enter into international trade. The only important exception to this general statement is the export of lean Lorraine iron ores from France to her near neighbours, Belgium, Luxembourg, and Western

Germany, who are members of the European Coal and Steel Community. The discovery and opening up of new ore fields, *e.g.* the Labrador, Venezuelan, Peruvian, Brazilian, and West African and Australian deposits, is rapidly changing the established pattern of iron-ore production. Some countries, such as the United States, the United Kingdom and Western Germany, though important producers of iron ore are unable to meet the demands of their great iron and steel industries and must perforce import appreciable quantities of ore. A country like Japan which has built up a large iron and steel industry but which has few home deposits of iron ore is greatly dependent upon foreign supplies of ore and scrap. Australia, with its extensive iron ore deposits—at Iron Knob and Yampi Sound and the more recently discovered ores at Paraburdoo and Northam—is becoming a major exporter to the Japanese market. The following table illustrates the dependence upon imported ore of some of the major iron and steel producers.

TABLE XXV

*Iron-Ore Imports*

| | *Ore imports (in million tons)* | | |
|---|---|---|---|
| | *1950* | *1961* | *1966* |
| U.S.A. | 8·2 | 25·8 | 40·0 |
| U.K. | 8·4 | 15·0 | 18·0 |
| West Germany | 8·4 | 32·2 | 36·0 |
| Belgium/Luxembourg | 8·1 | 20·2 | 20·0 |
| Japan | 1·4 | 20·7 | 40·0 |

The bulk of the world's iron is converted into steel which is an alloy of iron and carbon. Steel is of far greater importance nowadays in industry than iron; it is superior to iron in that it is far stronger, harder a little lighter, and less brittle. Many steel alloys are used these days; these are made by adding carefully measured quantities of other metals to steel. These alloy metals, the most important of which are manganese, nickel, chromium, cobalt, molybdenum, vanadium, and tungsten, impart special properties to the steel, such as resistance to rusting and high temperatures (required in the case of high-speed cutting tools) and increased hardness and toughness (needed for armour plating, etc.). The mechanical engineering, motor car, and aircraft industries are the chief users of alloy steels.

**COPPER**

Copper was second only in importance to iron until very recent years when it was displaced by aluminium. As we have already noted, copper was one of the first, if not *the* first, of the metals to be used by man. But,

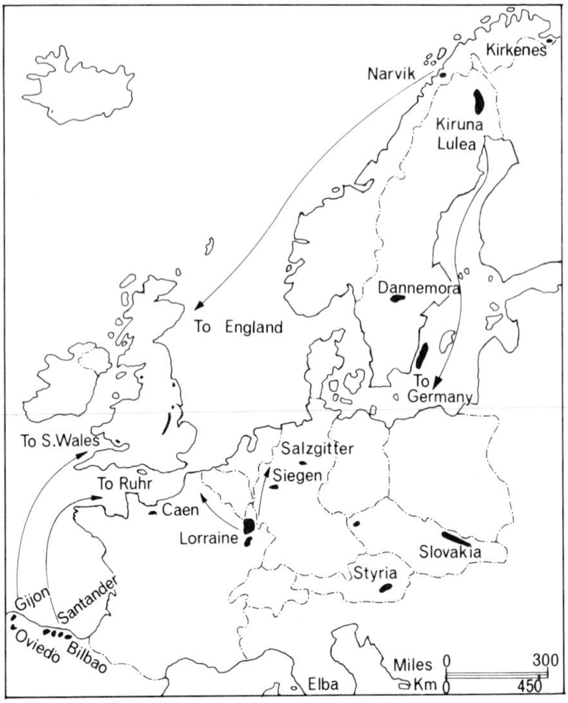


FIG. 88.—MOVEMENTS OF IRON ORE IN EUROPE

with the introduction of iron, copper lost its significance and did not really come into its own again until the development of electricity. Copper is used in large quantities today mainly because of its high electrical conductivity and the ease with which it can be drawn into wire. Accordingly, copper finds its chief use in the electrical industry which consumes about 60% of the total production; it is used for the making of generators, motors, cables, and radio and telephone equipment. The other chief use of copper is for the making of copper alloys and approximately 40% of copper is so used. The chief alloys are bronze (copper–tin), brass (copper–zinc), and cupro-nickel (copper–nickel).

Although copper occurs naturally in igneous rocks, the chief source of copper is from the sulphides of the metal. The United States, which has long been the world's premier producer and still produces roughly a quarter of the world's output, possesses very rich deposits of copper. Chile, too, has rich reserves and has for long been a major producer, but the most important feature of modern copper production has been the enormous output first from the Katanga copper belt of Zambia and the

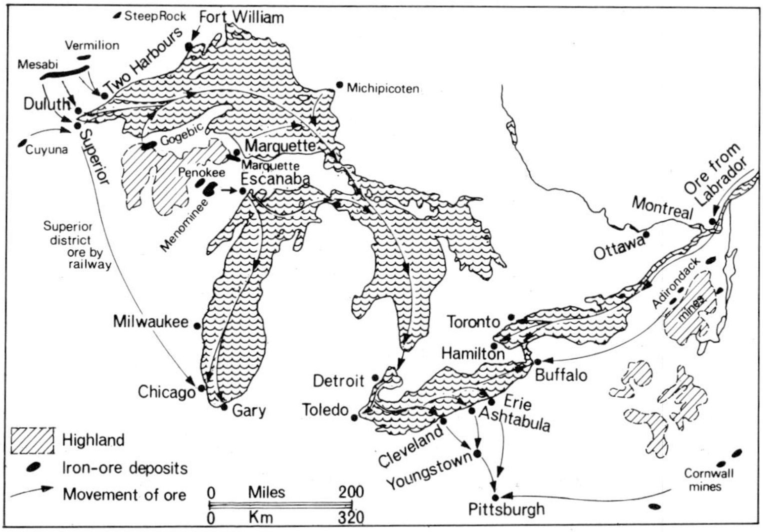


FIG. 89.—MOVEMENTS OF IRON ORE IN NORTH AMERICA

The rich iron-ore deposits in the Lake Superior region are transported chiefly by "lakers" to the lakeside iron and steel centres, *e.g.* Chicago, Cleveland; some of the ore goes directly by rail to Chicago. The newly-discovered iron-ore deposits around Knob Lake, in Labrador, are shipped upstream to the lakeside centres on Ontario and Erie.

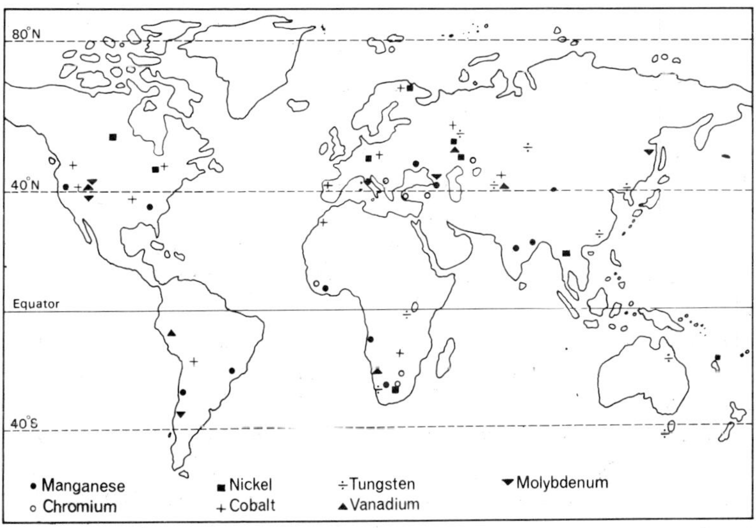


FIG. 90.—FERRO-ALLOY METALS

Congolese Republic (now renamed Zaire Republic) and, secondly, from Canada especially from the Sudbury and Noranda districts.

TABLE XXVI

*Copper Ore Production (thousands of tons copper content)*

| | *1960* | *1970'* | | *1960* | *1970* |
|---|---|---|---|---|---|
| U.S.A. | 1092 | 1560 | Peru | 50 | 206 |
| U.S.S.R. | 510 | 925 | Mexico | 66 | 128 (1968) |
| Chile | 587 | 668 | Japan | 98 | 120 |
| Zambia | 635 | 684 | Australia | 117 | 152 |
| Canada | 444 | 611 | South Africa | 52 | 148 |
| Zaire | 333 | 386 | | | |

Total world production 1970: 6,350,000 tons.
Pre-war production: 1,980,000 tons.

In spite of the United States' large production she imports additional supplies from Canada and Chile. The United Kingdom, France, West Germany, and Japan are important importers.

## LEAD AND ZINC

These two minerals may be taken together because they are found together; there is, in fact, only one known important zinc deposit in the world, that of Franklin Furnace in the state of New Jersey, U.S.A., which is not associated with lead. Because the two occur together, obviously they are mined together. Lead is used, in sheet form, for roofing material, for pipes, for cable sheathing, for accumulator plates, and for making solder. Lead oxides are much used in paint manufacture and for pottery glazes. A small but increasing amount of lead, in the form of lead tetraethyl, is used to improve the efficiency of motor spirit (anti-knock petrol).

Zinc, which is a non-corrodible metal, is one of the most widely used of the non-ferrous metals. The most important use of zinc is as a thin coating on sheet steel (galvanised iron) which protects the iron from rusting. But it is also used for the making of wire (wire netting), pipes, batteries, in the printing industry, in pharmaceutical products, and in the making of alloys. The chief alloy is brass (zinc–copper).

Considerably more zinc is produced and used than lead. World output of zinc has increased more than twofold since pre-war days, whereas lead production has increased by some 50%. While the demand for zinc has increased and could be met, the production of lead has been handicapped by the growing exhaustion of the ore bodies; the consequence has been a world shortage of lead which has not only put up its price but led to substitute materials being used in its place, *e.g.* copper piping in plumbing.

One or two of the famous lead–zinc mines in the world are worth

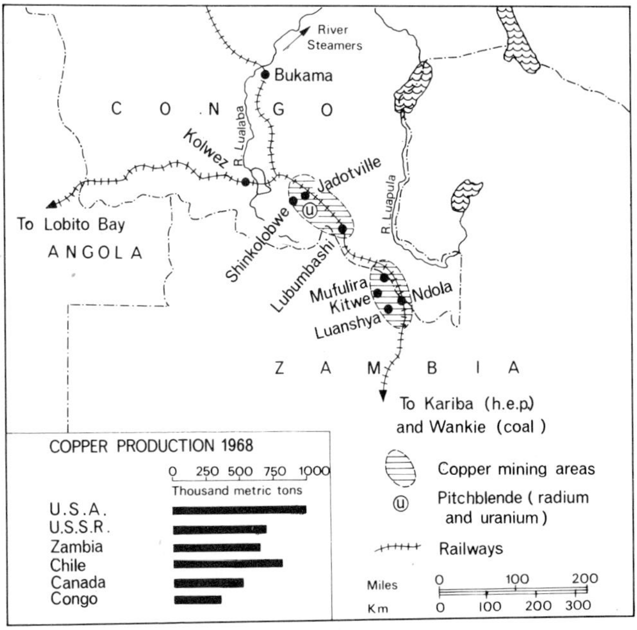


FIG. 91.—THE COPPERBELT

The Copperbelt of Central Africa extends some 250 miles (400 km) from the southern part of the province of Katanga, in the Congolese Republic, across the frontier into the northern part of Zambia. Although the area is especially rich in copper, hence the name "Copperbelt," other important minerals, such as cobalt and uranium, are found here. Although the exploitation of the mineral resources commenced in 1911, it was not until about 1930 that mining came to be seriously undertaken. Since that time the area has been completely transformed and now it is a modern industrial region whose importance as a mining area exceeds that of any other part of Africa except for the Witwatersrand region of South Africa.

The Copperbelt is one of the world's chief sources of copper: it produces some 980,000 tons or nearly a quarter of total world production. Nowadays very little of the ore is exported, the ore mostly being smelted on the spot, using electric power generated at Cornet Falls and other places and brought by transmission line from Kariba.

In the Congo, Lubumbashi (formerly Elisabethville) and Jadotville are the chief centres of copper smelting. Across the border, in Zambia, Ndola and Kitwe are the main copper centres. The refined copper, once exported via Dar-es-Salaam, Beira, and Cape Town, now goes mainly via Bukama to Port Francqui (on the river Kasai, a tributary of the Congo) or to Lobito in Angola (the shortest and most direct railway link). The new Tan-Zam Railway to Dar-es-Salaam will become a major outlet.

noting, especially the renowned Sullivan Mine, in British Columbia, Canada (the largest lead–zinc mine in the world), the well-known mines of Broken Hill in New South Wales, Australia; the Tri-state mines in west-central United States; the Santa Barbara–Hidalgo del Parral mines in Mexico; and Cerro de Pasco in Peru.

TABLE XXVII

*Lead Ore Production (thousand tons lead content)*

| | 1960 | 1970 | | 1960 | 1970 |
|---|---|---|---|---|---|
| U.S.S.R. | 360 | 440 | Peru | 144·8 | 157 |
| Australia | 325 | 457 | Yugoslavia | 101 | 127 |
| Canada | 205 | 357 | Morocco | 104 | 85 |
| U.S.A. | 244 | 519 | | | |

Total world production: 1960, 2,618,000 tons; 1970, 3,380,000 tons.
Pre-war production: 1,700,000 tons.

TABLE XXVIII

*Zinc Ore Production (thousand tons zinc content)*

| | 1960 | 1970 | | 1960 | 1970 |
|---|---|---|---|---|---|
| Canada | 410 | 1239 | Japan | 173 | 280 |
| U.S.S.R. | 360 | 570 | Mexico | 299 | 266 |
| U.S.A. | 432 | 484 | Poland | 159 | 242 |
| Australia | 273 | 489 | Italy | 122 | 110 |
| Peru | 138 | 323 | Zaire | 120 | 105 |

Total world production: 1960, 3,428,000 tons; 1970, 5,530,000 tons.
Pre-war production: 1,769,000 tons.

## TIN

The discovery of tin, somewhere about 3000 B.C., ushered in the Bronze Age, and until a hundred years ago its chief use was for the making of bronze. Though still used for bronze-making, it is also used for other alloys such as gun-metal, pewter, and solder. As a metal, tin has many advantages: it resists corrosion by air, water, and many acids, and herein lies its greatest utility, for it is used in the making of tin cans in which food is preserved. The idea of coating iron with tin to give it a protective covering apparently goes back to the fourteenth century and was first undertaken in Bohemia, but the process of coating sheet iron with a thin film of tin was first developed in the early part of the eighteenth century. The first United Kingdom tin-plate works was set up at Pontypool, South Wales, in 1720. The use of tin cans for food preserving led to the great development of the tin-plate and metal-box industries. An Englishman, Peter Durand, patented a process of enclosing food in containers of iron coated with a film of tin early in the nineteenth century. The tin-plate industry, which consumes about half the tin imported into Britain—some 25,000 tons a year—is concentrated in South Wales. Tin has been mined in Cornwall for several thousand years and until about a century ago Cornwall was one of the leading tin-producing areas. Although tin still occurs at depth, the accessible deposits have been practically exhausted.

Tin occurs either as cassiterite (tin oxide) in gravels or as metal in lodes. Most of the world's tin at present comes from alluvial or gravel deposits from which it is recovered by dredging or monitoring, *e.g.* in Malaya, but some, as in Bolivia, is mined from lodes. Tin is not very widely scattered; rather does it tend to occur in a few main areas. The richest deposits are found in south-east Asia where a tin zone runs from southern China,

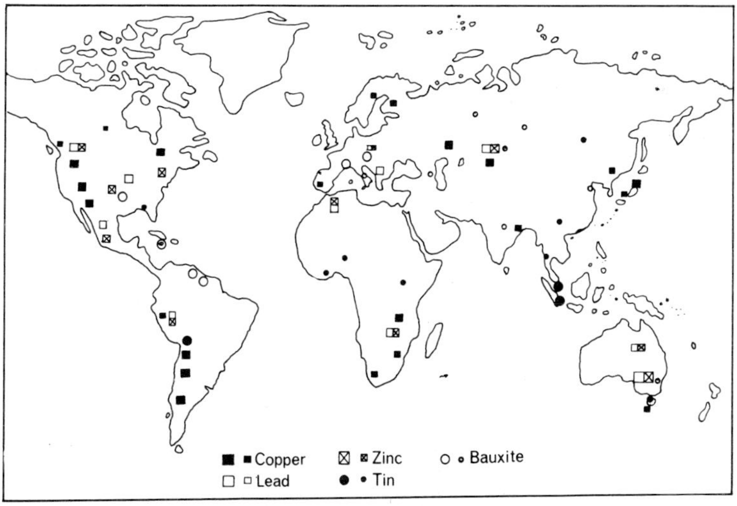


FIG. 92.—DISTRIBUTION OF NON-FERROUS METALS

through Thailand and the Malay Peninsula into Indonesia. Some two-thirds of the world total production comes from this region. Almost all the rest comes from Bolivia, Nigeria (the Jos Plateau), Congo (Leopoldville area), and the Soviet Union. Much of the tin from south-east Asia goes to the United States which is almost completely lacking in tin resources.

TABLE XXIX

*Tin Ore Production (thousands of tons tin content)*

| | *1960* | *1970* | | *1960* | *1970* |
|---|---|---|---|---|---|
| Malaya | 52·8 | 73·8 | China | 24·4 | 20·0 |
| U.S.S.R. | 25·0 | 27·0 | Indonesia | 12·2 | 19·1 |
| Bolivia | 19·7 | 30·1 | Nigeria | 7·8 | 7·9 |
| Thailand | 24·4 | 21·8 | Zaire | 19·0 | 6·4 |

Total world production: 1970, 234,000 (excluding U.S.S.R.)

Pre-war production: 167,000 tons.

## ALUMINIUM

Aluminium is the most abundant metal in the crust of the earth and accounts for as much as 8% of the rocky material. It is not often realised that up to a quarter of the substance of ordinary clay consists of aluminium. Clay is, in fact, an aluminium silicate but the metal is so intimately com-

[*Courtesy: Aluminium Development Association.*

FIG. 93.—BAUXITE MINING, GUYANA

This photograph shows many different operations: (i) bush-clearing (*top left*); (ii) a dragline stripping clay and sand overburden (*top centre*); (iii) blasthole drilling of the bauxite deposit below the overburden (*centre*); (iv) hydraulic stripping, *i.e.* breaking up bauxite by attacking it with water-jets (*foreground*).

bined with other elements that its extraction is not commercially practicable. Most of the aluminium produced comes from the ore known as bauxite (an impure aluminium hydroxide), which has a very wide distribution, and cryolite which is found naturally only on the west coast of Greenland.

Aluminium was first extracted from its ore about a century ago. During the present century the use of aluminium has steadily grown and at present more than five times as much is used as in pre-war days. It is one of the most useful and versatile of metals. A wide range of light but strong alloys, *e.g.* duralumin, magnalium, are now made and are used in aircraft, railway, and motor car construction. Aluminium is also used in the manufacture of domestic utensils, wire, foil, etc.

TABLE XXX

*Bauxite Production (thousand tons of ore)*

| | *1960* | *1970* | | *1960* | *1970* |
|---|---|---|---|---|---|
| Jamaica . . . | 6169 | 12,010 | Guinea . . | 1378 | 2642 |
| Dutch Guiana . | 3808 | 6011 | Yugoslavia . . | 1103 | 2098 |
| U.S.S.R. . . | 3858 | 5000 | Hungary . . | 1267 | 2022 |
| Guyana . . . | 2360 | 3211 | Greece . . | 1025 | 2283 |
| France . . . | 2246 | 2992 | Australia . . | — | 9338 |
| U.S.A. . . . | 2210 | 2562 | | | |

Total world production: 1960, 28,259,000 tons; 1970, 59,050,000 tons.
Pre-war production: 3,645,800 tons.

For many years British and Dutch Guiana were the leading producers of bauxite but during the past few years Jamaica has taken the lead. Most of the ore from Jamaica and Guyana goes to Canada for reduction. The United States, which possesses some bauxite of its own, imports additional supplies from Dutch Guiana. Both the U.S.S.R. and Australia are important producers. France has important deposits in the department of Var and she is the leading European producer.

## GOLD

Gold was one of the very first, perhaps the first, of the metals to become familiar to man. This was due, first to its attractive lustre and, second, to the fact that, unlike most other metals, it occurs "native." Gold, because of its beauty and rarity, has been the most highly esteemed of all the metals; and, because of its relative scarcity, it has always had a high value. Its chief use, because of its value, is for jewellery and, formerly, for coin. Being a very soft metal, gold is usually hardened by adding some alloy metal, such as nickel, copper, or silver.

TABLE XXXI

*Gold Production (kilogrammes)*

| | *1960* | *1970* | | *1960* | *1970* |
|---|---|---|---|---|---|
| South Africa | 665,105 | 1,000,417 | U.S.A. . | 52,248 | 56,390 |
| U.S.S.R. . | 328,125 | 195,000 (est.) | Australia . | 33,598 | 19,266 |
| Canada . | 143,151 | 76,736 | Ghana . | 27,340 | 21,892 |

Total world production: 1970, 1,480,000 kilogrammes.

Gold generally occurs either in alluvial deposits or placers (into which particles of gold have been washed from neighbouring veins of gold) or in quartz veins in which particles of free gold are embedded. In earlier

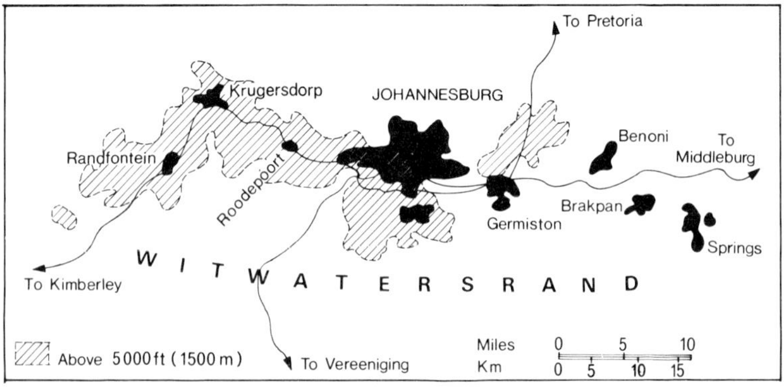


FIG. 94.—THE WITWATERSRAND

The Witwatersrand, or the "Rand" for short, is the name given to a ridge of high land running east to west in the southern part of the Transvaal. The Rand itself is about 50 miles (80 km) in extent but it is part of a 350-mile (560-km) long ridge. Its importance lies in its gold-bearing reefs. Gold was first discovered here in 1884 and it has been mined ever since. Today there are sixty-two mines working the precious metal in this "golden arc." The gold occurs in a hard, pebbly rock—quartz conglomerate—which is known as "banket." The banket is mined, crushed, and treated chemically to yield the gold. The early mines worked the surface rock but the reef, after being worked for nearly eighty years, is now worked at depth and some of the mines are 10,000 ft (3000 m) deep. One of the new mines is planning to mine at a depth of 13,000 ft (4300 m)—nearly 3 miles (4·8 km) below the surface! During more recent times, exploitation has taken place in the Far West Rand, in the Klerksdorp area of south-western Transvaal, and in the Orange Free State around the town of Welkom.

Some three-quarters of the world's output of gold (including the U.S.S.R., whose production is about 350,000 kg) comes from the Rand. Johannesburg is the great centre of the Rand and, with its population of over $1\frac{1}{2}$ millions, is the largest city in the Republic. It has become a big industrial centre with an emphasis upon engineering. Coal, needed by the mining industry and by the other growing industries of the region, is mined at Vereeniging some 25 miles (40 km) to the south of Johannesburg and to the east of the Rand. Water supplies, in great demand in this region, have necessitated the construction of a huge reservoir, the Vaalbank Dam, on the Vaal River.

Gold-mining depends upon a large labour force and some 300,000 Africans are employed in the mines. Mining is hard, difficult and dangerous work but great care is taken of the native workers and the casualty rate is less than two per thousand. Wages, though considerably higher than those paid to most African workers, are still low; but it is only by the employment of cheap labour that the gold can be profitably worked since each ton of banket yields only about 4 oz (113 grammes) of gold.

A point of interest is that uranium sometimes occurs with gold. Uranium has only become important with the development of nuclear power for which it is the "fuel"; hence many of the old mining dumps are now being worked over to recover the very small, but very valuable, quantities of this important mineral.

days much of the world's gold was obtained by panning river gravels and it was the discovery of placers which led to the great "gold rushes" of the nineteenth century to California, Alaska, and Australia. Nowadays, most of the gold mined is procured from veins worked at depth below the surface. Over half the total world production comes from South Africa. The Soviet Union is the second largest producer.

## SILVER

Silver is a lovely metal and, like gold, has long been held a precious metal. Its value has greatly declined, however, since the days when the Spaniards ransacked the Aztec and Inca Empires of the Americas for their silver treasures. Ductile and malleable like gold, silver resists corrosion although it tarnishes when it comes into contact with sulphurous fumes.

TABLE XXXII

*Silver Production (metric tons)*

| | *1960* | *1970* | | *1960* | *1970* |
|---|---|---|---|---|---|
| Canada | 1050 | 1383 | U.S.S.R. | 781 | 1150 (1969) |
| Mexico | 1385 | 1332 | Australia | 469 | 811 |
| Peru | 956 | 1217 | W. Germany | 450 | 758 |
| U.S.A. | 1146 | 1371 | Japan | 324 | 920 |

Total world production: 1970, 10,570 metric tons.

Silver sometimes occurs in its native state but it is often found in its sulphide form in association with the sulphides of copper, lead and zinc. In fact, a large proportion of the silver produced at the present day comes from the desilverisation of mixed ore bodies which are worked for other metals; in other words, it is very largely a by-product of copper- and lead-mining.

The Latin American countries of Mexico and Peru are still major producers. The United States and Canada are other major producers, most of the output coming from the Cordilleran states. A large number of countries produce small amounts of silver.

### EXERCISES

1. Stress the geographical distribution of the activities involved in the exploitation of copper under the following headings: (*a*) the location of copper ores, (*b*) the treatment of copper ores, (*c*) the utilisation of copper.

2. "Minerals provide a main incentive for man's interest in under-developed lands." Illustrate this statement by detailed references to examples drawn from several continents.

3. Choose any two minerals mined, quarried, or otherwise extracted on a large scale in Great Britain (except coal). Say, as accurately as you can, where they are worked and indicate briefly the methods of extraction and the uses to which they are put. (*Chartered Institute of Secretaries.*)

4. Give an account of (*a*) the occurrence, (*b*) the distribution, and (*c*) the trade in tin or lead.

5. Give an account of the distribution of the iron ore deposits in North America. Indicate the chief movements of ore to the consuming centres.

6. Examine the wealth and deficiencies of the United States in mineral resources.

7. Write a geographical essay on the location of iron ores in Western Europe indicating the chief movements of the ores to the major consuming centres.

8. What conditions are necessary for the economic exploitation of iron ore?

9. Chile is an important producer of (*a*) nitrates, (*b*) copper. Draw a sketch-map to show the locations of the main producing areas and describe the conditions under which the mining takes place.

## *Chapter XVIII*

# FUEL AND POWER

In the very earliest times all fetching and carrying, lifting and hauling had to be done by man himself—by human muscle-power. When, for example, the Ancient Britons built Stonehenge some three or four thousand years ago, the huge blocks of stone were dragged all the way from Pembrokeshire by teams of men and manhandled on the site. There were no steam-driven trains, no petrol-driven lorries, no motor-tractors, and no electric cranes to help move and mount the massive stone blocks. Man is a somewhat lazy creature and very soon he began to devise ways and means to ease human effort and toil. At first, by the institution of slavery, he got other men to do his work for him and most of the early civilisations were largely based on the energy provided by slaves. But slaves were not always to be had and sometimes they became scarce; hence man had to find alternative means of help. He turned to the animal world and soon began to use the animals he had succeeded in domesticating, such as the horse, the ox, and the camel, first to carry and then to draw loads.

Animate power, however, has certain disadvantages: there is a limit to the carrying and drawing capacity of an animal, the animal has to be fed and housed, and often it can work only within restricted environments. Fairly early on man sought to use inanimate sources of energy and his first success was with the wind. He found that by mounting a sail on a boat he could use the force of the wind. In the beginning he could only sail with a following wind but this relieved him of the necessity of plying oars. With increasing experience, he eventually learned how to sail against the wind, *i.e.* by tacking or following a zigzag course. This, clearly, was a tremendous step forward. On land, the wind was harnessed by the windmill, an invention, it is thought, of the Persians. About the time of Christ, the water-wheel was invented, an invention often attributed to the famous Roman engineer Vitruvius. Thus running water was used to turn wheels which provided the necessary energy for lifting water or grinding corn.

Wind and water remained the only important sources of inanimate energy used by man until about two hundred years ago. For a century or more several people had experimented with the use of steam as a source of power but not until James Watt perfected his steam engine was a satisfactory machine devised. Since Watt's day many new sources of energy have been developed and now man, with a great fund of scientific and technological experience behind him, is forever seeking and finding new sources of energy and power.

The amount of energy, inanimate power, used by a people or country is a pretty good guide as to the stage of its economic advancement. The United States, which is the most economically advanced nation in the world and whose peoples enjoy the highest standard of living in the world, is the world's greatest user of energy. About half a century ago, a well-known geographer, called James Fairgrieve, wrote a book, *Geography and World Power*, the underlying theme of which was that throughout history peoples had advanced and become powerful in relation to the energy resources they possessed, developed, and utilised.

## *THE DIFFERENCE BETWEEN FUEL AND POWER*

First, let us note the difference between fuel and power. These are terms which are often loosely and sometimes inaccurately used. Coal, for instance, is commonly said to be a major source of energy and although this may be allowed in a general sort of way, strictly speaking coal is merely a fuel which is burned to produce steam power or electric power. Moving wind and flowing water through their motion provide energy or power. Falling water, again, may have its inherent energy harnessed to drive a turbine which, in turn, provides a source of power—hydro-electric power. But coal, by itself, is not power; only indirectly is it a source of power.

The chief fuels which are consumed to produce power are:

Wood and associated vegetable material.
Coal and lignite (brown coal).
Peat.
Alcohol.
Petroleum oils.
Natural gas.
Nuclear fuels.

The chief sources of energy are:

Man-power.
Animal-power.
Wind.
Water.
Tidal-power.
Geothermal energy.
Solar energy.
Steam-power.

At the present time there are four main direct or indirect sources of energy: coal, oil, natural gas, and falling water. These account for 99% of all the power produced which is of an inanimate nature. It is important to note that the first three sources are non-renewable resources and once used they have gone for good. Falling water, on the other hand, is a non-wasting asset for the water can be used over and over again. All the fuels, with the partial exception of nuclear fuels, are exhaustible, though wood is capable of being grown and in this sense may be thought of as a renewable resource. There seems little doubt that atomic energy will supply most of the power of the future but for some time to come the more orthodox sources of energy will continue to provide the bulk of the world's power needs.

## *THE SIGNIFICANCE OF ENERGY*

In describing the significance of energy we are concerned with two matters: (*a*) the importance of power supplies in the modern world, and (*b*) the role played by the different sources of power.

In all technically advanced countries, modern life would be impossible were it not for the abundant supplies of power that are available. Think for a moment of what happens when there is an electric power cut: the lights go out, the radio stops—unless you have a transistor run by a battery—the oven stops cooking, the vacuum won't work, lifts remain stationary, etc. In other words, the rhythm in homes, in factories, in transport, and in communication grinds to a halt when power supplies run out. It is only when there is a breakdown that we realise how essential energy is to modern life. Like water supplies, we tend to take it for granted.

Although power is vital to many aspects of modern life, it is especially important in connection with industry. The growth of manufacturing industry has led to ever-increasing demands for inanimate energy and, since the world as a whole is becoming more and more industrialised and the pace of industrialisation, too, continues to increase, the demand for power is insatiable and, in fact, in some areas power requirements are greater than can, at the moment, be supplied. Power shortage has been a chronic problem in some of the South American countries, notably Argentina and Brazil, both of which have embarked upon programmes of rapid industrialisation.

The recognition by the so-called under-developed countries that industrialisation is one of the main keys to progress, power, and better standards of living has led in many cases to the energetic development of such power resources as they possess. Nowhere is this better exemplified than in the case of China. Until almost as recently as twenty-five years ago inanimate energy was looked upon very largely as the prerogative of the technically advanced nations but, increasingly these days, the under-developed countries are appreciating the need for, and seeking the development of, power resources.

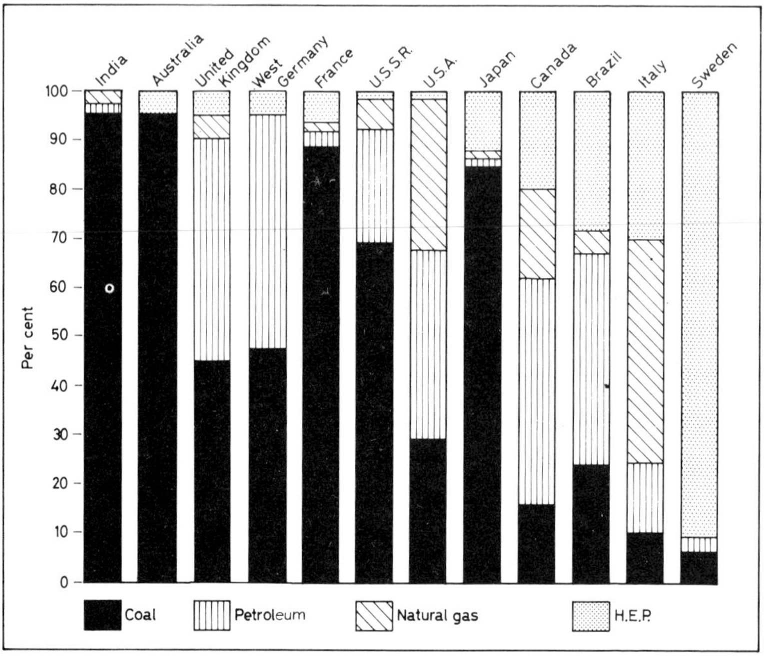


FIG. 95.—SOURCES OF INANIMATE ENERGY

Fig. 37 is an example of a simple bar graph. This graph is described as a *percentage bar graph*. It consists of a series of bars of uniform length, each representing 100%, each being subdivided on a percentage basis.

This graph attempts to illustrate the use made of four main sources of inanimate energy by a selection of countries. Carefully compare these different countries noting especially the different sources of power used by the United Kingdom, the United States, Brazil, Italy, and Sweden.

Some countries, such as the United States, Canada, the Soviet Union, the United Kingdom, Western Germany, and Japan, use large amounts of energy; others, such as Portugal, Turkey, Mongolia, Morocco, and Peru, use very little. The quantities used vary chiefly according to the degree of industrial development although it should be remembered that today, in the more developed countries, other productive occupations, including the service industries, are consuming large and increasing quantities of power. Think, for instance, of the modern farm which needs fuel for its tractors, machines and vehicles, and electricity for its milking machines, pumps, etc., or the modern hotel which needs fuel for heating and cooking, and electricity for its lighting, its lifts, its air conditioning system, its automatic dish-washers, etc.

Truly, then, the demands upon energy are almost unlimited and growing apace. This simple fact is well illustrated by the statement that more coal has been mined during the past 25 years than in the previous 250, and more petroleum mined during the past 10 years than in the previous 100. The phenomenal growth in energy demands during recent times seems likely to continue for a long time to come.

## ENERGY PATTERNS

One of the most interesting facts relating to the history of energy is the way in which one source of energy has become superseded by another. Each new source of energy, no matter how important or dominant it became in previous eras, has gradually declined in its significance as newer sources appeared and expanded. This wave-like pattern of growth and decline of energy sources seem to have been true ever since muscle power provided the first sole source of energy. The rise and fall of particular sources of energy is related to two conditions: the exhaustion of supply and new technologies which lead to the production and use of the superior or more efficient forms of energy.

In the beginning human and animal muscle power dominated the energy scene but early on man began to use wood as a fuel and then, somewhere about two thousand years ago, he discovered how to use running water as a source of energy. Coal was used occasionally for many centuries—"sea-cole" was shipped from Durham to London during the Middle Ages—but it was not until the Industrial Revolution that it came to be used on any scale; then, between 1800 and 1900 it quickly became the dominant source of fuel to provide steam power. From 1900 to 1950 coal maintained its supremacy but increasingly it began to be challenged by petroleum and at the present time slightly more oil is used than coal. But whereas the share of coal in world fuel consumption will decline rapidly in the next thirty years, the increased use of oil will, in all probability, continue to rise until about the end of the present century. Since the end of the Second World War there has been a rapid rise in the use of natural gas as a fuel but there are probably insufficient reserves of it to allow it to become a serious competitor of oil and it is believed that after about 1980 its share in world fuel consumption will begin to decline. Meanwhile, since about 1960, nuclear energy has entered the field and it is anticipated that, after a fairly slow start, it will be meeting 20% of the world's fuel consumption by the end of the century. Fig. 96 shows the pattern of world fuel consumption in the past hundred years and projects the shares of different fuels in world consumption during the next thirty years.

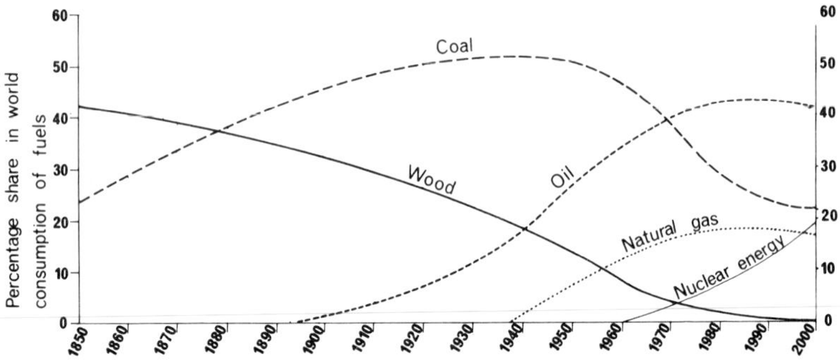


FIG. 96—ENERGY PATTERNS

At present (1972) the rate of consumption of energy in the world is approximately 7000 million tons of coal equivalent per annum. It is not easy to apportion this energy but it is believed that about one-half is used in manufacturing industry, about one-fifth in transport, approximately the same amount for domestic purposes, and the remainder for miscellaneous purposes. It is estimated that the current annual rate of increase is about 5%. The general opinion is that the demand for energy will double itself during the next generation and that by around 1980 the world

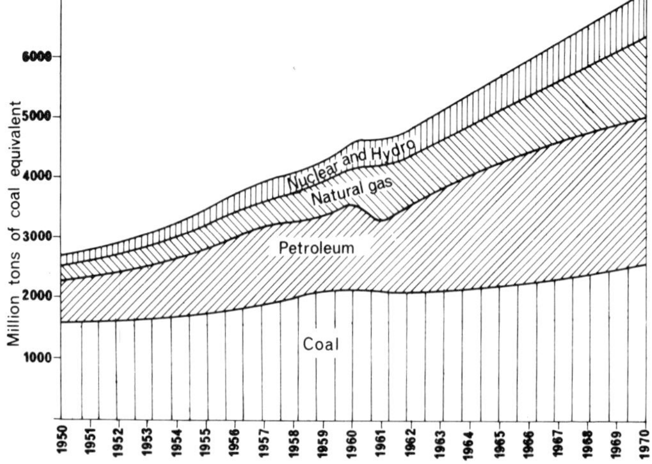


FIG. 97—WORLD PRODUCTION OF PRIMARY SOURCES OF ENERGY

will be using something of the order of 10,000 million tons of coal equivalent.

Coal, petroleum, natural gas, and hydro-electricity are the main sources of inanimate energy. Of this energy coal is responsible for about 37%, oil for about 40%, natural gas for about 15%, hydro-electric power for about 5% and nuclear energy about 3%. Since pre-war years coal has lost much ground, while petroleum and natural gas have forged ahead. It should be noted that in spite of the spectacular nature of the great hydro-electric power dams, water power is gradually declining in its relative importance to the other power resources. Figure 97 shows how the world production of primary sources of energy has grown since the end of the Second World War.

## *COAL*

Only fifty years ago coal was the direct or indirect source of nearly 90% of the world's power. Today, as we have just noted, it accounts for less than half, yet more coal is being mined in the world than ever before. Coal is, and is likely to remain, a major source of power for a long time to come.

### ORIGIN OF COAL

Coal is mineral matter found in layers or beds in the sedimentary rocks of the earth's crust. It is a hard, black, shiny substance, composed of varying amounts of carbon, oxygen, hydrogen, and nitrogen together with some impurities. Technically, coal is known as a hydro-carbon. Unlike most other mineral products, it is combustible and herein lies its importance. Coal is really fossilised vegetation. In earlier periods of the earth's geological history, especially in the Carboniferous Age—the name means "coal-bearing"—great swamp forests, very similar to those found at the present time in the Everglades region of Florida, grew in various parts of the world. Such forests were periodically submerged and covered with mud and sand. In due course, the vegetable matter partially decayed and was compressed by the weight of the layers of sediment laid down on top of it. An accumulation of vegetable matter 30 feet (9 m) thick would eventually become squeezed into a seam of coal only 2 or 3 feet (0·6–0·9 m) thick. Hundreds of millions of years later these coal seams came to be exposed at the earth's surface through the uplift and denudation of the rock layers. This explains why coalfields are often found on the flanks of highland areas, *e.g.* the various coalfields of Northern England are found on the margins of the Pennines.

## VARIETIES OF COAL

Coal is classified into various types according to the carbon content. For present purposes three main categories of coal may be distinguished:

1. *Anthracite*: a very hard, shiny, clean coal, containing over 90% of carbon, which burns with little flame and gives off little smoke but

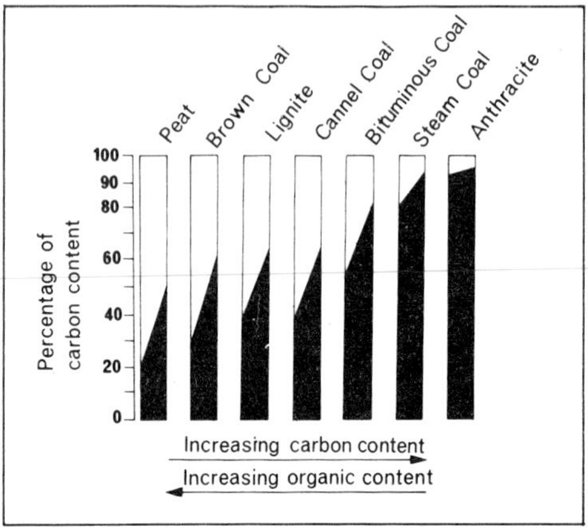


FIG. 98.—TYPES OF COAL

Another percentage bar graph showing the different kinds of coal and their respective percentages of carbon content.

which produces great heat and leaves little ash behind. Anthracite is in much demand for central heating plant, for certain types of modern kitchen cookers, and, formerly, was much used by steam-powered ships. Deposits of anthracitic coal are fairly scarce; the two most important anthracite fields in the world occur in the South Wales coalfield of the United Kingdom and the East Pennsylvania field of the United States.

2. *Bituminous Coal*: coals belonging to this group contain 70–90% of carbon. They are black, shiny, or dull, burn freely with a long, yellowish flame, and are smoky. They leave much ash behind, especially the poorer, lower-ranking types. The coals in this group are called bituminous because they all yield tar (bitumen) when heated; they also give off gas. It is for these reasons that bituminous coal is used in the making of coal gas and coke. Bituminous coals are often sub-divided into (*a*) coking coals, (*b*) gas coals, and (*c*) steam coals. Ordinary household coal is bituminous coal.

3. *Lignite or Brown Coal*: this type of coal has a low carbon content of between 45–65%. It is usually brown, but sometimes black, in colour, burns with a long, smoky flame, and leaves much ashy material behind; this is because it is of much more recent origin than anthracitic or bituminous coals and contains a higher proportion of vegetable matter. It

may be looked upon as being at the midway stage between peat and bituminous coal. Lignite is much used for steam raising, especially in the generation of thermal-electric power. It occurs in large quantities in Western and Eastern Germany and the Soviet Union where it is much used for making electricity.

FIG. 99.—COALFIELDS OF BRITAIN

The map shows the coalfields of Britain and the administrative divisions of the National Coal Board.

## COAL OUTPUT

The total annual production of anthracite, bituminous, and lignite coals in the world is around 2000 million tons. There are four major coal-mining regions—the United States, Western Europe, the Soviet Union, and China—and these account for over 90% of the output. As we shall see below, at the turn of the century Britain was overtaken as the world's leading coal producer by the United States which held this position until 1959. Now both China and the Soviet Union contest for first place. China possesses great reserves of coal, but until a few years ago she produced only about 20 million tons a year. When the Communists came into power, however, in 1947, they set out to industrialise the country and coal production has leapt up. The table below gives the output of the more important world producers.

TABLE XXXIII

*Coal Production (thousand metric tons)*

| | *1960* | *1970* | | *1960* | *1970* |
|---|---|---|---|---|---|
| U.S.A. . | 391,526 | 541,562 | South Africa. | 38,208 | 45,612 |
| U.S.S.R. . | 379,933 | 432,715 | Japan . . | 51,067 | 39,694 |
| China . | 420,000 | 360,000 (est.) | France . . | 55,965 | 37,354 |
| U.K.. . | 196,815 | 144,562 | Australia . | 22,933 | 49,547 |
| Poland . | 104,438 | 140,101 | Czechoslovakia | 26,214 | 28,053 |
| West Germany | 143,205 | 111,443 | Belgium . | 22,465 | 11,362 |
| India . . | 52,677 | 73,124 | | | |

Total world production, 1970: 2,126,000 tons.

## BRITISH COAL PRODUCTION

One hundred years ago Britain was the world's largest producer of coal and she maintained the lead until the turn of the century when she was outstripped by the United States. A high level of production continued until the end of the First World War when a rapid decline set in. The change-over in shipping from steam to oil propulsion, the rapid developments in the use of electricity, and more recently the use of natural gas have all been important contributing factors in the shrinkage of the British coal industry.

British coal production reached its peak in 1913 with a record output of 292 million tons. In 1970 output was approximately 175 million tons. Before the First World War Britain exported coal to the tune of some 80 million tons annually; today the export is negligible, some 5 or 6 million tons annually.

After the Second World War the British coal industry was nationalised and Fig. 99 shows the administrative divisions of the National Coal Board. Since 1957, when rationalisation of coal production really got under way, there has been a measure of progress in the industry. The labour force has decreased from over 700,000 to less than 280,000 and the number of mines in operation has dropped from around 800 to about 300, but the output has declined from its 1957 level by only one-quarter. Mechanisation, concentration on larger mines and more efficient underground working have marked the industry and while productivity has been increasing at an average rate of about 5% it must be recognised that this has been achieved through a substantial and increasing degree of protection.

While for a long time the industry was hampered by surplus production, since 1968 demand has been outrunning supply—an encouraging sign. Nearly three-quarters of the present production goes to supplying the demands of electricity generation.

## DISTRIBUTION OF COAL

Although some coal deposits may still remain hidden and unknown, it is probably true to say that most of the deposits have now been discovered and that even if we do not know the exact amount of the reserves in many

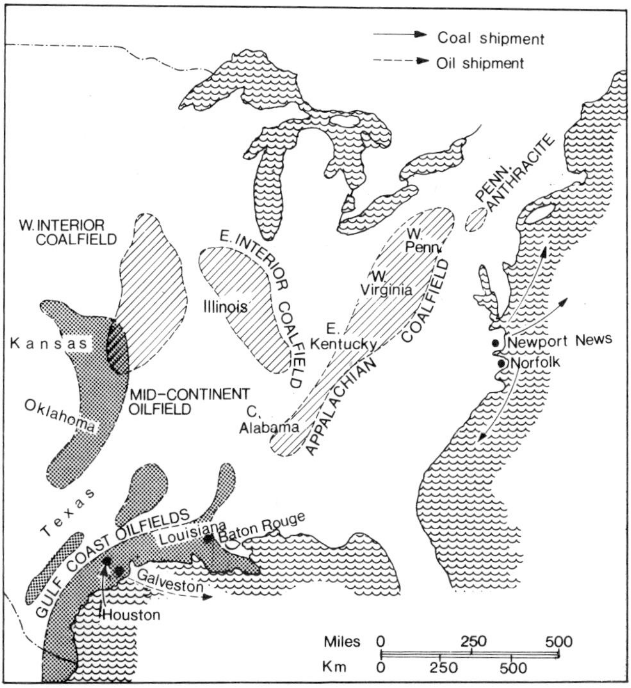


FIG. 100.—COALFIELDS OF THE UNITED STATES

cases the world picture is fairly complete. Two points with respect to the world distribution of coal need emphasising: first, the very uneven distribution of the earth's coal resources, and, second, the great concentration of the world's coal in the northern hemisphere. The latter fact is largely explained by the occurrence of very old rocks; the southern continents are mostly very ancient plateau areas of crystalline rocks in which coal does not occur. Only here and there are there sedimentary deposits in which pockets of coal are to be found.

A very large proportion of the world's coal deposits, perhaps about one-third of the reserves, are to be found in North America and especially in the United States which is bountifully endowed. The most important deposits are in the east of the United States, particularly on the western flanks of the Appalachian Mts (*see* Fig. 100). The fields of the interior, notably that in Illinois, also have an appreciable output, roughly a quarter of the total. There are small scattered pockets of coal in the Western Cordillera. Canada has significant deposits in Nova Scotia and the prairies.

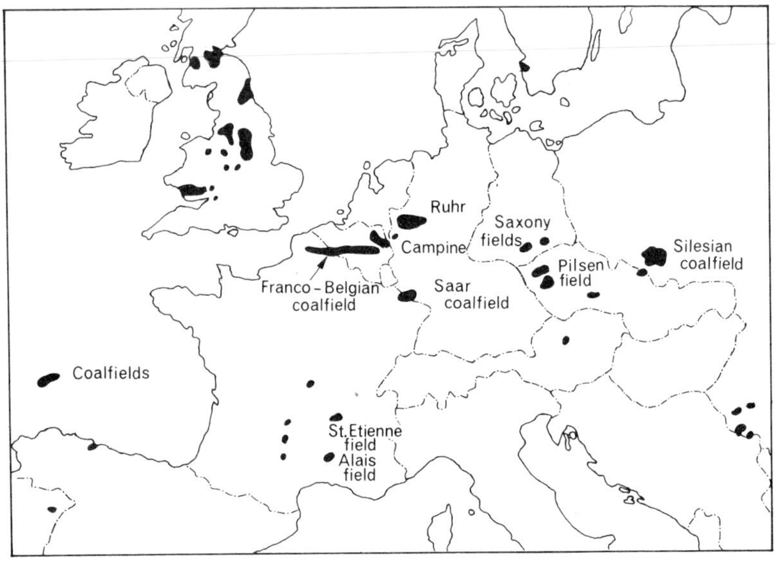


FIG. 101.—COALFIELDS OF EUROPE

In Europe a major coal trough lies along the northern flanks of the Hercynian mountain system. Discontinuous coalfields stretch from north-eastern France to Poland: the Franco-Belgian Coalfield, the Campine-Limburg Coalfield, the Ruhr Coalfield, the Silesian Coalfield. Outside this main coal belt are the important deposits in the British Isles, the small fields of Spain, the coal pockets of the Central Plateau of France, the Saar Basin, and the small fields of Czechoslovakia (Fig. 101).

The total coal resources of the Soviet Union are not yet fully known but there can be no doubting their extensiveness and richness: it seems very likely that the U.S.S.R. possesses the world's greatest reserves. The chief fields are the Donbas Coalfield in the Donets Basin in Ukraine and the Kuzbas Coalfield in the Kuznetsk Basin in central Soviet Asia. Other fields are the Tula field near Moscow, the Urals field, the Karaganda field in Turkestan, the Irkutsk field and the Far East field in the Amur valley. Recently the Russians have claimed the discovery of a vast new coalfield in the Lena Basin in northern Siberia.

The principal remaining coalfields occur in China. Extensive and rich deposits are found widely scattered throughout China proper and Manchuria; the chief deposits occur in south central Manchuria around Fushun where very thick seams can be mined by opencast methods, in the northern plateau region of Shansi and Shensi, and in the Red Basin of Szechuan province.

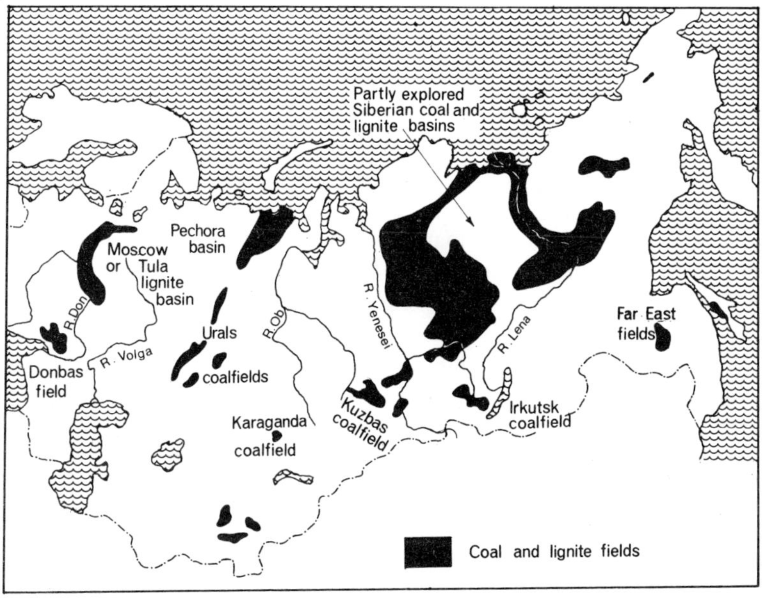


FIG. 102.—COALFIELDS OF THE SOVIET UNION

Japan has several small coalfields which yield fairly substantial quantities—practically enough to meet home demands—although she lacks good coking coal. In the Damodar valley, to the west of Calcutta, India possesses one of the few great coal deposits south of the Tropic of Cancer. Most of the output comes from the Jharria and Raniganj fields and the total production is gradually increasing.

Australia secures most of her coal from the large field centring on Sydney (Newcastle, near northern margin of field) in New South Wales but smaller fields occur in the vicinity of Ipswich in Queensland and near Collie in Western Australia. Exceptionally thick deposits of lignite occur in Victoria and these are now being worked. New Zealand has a number of small fields on the western coast of South Island but the coal is rather low-grade.

The large continent of Africa is very poorly endowed with coal. It is estimated that the reserves total less than 5% of the world total while the output is only about 2% of world production. The chief deposit occurs in South Africa on the borders of Transvaal, Natal, and the Orange Free State; the reserves here, though fairly large, are of relatively low-grade quality. In Rhodesia the Wankie Coalfield is of note. Small coalfields also occur at Enugu in Nigeria and at Colomb Béchar on the edge of the Algerian Sahara.

South America, like Africa, is deficient in coal. There is a small output of a few million tons all told from the small, scattered, low-grade deposits of southern Brazil, southern Argentina, central Chile, and Colombia.

## *OIL*

### THE ORIGIN OF OIL

Mineral oil, from which many different kinds of oil are eventually derived, is generally thought to be due to the decomposition of tiny marine creatures, minute plants and animals similar to the plankton which occurs in the oceans at the present time, which sank to the bottom of lagoons and seas and became intermingled with the muds on the lake and sea floors. Ultimately this dead organic matter became entombed and preserved in the sediments which were buried deeply, often to depths of several thousands of feet. This may have happened 100 to 200 million years ago.

How this organic matter imprisoned in the rocks came to be transformed into oil is something about which we are not very certain. It may well be that petroleum has originated in several different ways. Formerly, it was thought that great heat and pressure were necessary for the formation of oil but this old idea is tending to be discarded and modern research workers are suggesting the quick burial and quick transformation into oil without the need for great pressure exerted by the tremendous weight of overlying rock strata. This new idea does not rule out the earlier one of great pressure together with heat changing the trapped animal and vegetable matter into oil. As we have just said, several modes of origin may be involved in the formation of oil. The decomposition of the organic matter by bacteriological action, however, appears to be a necessary stage in the process.

During the time the dead organic matter was being entombed and transformed into oil, sediments continued to settle and accumulate on the lagoon or sea floors. As a result of the accumulation of the overlying loads of sediment, the deeper layers were squeezed, compacted, and hardened and turned into rock. Under such great pressure the muds in which the organic material was originally laid down became dense and impervious to fluids. In the sandstones, however, which were made up of larger mineral grains, pore spaces remained and in these tiny spaces water particles (for the sands had originally been laid down in sea water) continued to be present. The minute bubbles of oil and gas, products of

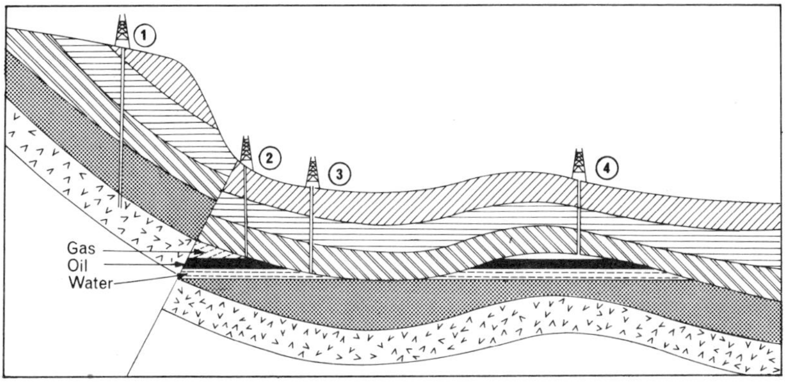


FIG. 103.—CROSS-SECTION OF AN OILFIELD

The diagram shows two kinds of "trap" in the rock strata of the earth's crust in which oil may be found. On the right, the oil is trapped in an upfold of rock or anticline; on the left, a fault in the rocks has led to a movement of the rock layers and an impervious layer has formed a barrier which dams up the oil. Well 1 hits a dry area; well 2 taps oil; well 3 draws water; and well 4 taps natural gas.

the alteration of the organic matter buried with the mud at the time it was originally laid down on the sea bed, were squeezed out of the muds as they were being compacted and moved into the adjacent sandstones whose pore spaces were able to accommodate them. Due to the gradually increasing pressure, the particles of oil and gas, as well as the sea water, migrated upwards until their movement was halted by an overlying layer of impervious mud.

The presence of an impervious "ceiling" meant that the oil and gas could now move only horizontally and so they tended to spread laterally along the sandstone stratum, floating on top of the water which permeated the pore space in the rock. As this lateral movement took place the minute oil particles gradually coalesced to form larger and larger bubbles. In due course, the oil bubbles accumulated in some dome-shaped concavity or irregularity in the strata where they lay trapped. In such "traps" they were confined by the hydraulic pressure of the brine on which they floated and it is these trapped accumulations of oil and gas that we call oil pools.

## EXPLORATION FOR OIL

Some kind of geological "trap" is necessary for oil to be present. Many kinds of trap occur in the rocks but the simplest, and a very common form of trap, is an anticline or upfold where the layers of rock have been gently arched upwards. In a trap of this kind, oil often floats on top of salt water while above the oil there is frequently natural gas. It should be noted that oil may occur without gas and that gas may be present without oil.

Oil is found fairly widely over the earth's surface but we should emphasise that it can occur only in sedimentary rocks. It is useless to attempt to search for oil in areas of igneous or metamorphic rocks for the tremendous heat and pressures involved in the formation of such rocks would destroy any oil if it happened to be present; and in the case of igneous rocks this, of course, would be quite impossible because they are not sedimentary in their origin.

Wherever there are areas of sedimentary rocks which have not been violently folded there is always the possibility of finding oil. Today the search for oil goes on far and wide and millions of pounds are spent every year by oil companies in the quest for oil. In spite of the increasing development of hydro-electric power and the growing importance of atomic energy, petroleum is still of primary importance and greater and greater quantities are needed every year.

The oil prospectors of half a century ago experienced no great difficulty in locating deposits of oil; they merely searched for areas where oil oozed out at the surface or where natural gas escaped, bored their holes and waited for the greenish-black petroleum to gush out. In the Middle East, for example, there were many pools of bitumen (petroleum) on the surface (remember Moses' ark was made watertight with bitumen) and natural gas found its way out of the ground in many places (the story of the burning bush is most likely linked with escaping gas which became ignited). Early prospecting was very much a matter of hit and miss but since the demand was still small sufficient oil could be found by such unscientific methods. When the demand for oil began to grow, especially with the development of the motor car and the rapid growth of motor transport, the search for oil had to be undertaken more seriously and scientifically.

The old-fashioned practice of "wild-catting," as the random drilling was called, has now been replaced by highly scientific methods of oil exploration. Where successful oil wells had been drilled, geologists carefully studied the rock structures and so were able to work out where oil was likely to be found. Petroleum geologists are still not able to do this with complete accuracy. For example, in 1947 a rich oilfield was discovered in the prairie province of Alberta, in Canada, but 30 years had been spent in finding it and 134 wells having a total depth of 160 miles (256 km) had been dug during the quest!

In an attempt to locate oil deposits, geologists make soundings with delicate instruments and these can give an indication of whether oil is likely to be found in a given place. Sometimes the geologists create artificial "earthquakes" by firing explosives; the "shocks" can be measured by instruments (similar to the seismographs which measure earthquake waves) and the readings help to indicate the arrangement, density, permeability, etc., of rocks which, in turn, indicate whether oil is likely to be found by boring.

## OIL DRILLING

When a possible source of oil has been located, derricks are mounted and drilling takes place. A derrick is a steel structure or tower, about 120 feet (36 m) high, which supports the machinery for the drill which bores down through the crust. The drill is tipped with a bit or cutter. As the drill bores downwards additional lengths are attached to it. The drill is in the form of a tube down the hollow centre of which mud is poured to lubricate and cool the rotating bit. As the drill penetrates downwards, the cores of rock are forced to the surface. As the drill probes progressively deeper, steel tubes of larger diameter are let down to form a lining to the bore hole.

The story of oil-drilling has often been told on the screen and on television: the optimistic beginning, the excitement and tension as the drill goes deeper and deeper, and the relief and jubilation when the treacly mineral oozes out to the surface or gushes out in a jet—or the great disappointment when, after great effort and expense, the boring proves to be useless and unrewarding. When the oil flows out of the ground of its own accord, the well is described as a "gusher"; this results from the great pressure exerted upon the oil underground. Often, however, pumps have to be installed to raise the oil to the surface.

One of the great hazards is fire. Sometimes a gusher ignites. When this happens, not only is there great danger but it is extremely difficult to extinguish the blaze. Recently, in the Sahara, one of the new wells caught fire and blazed for several weeks before it was eventually put out by using explosives, an extremely dangerous but often successful procedure.

The life of an oil well is largely unpredictable: some will continue to yield oil for many years—there are several which have continued to flow for over quarter of a century—but others may dry up in a relatively short time, sometimes within a matter of weeks.

## THE REFINING OF OIL

The oil which comes out of the ground cannot be used in its crude state; it must be refined or broken down into its so-called "fractions" which are groups of hydro-carbons (petroleum, like coal, consists of compounds of hydrogen and carbon) but which may be referred to as light, medium, and heavy oils. This splitting up of petroleum is called distillation. And the distillation of crude oil into its various fractions is the key to all petroleum refining. The distillation process is made possible

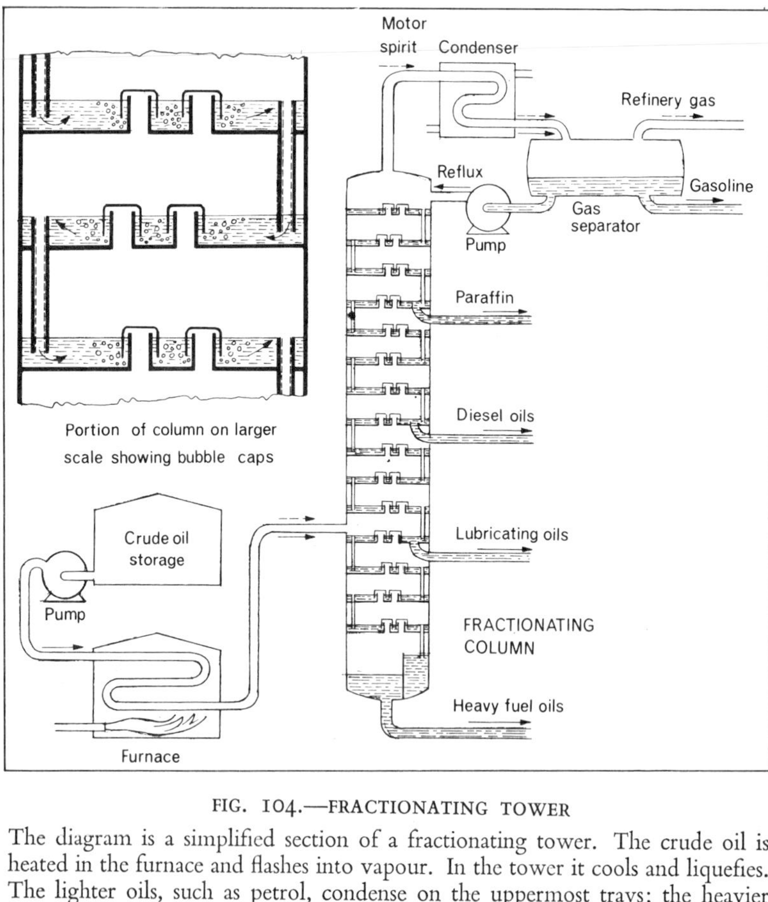


FIG. 104.—FRACTIONATING TOWER

The diagram is a simplified section of a fractionating tower. The crude oil is heated in the furnace and flashes into vapour. In the tower it cools and liquefies. The lighter oils, such as petrol, condense on the uppermost trays; the heavier fractions, such as lubricating oils, condense on the lower trays. Crude petroleum is thus split up into its different fractions by distillation.

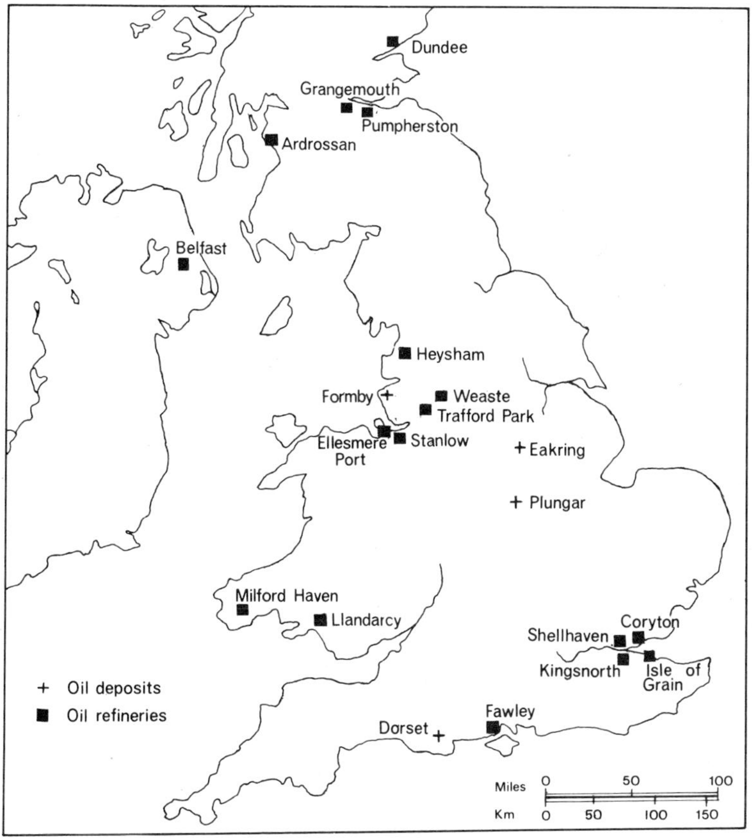


FIG. 105.—REFINERIES IN GREAT BRITAIN

Note carefully the location of the refineries: they are all near tidal water on flat ground. The refineries are linked by pipelines to the large consuming areas. The seventeen British refineries have a refining capacity of 70 million tons of crude oil per year.

by the fact that the various fractions boil and vaporise at different temperatures.

First, the crude oil is passed through pipes in a furnace which heats the oil to about 427° C (800° F). In the pipe still the crude oil is partially vaporised. It is then passed into the fractionating tower where it flashes immediately into vapour. As the vaporised oil rises in the tower it gradually cools, the constituent parts condensing out one by one. A series of horizontal trays are spaced all the way up the distillation tower and, as the various components of the vapour have different boiling ranges, one by one they condense and collect on the trays. Some of the liquid from

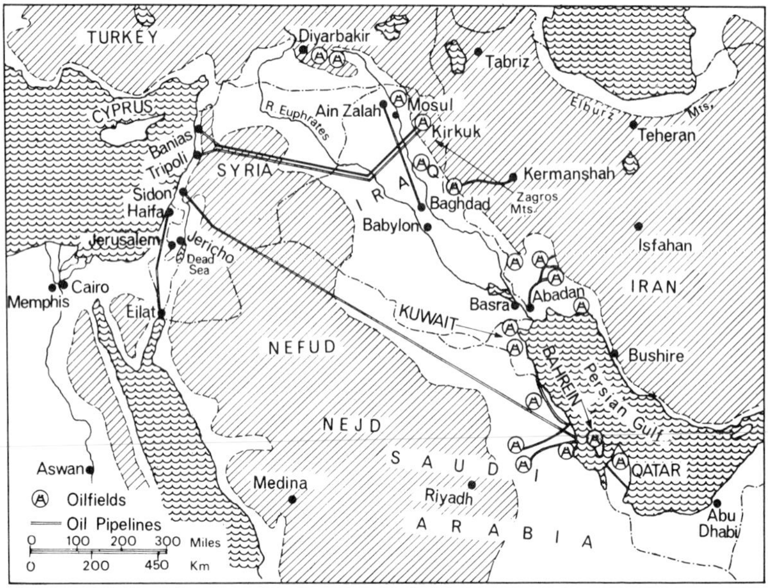


FIG. 106.—OIL IN THE MIDDLE EAST

Almost without exception the countries of the Middle East have rich deposits of petroleum. Of the world's confirmed reserves of "black gold" amounting to 42,000 million tons, the Middle East region is estimated to possess 25,000 million tons—more than half the total. The output from the Middle East countries is shown in Table XXXIV. Kuwait, Saudi Arabia, Iran, and Iraq are the four important producers but smaller quantities are also obtained from the sheikdoms of Qatar and Abu Dhabi and from the island of Bahrein in the Persian Gulf. Altogether, the Middle East accounts for some 500 million tons or about one-quarter of the total world production of 1923 million tons.

Although there are some oil refineries in the Middle East, as at Abadan in Iran, only 75 million tons are refined in the area. About three-quarters of the crude oil is despatched abroad for refining. Some of the oil is moved by pipeline to the ports on the Mediterranean coast whence it is shipped to Europe. The remainder is carried away by tankers, vessels specially designed to carry oil in bulk. The Middle East produces most of its oil for export since internal demand—a mere 40 million tons a year—amounts to a mere tenth of the output.

The development of the petroleum industry in the Middle East has been spectacular. Although oil was discovered in Iran before the First World War, the tremendous expansion of the oil industry in the Middle East has been mainly a post-Second World War development. While the Middle East's share of total world output was a mere 6% in 1938, it has now risen to about 25%. Iran nationalised her oil industry but the other backward and undeveloped producers of the Persian Gulf receive vast sums of money as royalties from their oil. The effect of this suddenly acquired wealth has brought many political, economic, and social problems.

each tray drops to the tray next below through overflow pipes. This liquid also has components having different boiling points, hence these also sort themselves out just as did the vapours on the way up.

Each component can be drawn off from the trays where it has collected as a liquid. Petrol, which has a boiling point of about 100° F (38° C), and which is the lightest of the vapours, collects at the top of the tower and the petrol vapour is led through a condenser where it is liquefied. Paraffin, which is another light fraction with a boiling point of about 149°C (300°F), also condenses on the upper trays. On successive trays below, heating oil, lubricating oil, and fuel oil condense out. The component with the highest boiling point collects at the bottom of the tower where the temperature is about 427°C (800° F), and is drawn off as a residue which is used as heavy fuel oil and from which asphalt and, sometimes, wax is obtained. Fig. 104 illustrates the distillation which takes place in a fractionating tower.

## FACTORS AFFECTING THE LOCATION OF OIL REFINERIES

In pre-war days most of the petroleum refining was undertaken on, or near to, the oilfields; this was especially true in the traditional oil-producing areas. In the past two decades or so oil-refining has tended to move from the sources of oil to the consuming areas—hence the great new refining centres at, for example. Fawley near Southampton, Rotterdam, Genoa, Marseilles and other places (*see* Fig. 105).

There are a number of factors or conditions which have influenced this change in the location of the oil-refining industry. Firstly, we see the strong influence exerted by market demand: the refining industry has tended increasingly to establish itself near to the areas of oil consumption. For instance, the largest oil-refining complexes occur in North-western Europe which, outside the United States, provides the greatest market for petroleum and its products.

Secondly, the problems of transporting crude oil have now been very largely overcome. The development of the giant tanker and the great growth in pipelining—think, for example, of the extensive pipelines in the Comecon Bloc—have made it possible to transport crude in bulk. Again, the deep-water harbours of North-western Europe have made the use of the super-tanker a practical possibility.

Thirdly, political factors have been at play. Nationalisation of the oil industry in many of the under-developed oil-producing countries has caused the Western powers to halt their investment in foreign countries. And, of course, the latter often lack both the capital to build refineries, or the expertise to run them.

## WORLD PRODUCTION AND TRADE IN PETROLEUM

Like most mineral deposits, oil appears to be irrationally distributed over the earth's surface. Since it can occur only in sedimentary rock deposits, petroleum is restricted to the sedimentary lowland areas, although it is not necessarily present in them all. There are six main areas where oil is present in large quantities:

(i) in the interior plains of North America;
(ii) in the Caribbean area of the Americas;
(iii) in the Ural–Caucasus region of the Soviet Union;
(iv) in the Tigris–Euphrates–Persian Gulf depression;
(v) in the region of South-eastern Asia;
(vi) in the Sahara region of north Africa.

The total confirmed oil reserves in the world have been estimated at some 46,000 million tons. Of this the Middle East has the lion's share with about 25,000 million tons. North America comes next with about 5000 million tons, then the Soviet Union with about 4500 million tons, then South America with just over 3000 million tons, and, finally, South-eastern Asia with approximately 1500 million tons.

Table XXXIV gives the world production and the output of the major producers. The United States continues to occupy the dominant position she has always had, although the combined output of the Middle Eastern countries now comes a reasonably close second. The Soviet Union is the third major producer, Venezuela fourth.

Oil is seldom found in the places where it is most needed. The bulk of the oil produced has to be moved to the consuming centres. This is

Table XXXIV

*Crude Petroleum (thousand metric tons)*

| | *1950* | *1960* | *1970* |
|---|---|---|---|
| U.S.A. | 271,081 | 347,121 | 475,346 |
| U.S.S.R. | 37,800 | 147,000 | 352,574 |
| Venezuela | 78,240 | 148,863 | 193,873 |
| Saudi Arabia | 26,179 | 61,500 | 176,850 |
| Iran | 32,259 | 52,050 | 191,740 |
| Kuwait | 17,291 | 81,863 | 137,398 |
| Libya | — | — | 161,708 |
| Iraq | 6427 | 47,500 | 76,448 |
| Canada | 3738 | 25,827 | 60,624 |
| Algeria | 42 | 8542 | 47,281 |
| Indonesia | 6414 | 20,592 | 42,102 |
| World total | 525,000 | 1,051,384 | 2,278,400 |

(*Source:* Statesman's Year Book.)

done in two ways: (*a*) by pipelines which frequently run for thousands of miles across deserts, forests, and mountains, and (*b*) by a tanker fleet of some 3000 vessels which make up about half of the world's mercantile tonnage.

Approximately 3 million tons of petroleum are consumed daily in the world but the consumption varies enormously between different countries. The United States heads the list of consuming countries with an annual consumption of some 500 million tons (more, note, than she actually produces herself); she is followed by the Soviet Union with 166, the United Kingdom with 85, West Germany with 80 and Canada with 65 million tons annually.

Oil has, therefore, to be moved to most of these countries and a map of oil movement would show six principal routes:

(i) from the Gulf of Mexico to the eastern seaboard of the United States;
(ii) from Venezuela to the United States;
(iii) from Venezuela to other South American countries and West Africa;
(iv) from Venezuela to Western Europe;
(v) from the Middle East to the Mediterranean and Western Europe;
(vi) from the Middle East to India, Australia, and eastern Asia.

## *NATURAL GAS*

The gas which is commonly found in association with petroleum is termed "natural gas" to distinguish it from coal gas. Methane, popularly known as marsh-gas, and ethane are the principal constituents of natural gas. Frequently tapped in oil-drilling processes, it was, formerly, allowed to escape and burn to waste. Nowadays, it is highly valued as a source of energy and its importance in post-years has grown by leaps and bounds; indeed, it now provides about 15% of the earth's total industrial power.

The United States, which possesses large resources of natural gas—the Texas gas-field is reputed to be the richest known deposit—was the pioneer in the use of gas. At the present time pipelines carry gas from the Central Plains, where most of the deposits occur, to all parts of the country; in fact, the United States has more miles of gas pipeline than she has miles of railway track! Canada also possesses very substantial supplies in Alberta and two pipelines have been constructed, one to carry gas from Princess in southern Alberta to Port Arthur and thence to Toronto and the industrial area of the Lake Peninsula, the other to pipe gas from Grand Prairie in west central Alberta over the Rockies to Vancouver.

Mexico, which is a substantial producer of oil, also has rich gas deposits and already considerable use is being made of this energy resource. Likewise, Venezuela has gas in association with her vast petroleum deposits; and she, too, is now beginning to utilise her wealth in gas. In Argentina an 800-mile (1286-km) pipeline carries gas from the Commodoro Riva-

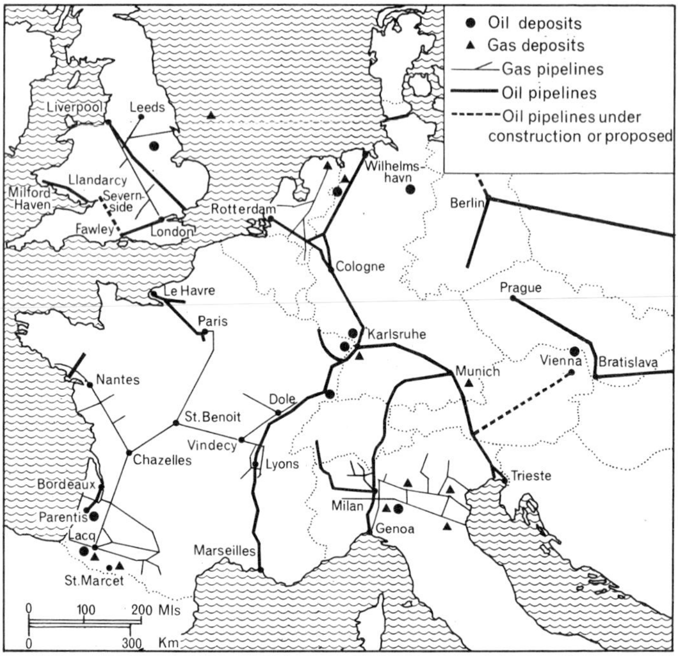


FIG. 107.—GAS AND OIL PIPELINES IN EUROPE

The gas pipelines in France carry the gas from the fields at Lacq and St Marcet in the Aquitaine–Pyrenean region to important consuming areas. The network of gas pipelines in the Plain of Lombardy, in Italy, serves the industrial towns of northern Italy. A number of gasfields occur in Lombardy. The gas pipeline linking London to Liverpool carries gas brought from Algeria. It is carried by special tanker in a liquefied form. The newly discovered gas deposits in Holland have resulted in a mesh covering that country. Without doubt pipelines will soon extend beyond the frontiers of Holland. Note the great oil pipelines running from Marseilles to Karlsruhe and from Rotterdam to Karlsruhe. The pipeline from Genoa to Milan has been carried across the Alps to Munich.

davia oilfields, in southern Patagonia, to Buenos Aires while another pipeline has been constructed from the Bermejo oilfield, in the extreme northwest, to the Plate estuary.

Use of natural gas in the Old World has tended to lag behind that in the Americas because: (i) there were major difficulties of transporting natural gas from the Middle East, and (ii) the discoveries of natural gas in Europe are of recent date. Remarkable developments, however, have taken place in the past decade. Italy was the first European country to make use of natural gas on any scale. Large gas-fields underlie the central part of the Plain of Lombardy. This gas has been tapped in a number of places and feeds an extensive gas grid which provides power for the northern industrial towns. France has located deposits in Aquitaine (at Lacq and St Marcet) and has a gas pipeline network linking Lacq, St Marcet, Paris, Lyons and Marseilles. In 1963, while prospecting for oil in northern Holland, a major gas-field was struck; preliminary estimates put the reserves on a par with the Texas gas-field. Germany and Rumania also have gas deposits. The Dutch gas grid has now been completed.

Largely as a result of the rich gas finds in the Netherlands, attention was drawn to the possibility of there being natural gas deposits under the bed of the North Sea. The first explorations were not very productive but more recently quite substantial finds have been made. By 1970 nine commercial gas fields had been located. These occur within the British, Dutch and Norwegian sectors. Both the Danes and the Germans have drilled a dozen wells each but none have proved to be commercially useful. Two pipelines carry gas from the British wells to coastal locations, *i.e.* at Easington in East Yorkshire and at Bacton in Norfolk. A third proposed line is to run to Theddlethorpe in Lincolnshire. Inland natural gas is being piped to Yorkshire, the Midlands, South Wales and the London Basin.

TABLE XXXV

*Natural Gas Production (million cubic metres)*

| | *1962* | *1970* |
|---|---|---|
| U.S.A. | 392,944 | 620,727 |
| U.S.S.R. | 73,525 | 197,945 |
| Canada | 27,058 | 64,465 |
| Mexico | 10,510 | 17,832 |
| Netherlands | — | 31,617 |
| Italy | 7150 | 13,171 |
| Venezuela | 5198 | 8990 |
| France | 186 | 6880 |

Under the sands of the Sahara Desert, gas has been found at Hassi R'Mel in Algeria and a 200-mile (320-km) pipeline carries it to Arzew on the Mediterranean coast. Some of this gas is being imported by Britain. In West Pakistan gas has been found at Sui in the north-west and it is now being piped over 300 miles (480 km) to Karachi, where it serves both industrial and domestic purposes. Finally, the Soviet Union has rich deposits in the southern part of Russia in Europe. The U.S.S.R. also boasts the world's longest natural gas pipeline which runs from the Gazli field in Uzbekistan to the Amu Darya river, crossing the Kizil Kum

Desert en route. When finally completed it will have a length of 1250 1250 miles (2000 km).

## *WATER-POWER*

Water-power has long been utilised by man to turn wheels. Who first invented the water-wheel is unknown, although some ascribe its origin to the famous Roman engineer Vitruvius. At any rate the force exerted by flowing water to move a wheel which in turn drove a simple mechanism has been employed in Britain and elsewhere for several hundred years. In earlier times running water was used in the corn-mills to grind flour, in the textile industry to power the fulling-mills, and in the iron industry to work the bellows. It was the swift streams of the West Riding which enabled that area to forge ahead of East Anglia and eventually displace it as the chief centre of the wool textile industry. Industries grew up in many areas where running water could provide motive power.

Nowadays water-power is utilised primarily as hydro-electricity. The transformation of water-power into electric power is a relatively recent development, dating only from the end of last century. The present-day large-scale transformation of water-power into electrical energy is done by directing water into pipes, called penstocks, and allowing it to fall on to turbines which drive dynamos that produce the electricity. The turbines are situated either at the foot of a waterfall or at the foot of a dam impounding water in a valley.

### FAVOURING FACTORS

The natural conditions favouring the large-scale development of hydro-electricity are: (*a*) heavy rainfall evenly distributed throughout the year; (*b*) a constant and uniform supply of water either from rivers or lakes; and (*c*) a mountainous relief to give steep slopes to provide a good "head" or fall of water (rapids and waterfalls, which are commonly found in mountainous regions, usually offer especially advantageous sites for water-power development).

While high, irregular relief and abundant precipitation are desirable (many rivers, *e.g.* the Volga, are now large sources of hydro-electric power though they flow through lowlands), a constant and regular flow of water is desirable to enable a hydro-electric station to work at full capacity. Few streams in actual fact have a uniform flow due to seasonal variations in rainfall, losses through evaporation, and freezing in winter. The occurrence of lakes, therefore, which act as natural regulators, is of great help in maintaining a constant flow. The lakes of Scandinavia, the Alps, and the Canadian Shield are of great value in this respect. Where river flow is

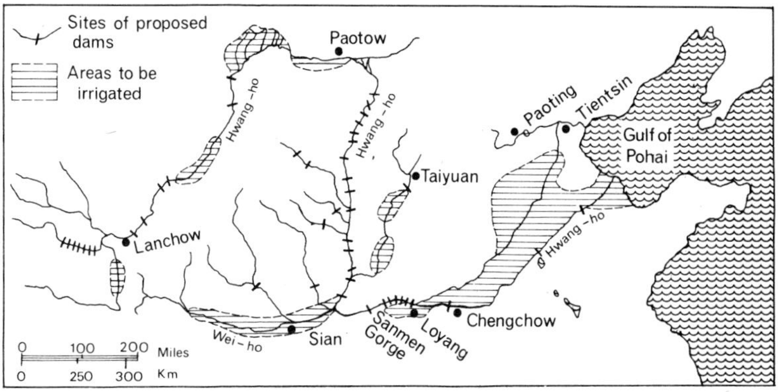


FIG. 108.—HWANG-HO: POWER AND IRRIGATION

The most gigantic multi-purpose river-control project envisaged up to press is the Chinese scheme for the taming and harnessing of the Hwang-ho. Northern China is prone to erratic rainfall and periodic drought, the Great Plain has regularly suffered extensive flooding resulting in great loss of life and property, and expanding industry will need additional power resources.

The Hwang-ho, popularly known as "China's Sorrow," because of its periodic disastrous flooding, flows in a great rectangular bend across the loess-covered plateaux of north-western China. The river has cut deeply into the plateaux surface and, for the most part, flows in a great gorge. The loess is a rich, fertile soil but it is very porous. Moreover, the north-western plateaux lie in an area of deficient rainfall. The cultivation of crops in this region is therefore largely dependent upon irrigation.

To provide irrigation water, to check flooding, and to provide power for northern China is the triple function of the Hwang-ho development scheme. The project, if ever finally completed, would result in some forty dams being built on the Hwang-ho which would have a total generating capacity of some 23 million kW; large areas in the Hwang-ho valley and in the valleys of its tributaries, the Wei-ho and Fen-ho, and on the Great Plain would be irrigated; while the age-old menace of flooding with all the misery and suffering it brings would be a thing of the past. A spectacular start was made with the construction of a 500-ft (166-m) high dam in the Sanmen Gorge, but further progress on the scheme has stopped.

The entire scheme would, of course, take decades and could not be completed before the turn of the century if, indeed, the plan was carried out to the full; for, whatever justification this grandiose scheme may have from the point of view of water supplies and flood control, it is more than doubtful whether the proposals for the lavish production of hydro-electric power are economic. Dams and plant are extremely expensive and since China has abundant coal supplies, it seems rather unrealistic to attempt to produce electric power in this way. One cannot see that this will be a cheap source of power. There can, however, be no doubt that the plan would bring great benefits to the Hwang-ho basin.

highly irregular, as in "Mediterranean" or "Savanna" lands, water must be stored in artificial lakes made by damming in order to maintain flow through the dry season.

Some of the difficulties attendant upon the shortage of water supplies or irregularity of flow have been overcome by the principle of pumped storage. Some of the generated power is used to pump water to storage reservoirs so that this stored water can be used to generate more power.

There are a number of other, largely non-physical, factors affecting hydro-electric power development: (*a*) there must be a demand for electric power, either for domestic, industrial or transport use; (*b*) the absence of other fuel supplies or the inability to produce electric power more cheaply from coal, oil, or other fuels; and (*c*) the accessibility of water-power sites; in other words, the sites must be near to consuming centres since there is a limit to the distance over which electricity can be transmitted.

## POPULAR MISCONCEPTIONS

Before we proceed to describe the potential resources and the developed water-power of the world, it will be useful to note and correct some popular misconceptions about hydro-electric power. It is true, of course, that whereas coal, oil, and other fuels are exhaustible and will one day be used up, water is inexhaustible and provides man with a perpetual source of power. It is because water is a permanent, inexhaustible resource that some of the popular misconceptions arise.

Broadly speaking, there are three chief erroneous ideas about water-power:

1. Since water is a free gift of nature, most people believe hydro-electric power is cheap. Although water is free and although the running costs of plants, once they have been built, are relatively low, it should be remembered that the initial capital outlay in building dams, creating reservoirs, installing plant, and building transmission lines, etc. is extremely heavy—for example, the new Aswan High Dam on the Nile, which has been constructed to provide hydro-electric power as well as water for irrigation, cost over £100 million.

2. Most people think that only pylons and cables are needed to carry electric power anywhere. In actual fact, the transmission of power is limited (at least at present) to a distance of about 400 miles (640 km); the Swedes, however, have been experimenting with the long-distance carriage of electricity and it is now becoming possible to convey current over much greater distances than has hitherto been feasible.

3. The importance of water-power is exaggerated. The great

dams and power plants capture people's imagination and they are inclined to think, as indeed happens in the United States, that all the electricity comes from water-power, yet in the latter country only about a quarter of the electrical energy derives from water-power, the bulk being generated by plants burning coal, oil, and natural gas. In spite of the tremendous hydro-electric power developments in various parts of the world, a mere 5% of the world's energy output is supplied by water-power.

## POTENTIAL AND PRODUCTION

Water-power resources, like most of nature's gifts, are very irregularly distributed throughout the world. The following table gives the estimated water-power potential for each of the continents:

TABLE XXXVI

*Hydro-electric power potential*

| *Continent* | *Potential water-power in millions of horsepower/kW* | |
|---|---|---|
| Africa | 272 | (202) |
| Asia | 151 | (112) |
| U.S.A. | 87 | (64) |
| Europe | 69 | (51) |
| South America | 55 | (41) |
| Australia–Oceania | 23 | (17) |
| | 657 | 487 |

It will be seen that Africa leads quite easily with over 40% of the world's potential water-power. Africa's resources lie mainly in the tropical belt where many great rivers carry a great volume of water. The Congo and Zambesi Rivers are especially rich. So far, however, there has been little development of Africa's resources partly due to the lack of demand for electricity and partly due to the great cost of schemes, which are beyond the pocket of most African countries. The growing exploitation of the rich mineral resources of the continent, however, is affecting hydro-electric development; for example, the Kariba project on the Zambesi was undertaken partly to supply the copper belt of Zambia with much needed power. The Volta scheme in Ghana has been associated with the development of the bauxite resources of that country.

Next to Africa, Asia has the greatest resources of potential power; these are greatest in the mountain zone where most of the continent's great rivers rise. Apart from Japan, where the water-power resources have long been developed and can be related to the country's mountainous terrain, heavy precipitation, numerous swift-flowing streams, and shortage of coal, there has been little hydro-electric power development until very recent years. Although development, in relation to the need, is still limited, several important projects have been, or are being, built in India and Pakistan, in China, and in Soviet Asia.

The United States and Canada early began to develop their rich resources of water-power and it is in these countries that the greatest progress has been made. Almost a half of the potential water-power supplies have been harnessed. The St Lawrence region and the Cordilleran region are the most important areas of development; this is partly due to favourable natural conditions—mountainous relief, abundant rainfall, swift rivers, numerous lakes, many waterfalls—and partly to the fact that these areas are remote from coal and oil. In the United States hydro-electrical power development has been closely linked with total river basin management; in other words, a river basin is developed as a whole and such matters as the generation of hydro-electric power, drainage, and flood control, navigation, irrigation, reforestation, and soil-conservation are carefully integrated. The well-known Tennessee Valley Project was the pioneer scheme for river basin management and its success has led to other TVAs not only in the United States but in other parts of the world.

Although great potential power resources occur in South America, particularly in the Andes and along the edges of the Brazilian and Guiana plateaus, there has been relatively little development until very recent times. Brazil, which possesses the lion's share of the South American potential, has gone furthest in developing her resources and she now has several great plants. The power demands of the great multi-million cities of Rio de Janeiro and São Paulo have greatly stimulated hydro-electric development, *e.g.* the Furnas and Tres Marias schemes.

Europe, like North and South America, is moderately well endowed with water-power resources. The hydro-electricity resources tend to be most highly developed in those countries where coal and oil resources are lacking, *e.g.* Norway, Sweden, Finland, Switzerland, and Italy. But power stations have been built in most of the mountain and highland areas of Europe while more recently attention has been turned to the possibilities offered by some of the major river valleys such as the Rhône, the Volga and the Rhine. In Britain, hydro-electricity plays only a very minor role; though there are numerous schemes in the Highlands of Scotland, the total amount of electricity generated is small. Britain's abundance of coal has militated against water-power development; at the same time there are no large catchment areas to feed great rivers that can service very big power stations. In sum, the countries of Europe which have developed hydro-electric power to the fullest extent are those which are mountainous, have appreciable industrial development, and are lacking in alternative sources of power.

Finally, at the bottom of the list comes Australia which has almost negligible resources of water-power. Only in the south-eastern portion of the continent where there are highlands and there is a reasonably high and reliable rainfall do possibilities exist. It is here that the Snowy Mountain Scheme is located. The islands of Tasmania, New Guinea, and New Zealand, with their high relief and heavy rainfall, have useful water-power resources and already considerable development has taken place, especially in North Island, New Zealand.

## THE USES OF HYDRO-ELECTRIC POWER

In many areas and countries the energy derived from hydro-electric power is passed into the transmission lines serving railways, factories, and domestic needs. This is what happens, for example, in Britain: the bulk of the power from the Scottish Highlands schemes is fed into the national grid. In some areas, however, hydro-electric power is produced to serve special industrial needs and three industries in particular are closely associated with, and dependent upon, hydro-electricity; these are: (*a*) the timber and pulp industries; (*b*) the electro-chemical industry; (*c*) the electro-metallurgical industry.

In Canada, Scandinavia, Finland, and parts of the Soviet Union, the exploitation of the coniferous forest belt has been made possible only by the development of the hydro-electric power resources, for it is the sole power resource available. The pulp and paper industries require large amounts of cheap power, and in the St Lawrence region, Scandinavia, and Finland, hydro-electric power has been especially developed to cater for the needs of these industries.

Hydro-electric power is used in the electrolytic refining of certain metals, especially aluminium. Bauxite, the ore of aluminium, is first reduced to alumina—this may be done with the help of coal—but the refining can only be done economically in an electric furnace since very high temperatures are required. Abundant supplies of cheap power are necessary. Hence the extraction of aluminium is most profitably carried on where power and bauxite are found in the same locality, as in the French Alps, or where electricity is so cheap that bauxite can be imported, as in Canada (*e.g.* Kitimat), and Norway (*e.g.* Rjukan). Again, in the United States, power from the Rio Colorado is used for the extraction of magnesium from dolomite.

In some countries, notably Norway, Sweden, and Canada, huge electro-chemical industries have been developed (*e.g.* the making of nitrates, fertilisers, calcium carbide, etc.) and these need abundant supplies of cheap electric power.

## *NUCLEAR ENERGY*

Hydro-electric power, unlike coal, petroleum, and natural gas, all three of which exist in finite quantities and will one day be used up, is an

FIG. 109.—NUCLEAR POWER IN BRITAIN

The nuclear power stations are located (*a*) near to river, lake, or sea water since they require vast quantities of water for cooling purposes—up to 20 million gallons an hour (90 million litres); (*b*) on firm ground which is necessary to bear the great weight of the plant, which may be up to $2\frac{1}{2}$ tons per square foot; and (*c*) away from centres of dense population, so far as is permissible, because of the possible danger from radio-activity.

inexhaustible source of energy. But hydro-electric power, by itself, could never meet all the demands for power, hence the scientists' search for some additional source of energy. Since the hydro-carbons are a wasting asset and since the world demand for power continues to increase, the advent of so-called atomic power appears to have come at a propitious time.

What is known as nuclear fission (the "splitting" of the atom) results in the release of tremendous amounts of energy—this is why atomic bombs are so horribly destructive. In an atomic explosion the energy release is uncontrolled. The idea behind nuclear energy as a source of power was the control of this released energy. In other words, the aim was to produce a steady flow of energy which could be harnessed for use. This has now been achieved through the medium of the nuclear reactor or "atomic pile." The first experimental commercial reactor was designed by British scientists and established at Calder Hall, Cumberland, in 1957. Energy produced by this reactor was fed into the National Electric Grid System. Britain now has several nuclear power stations in operation (*see* Fig. 109). Several other countries have followed Britain's lead, especially the more highly-developed industrial countries such as the United States, the Soviet Union, Canada, France, West Germany, Italy, and Japan, but some of the more backward, under-developed countries, such as India, are now following suit.

The "fuel" used in nuclear reactors is provided by radio-active metals, the most important of which are uranium and thorium. Only very small quantities of these are required; 1 oz (28 grammes) of uranium, for example, has the energy equivalent of more than 100 tons of coal. Although uranium ores are not especially common, the small amount of uranium used means there are ample supplies for a very long time to come. Most of the uranium is refined from two of its ores, pitchblende and carnotite. The ores, which are mined in a number of places, *e.g.* in northern Canada (Port Radium, Uranium City), in western U.S.A. (Central City, Marysville), in Australia (Rum Jungle, Mary Kathleen), Congo (Chinkolobwe), in South Africa (the Witwatersrand, where the ore is recovered from gold mine "tailings") and in Czechoslovakia and East Germany, are concentrated into uranium oxide from which uranium metal is finally produced.

The importance of nuclear energy lies, firstly, in the fact that countries lacking sources of conventional fuels can now develop industry and, secondly, that it will lead to a greater dispersion of industry. Such developments are not likely to come quickly, however, for a number of reasons. Nuclear power stations are very costly to construct, so this economic fact is bound to militate against their speedy adoption. Furthermore, a high degree of scientific and technological know-how is required to build and run these nuclear reactors, hence for some time they are likely to be confined to the technically advanced countries.

Finally, let us note that while tremendous developments may stem from this new source of energy, *e.g.* in the form of nuclear-powered transport and industrialisation, it will be a long, long time before the exploitation of "fossil fuels" is made unnecessary. In the foreseeable future nuclear energy is more likely to augment existing sources of energy than to displace and supersede them.

## EXERCISES

1. Give an outline of the power resources of the U.S.A., pointing out which industrial areas they serve and how they are transported to these areas.
2. "Coal is a magnet of industry." Discuss.
3. For either coal or oil or hydro-electric power (*a*) describe the physical conditions of occurrence, and (*b*) the particular value to industry of the form of power chosen. (*Institute of Bankers.*)
4. Write a geographical account of Britain's oil industry referring especially to (*a*) indigenous sources of oil, (*b*) the oil refineries, and (*c*) the uses of oil.
5. Describe the importance of hydro-electric power to *either* Norway *or* Switzerland. Name the different industries which are based upon water power. (*Northern Counties Technical Examinations Council.*)
6. On a map of the world locate precisely the main centres of oil production. What are the relative merits of oil refinery sites? Why are British oil refineries all sited on the coast?
7. What geographical factors favour the development of hydro-electric power? Illustrate your answer with reference to specific examples.
8. Describe very briefly the origin of petroleum. Name the principal producing areas in the world. Indicate the chief methods by which oil is transported and the chief movements of oil. (*Northern Counties Technical Examinations Council.*)
9. What are the main factors which appear to control the location of nuclear power stations in the British Isles? Briefly show the operation of these factors in an example known to you.
10. "World power demand is doubling every ten years." Comment on this statement and point out the changes that have taken place in the last sixty years in the different sources of energy used.
11. Two great modern schemes involving the harnessing of water power are the Aswan High Dam on the Nile and the Kariba Dam on the Zambesi. In each case give a brief account of the project and describe the uses to which the power will be put.
12. Electricity can be generated using either water power or coal. Locate *two* areas (excluding the British Isles) to illustrate the generating of electricity by (*a*) water, (*b*) coal, and describe the geographical conditions which encourage the production of electricity in each area.

# Chapter XIX

# MANUFACTURING INDUSTRY

## *THE MEANING OF MANUFACTURE*

WE have already referred to farming, fishing, forestry, and mining as the great primary industries; but here we are concerned with manufacturing industry. The term manufacture, strictly speaking, means to make by hand, but it has come to have a wider connotation. Simply put, manufacture may be defined as the processing and altering of materials to make new products to serve new ends. The materials used in manufacturing may be in their natural condition, *i.e.* the so-called "raw materials," such as ore, timber, wool, or they may be partly processed materials, such as steel, sawn timber, and leather, which are used to make other things. It will be clear that the finished product of one industry may be the raw material of another. For instance, timber is used to make wood-pulp, wood-pulp is used to make paper, and paper is the basis of the printing industry; in like manner, wool is the raw material of the woollen textile industry and textiles the basis of the tailoring industry.

A principle of industrial manufacture is that the more processing and altering a material is subjected to, the more valuable it becomes; in fact, one way of defining manufacture might well be "the processing of raw material to enhance its value." A good example of this is provided by a watch: the amount of raw material used in its making is small and costs very little, but the finished product is expensive. The difference in cost between the "raw material" and the "finished product" is "the reward due to management and labour, the payment for rent and maintenance of premises, local and national taxes, and costs of transport, advertisement, and marketing."*

It is the task of the producer to reduce manufacturing costs to as low a level as possible consistent with satisfactory workmanship. This can be done in a variety of ways but chiefly by: (*a*) mass production: usually the greater the number of articles produced, the smaller is the unit cost of production; (*b*) mechanisation: the machine can increase the numbers and speed of output of articles; (*c*) the reduction of labour costs: the cost of labour is normally the largest single production cost; (*d*) selecting an economic factory site: a site which will give economies with respect to raw materials, fuel, transport costs, etc.

* POUNDS, N. J. G., *An Introduction to Economic Geography*, p. 143.

## *TYPES OF MANUFACTURE*

Manufacturing falls broadly into three types or classes: craft manufacture, domestic manufacture, and factory industry. The first two share many features and there is a certain amount of overlapping between them. Factory industry is distinct from the other two in many important respects.

The craftsman was the earliest industrial worker. He appears as soon as civilisation appears, that is, as soon as man commenced to live in organised communities, for it was this that made specialisation of labour possible. Potters, wood-workers, metal-smiths, boat-builders, etc., appeared early on the scene; in other words, individuals earned their livelihoods by following specific crafts. Even in Britain, where the craftsman, as an agent of manufacture, has long since been displaced, a few still linger on, *e.g.* cabinet-makers, silver-smiths, textile workers, together with a few quaint survivals like thatchers, basket-makers, brush-makers, coracle-makers, etc. The craftsman is still a common feature of the pre-industrialised and less industrialised societies as, for example, in Egypt, India, and the Central American Republics. The chief characteristics of craft industry are that the craftsman finds or buys his raw materials, fashions his articles, and sells them himself.

Domestic, sometimes called cottage, industry may be said to have grown up out of the difficulty which the craftsman had in disposing of his goods. In domestic industry the merchant comes into the picture. He delivers the raw materials to the worker and then collects the finished products at regular intervals. Thus the merchant, in effect, controls the output of the worker but not the premises in which he works. As with craft industry, production is by hand and is carried on domestically. The wool and silk industries in Britain were originally organised on this basis as is much of the present-day silk industry in the Lyons area in France. Another good example of the domestic industry is provided by the needleworkers of Puerto Rico; New York textile firms distribute cloth among the islanders which is decorated by the womenfolk and then collected and despatched to New York for sale in the shops.

Factory industry differs markedly from craft and cottage industries. Although factories vary in size, employing between as little as a dozen or as many as several thousand, the distinguishing feature is the employment of groups of people on an organised basis using powered machinery. Mechanisation is the distinguishing feature of factory industry. Today, with the introduction of automation, where machines are used to control machines, factory industry has advanced a stage further. However, at this point, before we proceed to study modern factory industry, it will be helpful to look at the causes of, and the effects of, the Industrial Revolution.

## *THE INDUSTRIAL REVOLUTION*

We are all familiar with the term the Industrial Revolution but what, exactly, does it mean, where did it begin, and what were its causes?

The Industrial Revolution may be said to have had its take-off in the 'seventies and 'eighties of the eighteenth century. It should not be thought that it occurred with dramatic suddenness; on the contrary, it had been developing for at least a century beforehand, but towards the end of the eighteenth century a combination of circumstances suddenly, as it were, let it loose and rapid developments followed.

The Industrial Revolution brought a change in industrial methods and organisation: a change from handwork to machine work, from the domestic system to the factory system, and from wide dispersion to the centralised location of industry in particular areas. A new relationship between men, machines, and resources was the central and distinguishing feature of the Industrial Revolution. The traditional technology, which had been based upon wood, water, and wind, was supplanted by a new technology based upon coal, iron, and steam.

The Industrial Revolution began in Great Britain and Britain led the world in industrialisation until the beginning of the present century. Britain is still a major industrial nation, ranking third in the world after the United States and the Soviet Union, but relatively she is much less important industrially than formerly. It is impossible to interpret the Industrial Revolution in terms of a single factor but one very important factor was that, with the growth in the numbers of people, the domestic system was incapable of meeting the growing demand for goods. Britain, too, unlike her continental neighbours, had enjoyed a long period of peace and internal development; moreover, she had acquired colonies which supplied her with raw materals and provided her with expanding markets. Thus political and economic conditions conspired to foster the early industrial growth. Finally, Britain had abundant supplies of coal and iron ore, the two prerequisites for this industrial development, while a whole series of technical inventions in the eighteenth century made possible shaft coalmining, the large-scale production of iron and steel, and the mechanisation of textile manufacture.

## *THE LOCATION OF INDUSTRY*

Have you ever thought why pencils are made at Keswick, liquorice sweets at Pontefract, biscuits at Reading, marmalade at Dundee, motor cars at Oxford, chemicals at Billingham, or iron and steel at Middlesbrough? Although it is true some industries have grown up by chance, as it were, very often there is a reason (or reasons) why an industry happens to occur in a particular place. This is more especially true of the

FIG. 110.—INDUSTRY IN EUROPE

The dotted line encloses the major zone of industrial activity in Europe: it is sometimes called the axial belt of industry. Compare the distribution of the industrial areas with Fig. 101 which shows the coalfields of Europe. Note the two important outlying industrial areas of Central Sweden and the Plain of Lombardy.

older, long-established industries. Many factors—economic, historical, human, political, and geographical—are involved in the location of industry. In the past there has been a tendency to emphasise, perhaps sometimes to over-emphasise, the role played by purely geographical factors in locating industrial development. While there is no doubt that geographical factors have sometimes played a major part in industrial location we must be careful not to interpret the growth and siting of industries always and only in terms of geographical influences.

However, first, let us look at the geographical factors which are commonly cited to explain the location of individual industries; these are: (*a*) raw materials, (*b*) power resources, (*c*) labour supply, (*d*) markets, and (*e*) transport facilities.

## RAW MATERIALS

Among the various factors influencing the location of an industry, the occurrence of necessary raw materials is one of the most obvious. The old-established pencil industry of Keswick owed its development to local supplies of graphite and timber, the two chief raw materials required in the making of pencils. The leather and footwear industries of such Midlands towns as Northampton, Kettering, and Rushden, owed their origin to the availability of hides in a cattle-rearing area. The sweetmeat industry of Pontefract owes something to the former cultivation of the liquorice plant (originally by the monks) in the neighbourhood. Industries which are based upon a few heavy, bulky raw materials, *e.g.* iron-smelting, brick-making, cement-manufacture, are frequently located near the supply of raw materials. But even in cases of this kind we should note that indirectly the factor which is often involved is the deterrent effect of transport costs rather than the positive attraction of the raw material. Where raw materials are of the lighter kind, less bulky, and more valuable, the manufactured products are better able to stand transport costs. Finally, such light industries as radio engineering, the making of pharmaceuticals, tailoring and dress-making, which use varied, valuable, and relatively small amounts of raw material, are much more mobile and are readily attracted to large towns which at once afford a ready supply of labour and a big consumer market.

## POWER RESOURCES

All modern manufacturing industries are dependent upon some source of power. The particular nature of an industry may affect the part played by power as a factor in the location of industry; for example, the heavy iron and steel industries which need large quantities of coking coal, are frequently tied to the coalfields, although the growing need to import ore in large amounts is causing many of the newer plants to move to coastal locations; again, the need for abundant supplies of cheap electric power required by the electro-chemical and electro-metallurgical industries explains why such industries are located in rather inaccessible mountainous areas such as the French Alps, the Norwegian Plateau, and the Highlands of Scotland. Another industry closely associated with hydro-electric power is the wood-pulp-newsprint industry which requires large amounts of power. The development of electricity and its ease of transmission over long distances has made possible a greater dispersion of industry, especially of the lighter type. The growth of manufacturing industries in Essex and the home counties generally has been due primarily to two factors, the development of the electric grid and of motor transport.

## LABOUR SUPPLIES

A labour supply is important in two respects: quantitatively and qualitatively. Modern industry requires large numbers of workers notwithstanding increasing mechanisation and even automation. In many industries, especially where machines do most of the work, little skill is required. Repetitive jobs based upon a machine, *e.g.* packaging and assembly work, demand no special skills. Industries of this nature are often attracted to large urban centres. In such cases, the availability of a labour supply is often the only predisposing factor. Formerly, especially in the days of domestic industry, skill in a particular craft or trade was important. The so-called traditional skill in an industry was often an operative factor, *e.g.* in the woollen textile industry where, in fact, it still counts for much. There are cases where the tradition of an industry and an intimate knowledge of an industry—for there is no such thing as traditional skill: all skills have to be learnt—have accounted for the continuance of an industry or the establishment of an associated industry in a locality where the original advantages it possessed have long since disappeared.

## MARKETS

Clearly there is little point in producing goods unless there is a market for them. A market is dependent not merely on numbers but also, and this is the more important of the two, on the purchasing power of the people. China, for example, with a population of some 750 million, ought to provide an enormous market, but because of the low living standards the country's market value is very small. However, we should note that its *potential* value as a market is tremendous. One of the reasons why London has become a great light-industrial centre is simply because there is a large market to hand—some 10 million people with the capacity and the willingness to spend freely. Earl B. Shaw cites an interesting example of the importance of the market with respect to the manufacture of bread and confectionery.* Bread, cakes, and pies, which are known as soft bakery goods, need to be consumed in fresh condition hence bakeries lie close to the consuming areas. On the other hand, the hard bakery factories, producing such commodities as biscuits, which will keep for long periods without deterioration, are much less dependent upon immediate markets; indeed, some hard bakery products may enter into the foreign export trade.

* *World Economic Geography*, John Wiley & Sons Ltd., 1955.

## TRANSPORT

Facilities for adequate transport, whether by land or water, are necessary for industrial development; obviously they are required for the assembly of raw materials and for the transport of the finished products. But it is not easy to estimate how much a particular industry owes to the original transport facilities for there is a distinct tendency for industrial development to foster and further improve communications facilities. For example, the growth of the West Riding woollen industry greatly stimulated the development of communications, particularly of the canals, but at the same time the improved transport facilities greatly helped industrial growth; this is particularly well illustrated in the case of Brad-

[*Courtesy: Japanese Embassy Library.*

FIG. 111.—INDUSTRY IN JAPAN

This industrial scene, with its factories, chimneys, and smoke, is very like one of Britain's great industrial areas. The view is part of Yawata, in northern Kyushu, one of Japan's greatest heavy-industrial centres.

ford where the construction of a canal, linking the town to the River Aire, led to an immediate and rapid growth of the city and its woollen trade. Another example is provided by the Manchester Ship Canal which has led to numerous industries growing up along its banks. Railway development can show similar responses; for instance, most of the industrial development in Soviet Siberia clings to, and is largely dependent upon, the great communications line of the Trans-Siberian Railway. Partly because of transport problems and costs many articles are often made in the consuming areas, *e.g.* agricultural machinery in farming areas, ships' engines and boilers in the ship-building areas. While the factor of transport facilities is seldom a dominant one in industrial location and, in fact, has become of less significance with the development of modern communications, the detailed siting of a factory or plant may be determined by the transport facilities available.

## OTHER GEOGRAPHICAL FACTORS

Sometimes, in the past, a specific geographical factor has determined the location of an industry. Oft-quoted examples are the Lancashire cotton textile industry in relation to the humid atmosphere which assisted spinning, or the soft water of the Pennine streams which was indispensable for the washing of wool and woollen fabrics, or the peculiar character of the water supply derived from the Keuper rocks which greatly assisted the brewing industry of Burton-on-Trent. These, without doubt, were important factors in the early development of these industries but nowadays they are merely of historical significance since it is an easy matter and relatively cheap to humidify the air and to soften water.

Because of modern scientific developments, the ubiquity of communications, and the ready availability of electric power, not to mention the mobility of labour these days and the possibility of using alternative raw materials, one is tempted to think that geographical factors are no longer of much importance. The significance of many, it is true, has been much reduced, sometimes completely nullified; yet, in some respects, their importance is still as great as ever and man cannot afford to ignore them. Perhaps one of the best examples is provided by the nuclear power industry; the generation of power in nuclear stations demands vast quantities of water for cooling purposes, hence the location of nuclear power stations is tied very closely indeed to sites affording abundant water supplies.

## NON-GEOGRAPHICAL FACTORS

We have been at pains to emphasise the purely geographical factors which have influenced industrial location but we should remember that there are other factors of an historical, economic, political and human character which have exerted an influence. The availability of capital for investment in industrial enterprises is clearly an economic factor of tremendous importance. It is largely because capital is available in capital cities such as London, Paris, and New York that these cities have become major centres of industry. At the present time much of the industrial development that is taking place in such countries as India and Brazil is only possible because of investment by foreign countries. Again, governmental

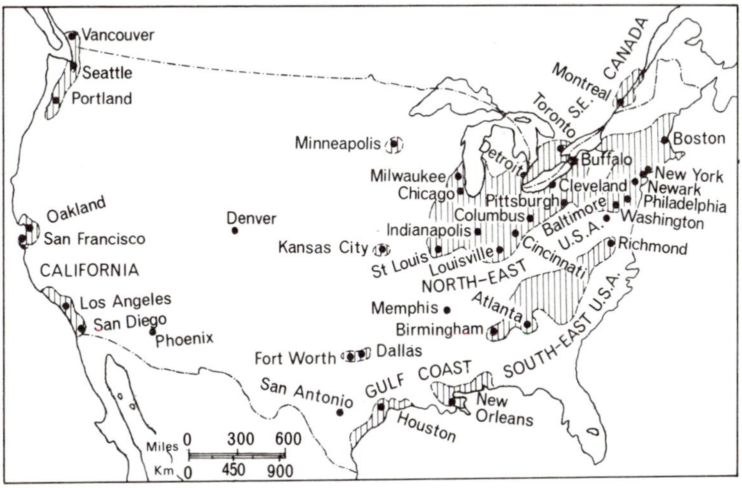


FIG. 112.—INDUSTRIAL REGIONS OF NORTH AMERICA

interference, direction, or control may be responsible for industrial location and development. An obvious illustration of this is provided by the Trading Estates which have been set up in the United Kingdom. Political considerations such as the desire to attain national self-sufficiency or at least a more nicely balanced economy may lead to the establishment of industries for which there are no obvious natural advantages. Purely historical factors, too, may play a part; for example, the settlement of refugees in a country may lead to the introduction of a new industry and the mere fact that a particular area has a long tradition of industrial activity may result in newer industries being attracted to that area.

Finally, human factors—maybe purely human whims—have been known to be instrumental in the location of industries. As the *P.E.P. Report on the Location of Industry, 1939,* suggested, the siting of a factory may be related to the managing director's addiction to golf. It is sometimes said that Lord Nuffield chose Cowley, near Oxford, as the site for his great motor works simply because he was born in Oxford, although it is equally likely that he chose this location because of the great surplus of rural labour which was available and which he foresaw would be needed. But the classic example of the human factor may be said to be the great shirt and underwear industry of Albany in New York State, U.S.A., which owes its origin to the simple fact that a local parson invented the first shirt to have a detachable collar!

## *FEATURES OF MODERN INDUSTRY*

Let us conclude this general consideration of manufacturing industry by emphasising some of the more important features of modern industry.

First, modern industry has spread to most parts of the world. Less than a century ago practically all the factory industry in the world was concentrated in Western Europe and most of this was in Britain. Now, many countries, formerly engaged almost exclusively in the production of primary products, *e.g.* Argentina, Mexico, South Africa, Australia, and New Zealand, possess manufacturing industries, industries which sometimes are dependent upon imported raw materials and fuels. The reasons for this industrial development are mainly twofold: first, during the two World Wars such countries were often cut off from supplies of manufactured goods and this meant either they had themselves to make the manufactured articles they required or do without them; and, secondly, many countries see in industrialisation a means of making themselves more self-supporting and, also, of absorbing their surplus populations and raising standards of living. It should be noted that many of the so-called under-developed countries are developing industry and looking to industry to help them solve some of their economic and social problems.

The second feature to emphasise is the tremendous expansion of the secondary, light, and consumer goods industries. Although some countries, *e.g.* the U.S.S.R. and China, give the basic or heavy industries priority, most countries beginning an industrialisation programme develop the lighter type of industries, *e.g.* textiles, food processing, first. In the countries which have well-developed industrial manufacture and high living standards the demand for a wider range of products and for more sophisticated products grows year by year; also, luxury, as distinct from utilitarian, goods assume increased importance. The variety, output, and value of the secondary industries now reach staggering proportions; among these industries are electrical and radio goods, furniture, pharmaceuticals, pottery, glassware, clothes, footwear, food-packing, confectionery, beverages, tobacco, printing, and publishing.

The third factor is the vast size of modern industrial plants—think of the Ford motorworks at Dagenham or the giant steelworks at Margam—and modern methods of mass production. Large, integrated plants are more economic to run than smaller ones, for production can be streamlined, large-scale mechanisations can be applied, standardisation can be adopted, by-products can be used, and big concerns can employ their own research and training staffs. The logical development of such major-scale mass production is automation. Automation may be defined as the use of machines to control machines; in other words, electronic devices supersede man in the supervision and control of machines. In automation, the bulk of the manufacture is done by automatic processes, and the need of human intervention and control is reduced to an absolute minimum. The future will see an increasing application of automotive techniques but how quickly and to what extent, as well as with what results, it is impossible to forecast. All that one can say is: automation is bound to have far-reaching effects upon industrial production, organisation, and employment.

## *THE INDUSTRIALISATION OF BACKWARD COUNTRIES*

Most of the backward or so-called "under-developed" countries of the world are beginning to industrialise themselves. A number of reasons help to explain this development: (i) in the twentieth century, industry is a primary determinant of national power and no country can be powerful or politically secure without a well-developed industry; (ii) the simple exchange of primary products for foreign manufactured articles savours too much of the "colonial" stage of development and psychological motives of this kind have helped to promote industrialisation; (iii) industrialisation offers one avenue of absorbing excess population, a problem which faces many of the under-developed countries; (iv) industrialisation is seen to be a means of raising low living standards, an acute problem in all of the under-developed countries; and (v) industrialism is a means of diversifying grossly unbalanced economies and assisting national aims of greater self-sufficiency.

It is questionable whether industrialisation is always to the good. Many of the "have-not" countries have tended to regard industrialisation as an essential and unquestionable means of solving their problems of under-development and rectifying their economic ills. For many countries, large-scale industrialisation cannot be regarded as a panacea for their under-development nor for their over-population problems. Even in those cases where a measure of industrialisation is feasible, it cannot provide a remedy unless there is a firm foundation on which to build up an industrial superstructure. Successful industrialisation implies not only the possession of adequate power, mineral, and raw material resources but the acceptance of new values, an interested and mobile society, the availability of capital, industrial expertise, technological know-how, and an educated labour force. Without these, which usually necessitate radical reforms in the economic, social, and institutional structures of a country, industrialisation may well create more problems than it solves.

All this, of course, is not to deny that most of the under-developed countries would benefit from some measure of industrialisation. Indeed, it is desirable that some industrialisation should take place. It would, in fact, be economic good sense to assist the under-developed countries in achieving this goal. Such help, granted, may result in the introduction of new trade rivals in an already keenly competitive world, but it would, at the same time lead to the creation of new markets through the demands by the new wage-earners for consumer goods. Thus Western standards of living would not necessarily suffer from the increasing industrialisation of the under-privileged countries. The manufacturing monopoly of a relatively few highly industrialised countries would naturally come to be broken but such industrially advanced countries are constantly developing new techniques of production and new products so that their continued superiority would be maintained and their present high standards of living ensured.

Summing up, it would be desirable for the under-developed countries to adopt a measure of industrialisation where this is a feasible proposition. Industrialisation would help to promote better standards of living; these, in turn, would affect social aspirations which, as past experience has shown, help to check population growth. But these countries need the help, experience, and financial assistance of the West to guide them in their efforts to industrialise, to promote their economic development, and to raise their living standards.

## EXERCISES

1. What are the main factors which influence the location of manufacturing industries? Select examples from old and new industries to show the relative importance of the various factors.
2. Explain the geographical factors which have helped to make the U.S.A. the world's greatest industrial power.
3. Select *one* large area noted for modern manufacturing industries and examine the conditions that have favoured this development.
4. Explain why and how the industrial activities of Greater London flourish without the presence of either a coal or an iron-ore field. What are the advantages and disadvantages of a heavy concentration of population in the region?
5. Show by examples that a manufacturing region rarely, if ever, owes its origin and growth to one favourable factor, but usually to a combination of advantages.
6. Discuss the problems involved in the industrialisation of backward countries.

# Chapter XX

# THE IRON AND STEEL INDUSTRY

## *THE IMPORTANCE OF IRON AND STEEL*

THE making of iron and steel is the most important of all the manufacturing industries for iron and steel provide the basic materials for a large number of secondary industries. Moreover, the production of steel has long been a pre-requisite of military might and it has been said that "steel is power." Again, the production of steel forms a pretty good yardstick of the manufacturing capacity of a country.

Since iron seldom occurs in its native state, it must be obtained by the reduction of its ores. This is done by mixing the ores with coke (or occasionally charcoal, as is sometimes the case in Sweden) and limestone, and firing or smelting them in a blast furnace. The high temperatures required to melt the ore are secured by blasts of hot air or, more recently, oxygen and oil—hence the name blast furnace. Under the great heat produced, the carbon (supplied by the coke or charcoal) unites with the oxygen in the iron ore to liberate the iron; at the same time the impurities in the ore combine with the limestone to form slag. The molten iron, being heavy, sinks to the bottom of the blast furnace while the scum or frothy slag rises to the top of the liquefied metal. Upon completion of the smelting process, the duration of which differs according to the process used, the molten iron is either run off into moulds where it cools and solidifies to form pig-iron or transferred directly to steel furnaces where it is further treated to form steel. Two points in this smelting process are worth noting: first, it requires about 1 ton of coal (*i.e.* 15 cwt of coke) to produce 1 ton of pig-iron and, secondly, the limestone is termed a "flux" because it is an agent facilitating the reduction of the iron ore.

Pig-iron is the raw material used in the making of cast iron, wrought iron, and steel. Cast iron is the term used for such articles as are "cast" from molten iron; in other words, the pig-iron is re-melted and poured into moulds of the shape of the articles required. Cast iron is hard and strong but brittle; this is because the pig-iron contains many impurities, *e.g.* about 4% of carbon and smaller quantities of such minerals as silicon, sulphur, and phosphorous. Cast iron, however, is relatively cheap. Wrought iron, which is the purest form of iron, is produced by re-melting and removing the impurities from the pig-iron, especially the carbon. Refined in this way, the iron loses its brittleness at red heat and it can be welded or hammered into any desired shape. Wrought iron is very strong and tough and resists rusting. It is, however, expensive to make—nowadays it is dearer than steel—and tends to have rather limited uses. It is mostly used in craftwork, *e.g.* in the making of ornamental gates. Steel is made from pig-iron by first removing all the impurities to produce pure iron and then adding a small but precisely calculated amount of carbon, which may vary from 0·3% to 2·2%. The steel takes on a special hardness through the addition of the carbon. Whereas cast iron is brittle and incapable of being forged, rolled, or drawn, steel is very versatile and can be worked in various ways to produce such widely diverse products as girders, rails, plates, sheet steel, tubes, and wires.

Moreover, by the addition of alloy-metals, various special steels, having particular properties which can be used for special purposes, can now be made and are, in fact, widely used. The more important of the alloy-metals are manganese, nickel, cobalt, chromium, molybdenum, tungsten, and vanadium. Chromium is used in the making of stainless steel which does not rust; chrome steel is also very hard and is used in the manufacture of armour-plating. Nickel also imparts a special hardness to steel and nickel steel is used for armour-plating, for turbine blades, and for sheathing submarine cables. Cobalt steels are very resistant to high temperatures and hence are much used in the making of jet engines. Manganese steel is extremely hard; molybdenum is also used for hardening steel. Tungsten and vanadium are used to produce ferro-alloys used for making high-speed cutting tools.

## STEEL-MAKING PROCESSES

There are a number of methods of producing steel; the one selected depends largely upon the quality of the pig-iron and the character of the steel required.

*The Bessemer Process.* This was the first method of large-scale steel production. It was invented in 1856 by Henry Bessemer. At first the Bessemer converter could take only non-phosphoric ores, but the basic process discovered in 1878 by Thomas and Gilchrist made it possible to use the iron ores which had a high phosphorus content (such as the "minette" ores of Lorraine in France). The basic process involved the addition of lime to the pig-iron and lining the converter with "basic" refractory bricks made of dolomite, which, with the limestone, combined with the phosphorus to remove it from the iron. The Bessemer converter usually has a capacity of about 25 tons. In the process the pig-iron is melted; all the impurities are extracted, including all the carbon, by blowing air through the molten metal; then the requisite percentage of carbon is added. The basic Bessemer Process is much used on the Continent, especially in France, Belgium, and Luxembourg where the phosphoric minette ores are used. Great strides have been made recently in the United Kingdom with rotating converters using oxygen in place of air. It is anticipated that before very long oxygen converters will be supplying the bulk of British steel.

*The Open-hearth Process.* This process, alternatively called the Siemens–Martin Process after the two men who invented it at the same time, differs from the Bessemer Process in that a certain proportion of scrap iron (up to 50%) may be used with the pig-iron. The iron is melted in a shallow hearth and gas is burned over it. As a result of the action of air and the heated gases, the impurities are eliminated or chemically combined to produce a slag. The processing of the molten mass is continued until the metal contains the desired percentage of carbon, when it is stopped. The furnaces vary in size, but the larger ones take up to 350 tons. The steel-making process takes longer (from 8 to 20 hours) by the open-hearth method than by the Bessemer (from 12 to 15 minutes). The Siemens–Martin process is the method most widely adopted at the present day in the United Kingdom and the United States.

*Other Processes.* The cementation or crucible process, invented by Huntsman in 1740, and the modern electrical processes are chiefly used in the making of alloy and special steels. The crucible process, though old and expensive, produces the finest steel and is sometimes used when especially high quality steel is required as, for example, in the making of cutlery.

In the more technologically advanced countries oxygen processes are increasingly being used in steel-making. Of particular note are the L.D., the Kaldo, and the Rotor converters which are used. In these new pneumatic or converter processes use is made of an oxygen/steam blast in bottom-blown converters, or oxygen is injected at high speed on to the surface of the liquid iron, *i.e.* from above instead of from below. In Britain, for instance, the use of oxygen in steel-making processes has increased more than tenfold since 1958 and while in 1961 only 16·8% of the total crude steel production came from processes other than open hearth furnaces today over 45% of our steel is made by these new techniques.

1968 saw steel production attain new records almost everywhere, and the total world output was 528 million tons. Compare this with around 350 million tons in 1961. The largest producers of steel in 1968 were: the United States (119·2 million tons), the Soviet Union (106·5 million tons), Japan (66·9 million tons), West Germany (41·1 million tons), the United Kingdom (26·2 million tons), France (20·4 million tons), Italy (16·9 million tons) and China (an estimated 16 million tons).

## *THE UNITED KINGDOM*

In Britain the smelting of iron ore by charcoal goes back to Roman times. Until the eighteenth century the production of iron was small, but with the arrival of the Industrial Revolution the demand for iron increased substantially. The traditional method of charcoal-smelting was quite incapable of meeting the increased demand. Fortunately, at this juncture, a method of smelting iron ore by using coke was discovered. The method was perfected by Abraham Darby and by 1800 considerable quantities of iron were being turned out by the British ironmasters.

The iron-smelting industries, which prior to this time had been located in the Weald, the Forest of Dean, and in the Midlands, now moved on to the coalfields where iron ore from the coal measures and coking coal were found together. Often, too, there were supplies of limestone near at hand for flux. Thus the Yorkshire, Northumberland and Durham, South Staffordshire, South Wales, and Lanarkshire coalfields became the seats of important iron, and later steel, industries. The coming of the railway and later of the iron ship, together with the growing demands for machinery, led to a rapid expansion and Britain became the leading world producer of iron and steel, a position she held until the last decade of the nineteenth century when she was successively passed by the United States (in 1890) and Germany (in 1893).

The present-day British iron and steel industry is widely dispersed on inland and coastal sites and all have ready access to supplies of coal. As the home supplies of iron ore became increasingly depleted, greater dependence came to be placed upon imported ores, first from Spain and Sweden and, more recently, from North Africa and Canada. One result of the necessity to import foreign ores has been the movement of many smelting centres from interior to coastal locations. Wartime demands led to an upsurge in iron and steel production after the stagnation of the inter-war period and this increasing output has not abated. Today the United Kingdom produces 26 million tons of steel, as compared with about 7 million tons in pre-war years. It should be emphasised that most of Britain's rich ore deposits are now exhausted and that over 90% of the home produced ores consist of the "lean" or low metal-content ores derived from the Jurassic escarpment which runs from Oxfordshire, through Northamptonshire, to Lincolnshire. The United Kingdom is dependent upon foreign imports for half its iron ore requirements.

The chief iron and steel centres are:

*South Wales.* This is the most important region, with an annual production of 5–6 million tons. The industry, based originally on the ore from the Brecon Beacon and the coal of the South Wales coalfield, first grew up in north Glamorganshire with Merthyr Tydfil, Aberdare, and Tredegar as the chief iron-smelting centres. As the local ores gave out, higher-grade ores came to be imported from Spain and North Africa.

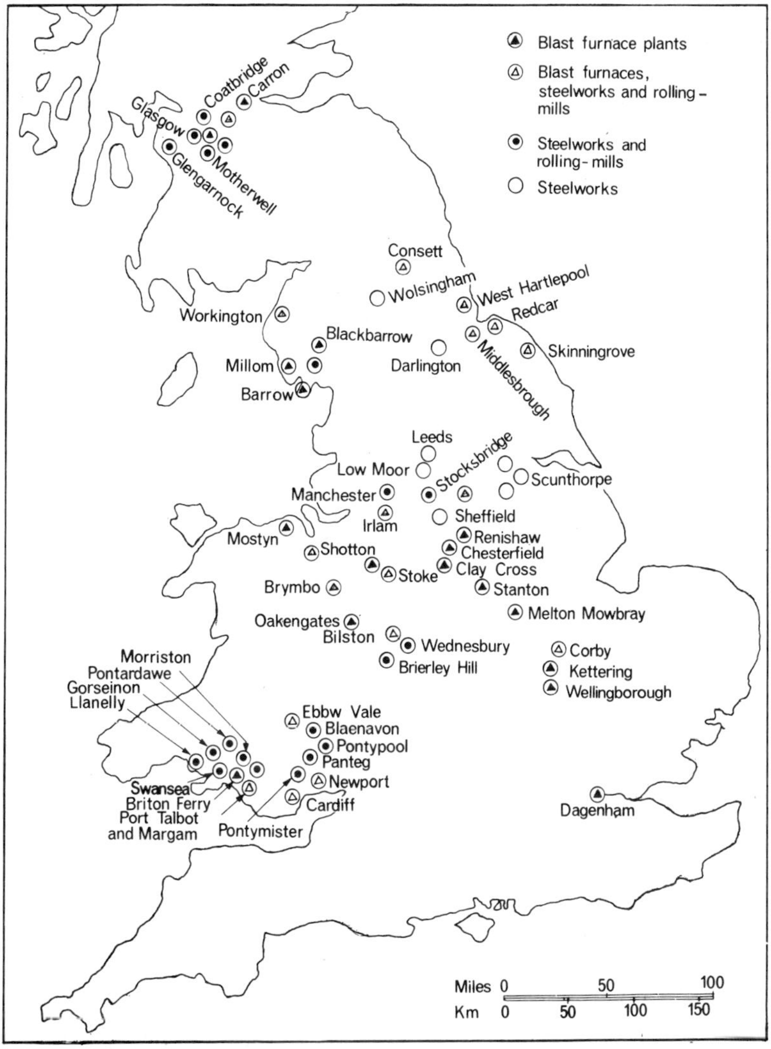


FIG. 113.—IRON AND STEEL IN BRITAIN

The industry grew up first on the coalfields using coal-measure ores; then, when the industry began to depend to a large extent upon imported ores, it tended to gravitate to the coast. Note the newer centres based on the low-grade Jurassic ores and the movement of these ores to the older centres. Rationalisation is leading to the closure of many of the smaller uneconomic plants.

At Cardiff and Swansea, the ports handling the imported ores, important iron and steel works grew up. Gradually the older-established centres of the interior declined and now, with the exception of a major works at Ebbw Vale, the industry has become concentrated on the coast. At Margam, near Port Talbot, a vast new works was opened in 1953 and here is one of the world's largest "continuous process" steel plants. Another vast plant has even more recently been opened at Newport. Rolled and sheet steel (for ships' plates, motor car bodies, and tin-plate) is especially important.

*North-east Coast.* The north-east coast, and especially Tees-side, ranks second with a production of 4·5 million tons. The chief steel-making

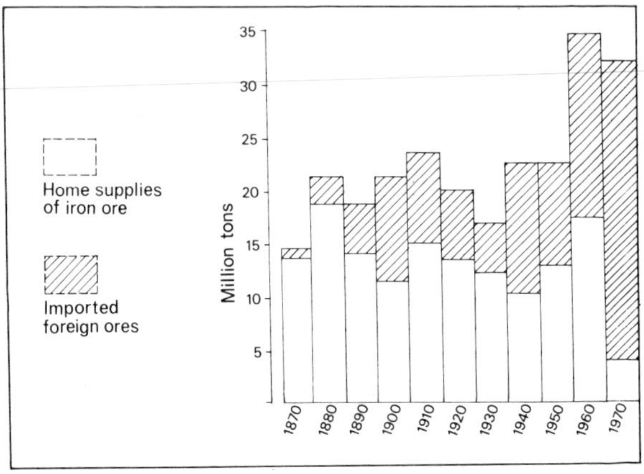


FIG. 114.—IRON ORE IMPORTS

The graph shows (i) the total quantity of iron ore used by the British iron and steel industry over the past hundred years; (ii) the proportion of the ore used accounted for by home supplies; and (iii) the percentage of the total ores supplied by foreign imports.

centre is Middlesbrough, which developed rapidly during the nineteenth century. The industry grew up on the nearby iron-ore deposits of the Cleveland Hills, the excellent coking coals of the South Durham coalfield, and the limestone available in mid-Durham to the west. The Cleveland ironstone is now almost worked out and, in fact, the last mine closed in 1963, but ore from Sweden can be easily imported by sea via Narvik in Norway and this is mixed with the low-grade home-produced ores brought north from Lincolnshire and Northamptonshire. Most of the pig-iron produced is turned into ships' plates (for the ship-building of the Tyne, Tees, and Wear estuaries), steel girders, and railway lines. Other steel centres in this region are Hartlepool, Stockton, and Darlington, and farther inland, on the coalfield, Consett. Other works are located at Redcar and Skinningrove (established 1855). Here is an example of geographical inertia.

*South Yorkshire.* Sheffield was the original home of the South Yorkshire iron industry. Here were local supplies of iron ore, forests providing charcoal, ganister for the furnaces, gritstone for grind-stones, and fast-flowing streams which worked the bellows and powered the hammers in the early days of the industry. None of these factors is of any significance at the present day, but the industry continues to flourish in Sheffield and the surrounding area, providing a good example of the principle of geographical inertia. This may be defined as the tendency for an industry to remain in a locality and to continue to prosper after the conditions which originally determined its establishment have either appreciably altered or completely disappeared. Nowadays, the industry is dependent upon scrap and pig-iron from Middlesbrough and Scunthorpe, for no smelting is done in Sheffield itself, but there are two blast furnaces at Rotherham. Coal from the Yorkshire coalfield supplies both the power and the carbon for smelting. Sheffield is famous for its cutlery industry but it also produces "heavy" products such as castings and forgings and makes alloy-steels. Chesterfield is an important iron and steel town with eight blast furnaces. Other centres in the Sheffield area are Rotherham (heavy steel castings) and Stocksbridge (steel), which largely uses scrap iron.

Farther north, and outside the Sheffield area, are Leeds, which has important blast furnaces and iron and steel works, and Low Moor, near Bradford, which specialises in alloy-steels.

*Lincolnshire.* The north Lincolnshire area, producing 2·5 million tons of steel annually, is one of relatively recent development. The industry is centred upon Scunthorpe and Frodingham. It has grown up on the thick beds of low-grade iron ore which can be easily mined. Such "lean" ores are costly to transport and so, instead of the ores being moved to the coalfield, the coal goes to the ironfield. Plentiful supplies of good coking coal are available from the nearby Yorkshire Coalfield. Here, in north Lincolnshire, are to be found some of the largest and most efficient blast furnaces and steel plants not only in Britain but in the world. The region accounts for about 15% of the British output of pig-iron and 10% of the steel.

*Central Scotland.* Not far behind Lincolnshire comes the Central Valley of Scotland with a yearly output of 2·4 million tons of steel. Originally the industry in the region had two special assets: the so-called blackband ironstone mined in Lanarkshire contained sufficient carbonaceous material to allow it to be smelted without the need for coke; and the type of coal found there could be used in its natural state without the need to process it into coke. These special assets no longer obtain because the ores have been exhausted. The region has maintained its importance, however, owing partly to the local skill in metal-working and partly to the demands of the important engineering and ship-building industries. The industry is dependent upon imported ore. The chief iron and steel towns are situated in the Clyde valley and upon the Lanarkshire coalfield, *e.g.* Glasgow, Coatbridge, Airdrie, Motherwell, and Wishaw. This is essentially a "heavy" industry producing steel girders, frames, plates, and bridge pieces. Falkirk, to the north-east, an old iron centre, produces light castings. There is a new integrated steel works at Ravenscraig, near Motherwell.

*The Midlands.* The West Midlands region, more commonly known as the Black Country, is one of the older iron centres. In the days before the Industrial Revolution local iron ores were smelted with charcoal from the Forest of Arden and Cannock Chase. Later when coal came to be used the industry persisted. Blackband ironstone was mined along with coal in the South Staffordshire Coalfield and limestone was near at hand. At one time South Staffordshire was the chief smelting area in Britain. Today the local ores and even the best coals are largely exhausted, the blast furnaces have almost disappeared, and the steel industry is based upon pig-iron brought in from the East Midlands. Bilston is the chief steel producing centre. Numerous towns on or near the South Staffordshire and Warwickshire Coalfields are concerned with the manufacture of a great variety of iron and steel goods, anything, it is usually said, "from a pin to a steam engine."

*Northamptonshire.* In the Jurassic or Oolitic limestone belt of the East Midlands are found abundant supplies of low-grade iron ores; though the iron content of the ore is small, the ore can be easily quarried or mined—often by open-cast techniques—hence a number of important blast-furnaces and steel-making centres have grown up at Wellingborough, Kettering, and Corby. The last named is a new steel town which has grown up during the past twenty-five years. The Stanton ironworks at Melton Mowbray (Leicestershire) has three blast furnaces. This region, like north Lincolnshire, provides an example of coal moving to the ore instead of vice versa. Over 1 million tons of steel are produced each year. Much of the steel produced at Corby goes to the Black Country to feed the engineering industries of that region.

*South Lancashire and Deeside.* The chief centres in this region are Irlam on the Manchester Ship Canal and Mostyn and Shotton on the Dee estuary. There is no local ore available and the industry is completely dependent upon imported supplies, although ores can be easily brought in by water transport. Locally produced coke from the Lancashire and

North Wales coalfields is used for smelting. Their big advantage is their proximity to major centres of industry and population in South Lancashire and the Midlands. The area produces some 9%, or approximately 2 million tons, of steel annually. Much of it is used by the ship-building industry of Birkenhead and the remainder for the production of such things as axles, wheels, sheet steel, light rolled products, and wire.

*The North-west Coast.* The iron and steel industry of the Furness district of North Lancashire and the Cumberland coast grew up originally on the rich hematite ores mined at Egremont, Beckermet, etc. These ores are rapidly running out and the industry is mainly dependent upon ores imported from Spain and North Africa. The coal from the Cumberland field is unsuitable for smelting, hence supplies of coke have to be brought from Durham and South Lancashire. Smelting is carried on at Workington and Barrow. The steelworks at Workington supply the ship-building industry of Barrow with some of its steel requirements; they also turn out steel rails. The area, however, produces only about 1% of the steel manufactured in Britain.

## THE FUTURE OF THE BRITISH IRON AND STEEL INDUSTRY

The nationalised British steel industry is currently losing money. The industry, nationalised in 1967, inherited many small and uneconomic plants and, although it has attracted heavy investment, it is unable to compete with the cheap steel produced by countries such as Japan. The British Steel Corporation, however, has plans which will, if they go through, drastically change the whole industry. The plans involve the closing down of 30 of the steel-producing plants out of the 39 inherited on nationalisation, although some of these sites will continue to carry on finishing processes. New major steelworks are to be built. A 12-million-ton-a-year steelworks is planned for completion in the late 'seventies. Just where this new complex will be situated is not yet decided but both the Thames estuary and the Southampton–Portsmouth areas are being seriously considered. Such sites have the advantages of easy access to the markets of the home counties and nearby continental Europe. It is estimated that the giant works will cost £1000 million. Another 6-million-ton-plant is to be built at Redcar on Tees-side; this will replace the old Redcar works. The intention is to concentrate steel-production in a few great plants, in effect to nine within the present decade. The new plants will be streamlined and largely automative. The changes will involve the loss of over 50,000 jobs in the industry but the capacity of the plants will rise to 43 million tons a year by 1980 and allow British steel to become competitive in the world market.

## *THE UNITED STATES*

The United States is the world's largest producer of steel, accounting for approximately one-quarter of the total world output. She has led the world for over half a century and there seems to be little chance of her nearest rival, the Soviet Union, overtaking her, at least for several years to come. Like the Soviet Union's, the United States' iron and steel industry is faced with the disadvantages of widely separated basic resources and the need for the long haul of either coal or ore. This great drawback has largely been surmounted and offset by (*a*) the ease with which both coal and ore can be mined and, in the case of ore, imported; and (*b*) the relatively cheap and efficient methods of transport, notably water carriage. Most of the United States' iron ore requirements (some 60%) are still met from home supplies, notably from the hematite deposits mined in the Mesabi, Marquette, Menominee, Gogebic, and Vermilion Ranges to the west and south of Lake Superior, though these resources have been exploited and are gradually being exhausted. It is now becoming necessary to import substantial quantities of iron ore from Labrador, Venezuela, Chile, and West Africa. The chief supplies of coking coal come from the extensive deposits of the Appalachian Coalfield.

The iron and steel centres of the United States can be grouped into six main areas:

1. On or near the Pennsylvania coalfield focusing on the *Pittsburgh–Youngstown district*, now dependent upon ore brought from the Lake Superior ironfields; this area has long dominated United States steel production, but though it still accounts for about one-third of the total output (approximately as much as the total West German output) it is gradually becoming relatively less important.

2. In the middle west at *Chicago and Gary*, not far removed from ore and coal, the former being brought by rail and boat from Lake Superior and other Lake ports, the latter by rail from the Appalachian and Eastern Interior fields; the industry has been stimulated by the large local demands of the engineering, railway, and shipbuilding industries and the area has become Pittsburgh's leading rival.

3. *The Lake Erie centres, e.g.* Detroit, Cleveland, Buffalo, where ore can be easily and cheaply shipped by water, either from Lake Superior or Labrador, and coal brought by rail; with excellent transport facilities and a location near to big consuming centres, the importance of the lakeside centres has grown rapidly during the present century and, producing nearly 20 million tons of steel a year, they rank third among the producing areas.

4. *The mid-Atlantic seaboard*, where the chief centres are Bethlehem (near Philadelphia), Sparrows Point (near Baltimore), and Morrisville (near Trenton), is an important and growing steel area; at or near tide-water, these plants depend mainly upon high quality imported ore from Labrador, Venezuela, Chile, etc., smelted by coke brought from Pennsylvania.

5. The *Birmingham area* of Alabama, where local supplies of ore, good coking coal, and limestone are fortuitously in close proximity, was originally presumed to hold out great possibilities and it was prophesied that it would become a second Pittsburgh, but its growth has been greatly handicapped by the lack of an industrial market; moreover, the high costs of producing the local ore have led to substantial imports from Venezuela.

6. The centres in the *Western States*, *e.g.* Denver, Tacoma, Pueblo, San Francisco, and Los Angeles, primarily serve local and limited markets (railway, mining, agricultural, engineering needs); the industry is mainly a modern development (though the first steel plant in the west dates back to 1882) and though the output has trebled since 1940 the total production of about 7 million tons amounts to only approximately 6% of the total national output.

## *THE EUROPEAN COAL AND STEEL COMMUNITY*

Western Europe is roughly comparable in population, resources, and productivity with the two economic giants of the United States and the Soviet Union. Unfortunately, the countries of Western Europe were individual nation states divided one against the other. The end of the Second World War left them shattered and weakened, but it also brought home to them the need for a greater degree of co-operation than had ever existed in the past, especially in view of the economic threat from the United States on the one hand and the Soviet Union on the other. As a result, six of the countries of Western Europe (France, West Germany, Holland, Belgium, Luxembourg, and Italy) joined together to form the so-called European Coal and Steel Community.

This association came into being in 1951. The idea behind this co-operative effort was to lessen the hindrances and obstacles to iron and steel production by promoting the free flow of iron ore and coking coal across the national boundaries. Individually the countries suffered from shortages of one or the other; united, they were almost self-sufficing. The establishment of the Community welded the various national industries into a single organisation which has enabled it to compete with the giant producers of the United States and the Soviet Union.

Here we cannot go into all the details of European iron and steel production and we must be content with drawing attention to the more important producing centres.

*The Franco-Belgian coalfield* is the seat of an important iron and steel area. Dunkirk, Anzin, and Valenciennes in France and Liège and Charleroi in Belgium are the chief centres. The industry, which was originally based upon local ores, now has to rely upon ore from Lorraine, Luxembourg, and Normandy. Much scrap steel is used here since the area lies in the midst of a great industrial region.

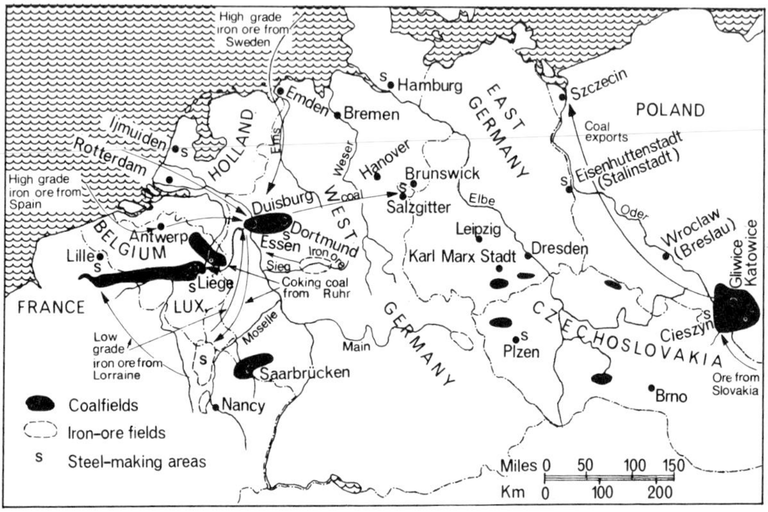


FIG. 115.—IRON AND STEEL IN EUROPE

In earlier times the iron and steel industry was primarily coalfield-based using coal-measure ores. Today many areas lie off the coalfields (*e.g.* Hamburg, Salzgitter, Ijmuiden); they are located on iron-fields or in ports where ore can be easily imported. Note the interrelation of coal and ore movements amongst the members of the Coal and Steel Community.

*The Lorraine–Luxembourg–Saar area.* France is fortunate in possessing in Lorraine, one of the greatest iron-ore reserves in Europe. These lean, oolitic, limonite ores occur in vast quantities and can be worked very easily by open-pit methods. A valuable coalfield exists in the Saar basin, but the Lorraine steel industry draws heavily upon the coking coals of the Ruhr which can now be easily brought by water carriage along the newly-canalised Moselle. The chief iron and steel centres are Metz, Briey, Nancy, Thionville, and Longwy. There are several locations in southern Luxembourg, the largest of which is Esch-sur-Alzette.

*The Ruhr in West Germany* has been the most important steel area in

Europe for almost a century. Possessing excellent coking coal and local supplies of ore (some ore is still supplied from the Sieg valley) the Ruhr made Germany the world's number two steel producer in the early years of the present century. The bulk of the ore now used comes from Sweden. The Ruhr produces most of West Germany's output of steel (some 45 million tons) and accounts for approximately half of the European Coal and Steel Community's production.

Smaller steel producing centres in the Community are: Ijmuiden in Holland; Salzgitter, Hanover, and Bremen in West Germany; and Genoa and Milan in Italy.

TABLE XXXVII

*Steel Production (million metric tons)*

| | *1961* | *1970* |
|---|---|---|
| West Germany . . . . | 33·5 | 45·0 |
| France . . . . . . | 17·6 | 23·8 |
| Italy . . . . . . | 9·1 | 17·2 |
| Belgium . . . . . | 7·0 | 12·6 |
| Luxembourg . . . . . | 4·1 | 4·5 |
| Holland . . . . . | 2·0 | 5·0 |
| Total | 73·3 | 108·1 |

## OTHER EUROPEAN PRODUCERS

Most of the other states of Europe are only minor iron and steel producers.

*Norway* has a very small industry based upon local iron ore deposits and using hydro-electric power.

*Sweden* has abundant ores but little coal, and imports it from Britain and Poland. Steel is produced around Dannemora and Grangesberg in central Sweden. This country produces 4 to 5 million tons annually.

*Poland* has an important iron and steel industry based on the Upper Silesian Coalfield, notably at Katowitze. Poland relies on Sweden for much of its iron ore. 11 million tons is produced annually.

*Czechoslovakia* uses coking coal from the Teschen coalfield. Steel-making is carried on at Ostrava and near Plzen at the famous Skoda works. 10 million tons is produced annually.

*East Germany* has built up a new steel centre at Eisenhuttenstadt (Stalinstadt) on the River Oder.

*Austria* has steel centres at Donawitz, Linz, and in the valleys of the Mur and Murz.

*Spain* possesses ironfields in the Biscayan coastal region along with coal deposits. Spain has no great iron and steel industry, though a large modern integrated works has been set up at Aviles.

## *THE SOVIET UNION*

The U.S.S.R. is richly endowed with iron ore and coal resources and has built up a vast iron and steel industry. In 1970 she produced slightly more than 116 million tons of crude steel (as against the U.S.A.'s 119 million). This large output is all the more remarkable since it has been achieved within a relatively short space of time; in 1928 (the year which saw the introduction of the First Five-Year Plan) steel production was about 4·5 million tons, in 1953 it was 37, in 1960 it was 65·3 and in 1968 106·5. The Soviet Union is now the world's second largest producer of steel. The Soviet government believing, rightly, that a large iron and steel industry was necessary for military strength and industrial development, pushed ahead with its development in a ruthless but efficient manner. Steel had first priority.

The Soviet Union's iron and steel industry has many points of resemblance to that of the United States. Both countries have the basic requirements for the industry in abundance; both find their major markets within their own frontiers; and both are faced with the problem of long-haul in assembling the raw materials. But the Soviet Union's iron and steel industry differs from that of the United States, and of Western Europe, too, in two major respects: first, it is a State-owned, State-run industry and, secondly, the industry tends to be rather far removed from the great consuming centres. Because of this latter circumstance, the Soviet Union has in more recent years begun to establish plants outside the major iron and steel areas to serve regional markets.

In the early years of the industry's development the Soviet planners set up a few large-scale plants, thereby achieving low-cost production; but this was achieved at the expense of high transport costs for the assembly of raw materials and the carriage away to the consuming centres. Immediately prior to the Second World War the Soviet government changed its policy and began to adopt a more rational location of the industry in relation to both raw materials and internal markets. This has led to the wider dispersion of the industry and the building of smaller plants.

The chief iron and steel centres in the Soviet Union are as follows:

*The Donbas region.* This, the U.S.S.R.'s greatest iron and steel region, is based upon the Donbas* coalfield and the ironfield of Krivoi Rog and the Crimea. Krivoi Rog is the traditional centre of the iron and steel industry in the Ukraine, but the greatest steel centre is Donetz (formerly Stalino), where the industry was founded in 1871 by a Welshman named

* The term Donbas is a shortened form of *Donetz Basin*. Compare Kuzbas derived from *Kuznetsk Basin*.

Hughes who set up the first blast furnace. Coal and iron ore are now shuttled between the Donbas coalfield and the Krivoi Rog and Kerch ironfields with the result that the whole area forms a vast integrated industrial "combine."

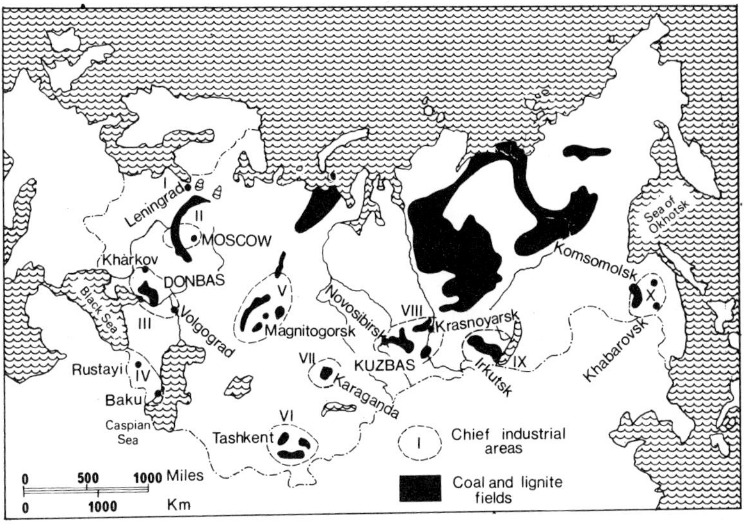


FIG. 116.—INDUSTRIAL REGIONS OF THE U.S.S.R.

Note: (i) how most of the industrial areas are based on coal deposits (the chief exceptions are the Leningrad and Caucasus areas); and (ii) the well-scattered character of the industrial regions.

*The Ural region.* The industry first grew up at Zlatoust where local deposits of iron ore were smelted with charcoal. Today, however, the industry is located on the eastern flanks of the Ural Mountains, chiefly at Sverdlodsk, Magnitogorsk, and Nizhny-Tagil. The local coal deposits are of little, if any, value for coking, hence coal has to be brought from the Karaganda coalfield (some 600 miles (960 km) away) or from the Kuznetsk coalfield (some 1400 miles (2240 km) away). Much of the high-grade ore is still smelted with charcoal and the presence of ferro-alloy metals, such as manganese, nickel, and chromium, enables the Ural region to produce high-quality steels.

*The Kuzbas region.* Since coal was taken to the Urals, it was found expedient to bring ore from the Urals ironfields to the Kuznetsk coalfield where an iron and steel industry was set up. This coal-ore exchange made the carriage more economic, although a long-haul of 1400 miles (2240 km) was not very satisfactory. The discovery of the Karaganda coalfield, some 800 miles (1280 km) to the west of the Kuznetsk basin, eased the problem.

Furthermore, the discovery of iron ore in the Kuznetsk basin lessened the latter's dependence upon Ural ore. Novokuznetsk (formerly Stalinsk) is the seat of a great integrated steel plant in the Kuzbas region.

*The Caucasian region.* The Caucasus, formerly known mainly for its oil industry, has in recent years experienced a surge of industrial development and many new heavy industries have been set up. With iron ore from Chiatura, in the Rion Basin, smelting and ferro-alloy production is carried on at Zestafoni. There are also iron and steel works at Dashkezan and at Rustavi in Transcaucasia. Although there are coalfields in the region, some of the smelting is carried out by hydro-electric power.

*Other areas.* During more recent times new plants, outside the major iron and steel areas, have been built mainly to serve regional markets. Some of these are: at Cherepovets, 250 miles (400 km) south-east of Leningrad, which draws iron ore from the Kola Peninsula and coal from Vorkuta in the Pechora Basin; at Solonichki on the Karanganda coalfield; at Komsomolsk in the Amur valley; and at Vladivostok at the Pacific terminus of the Trans-Siberian Railway.

## *MONSOON ASIA*

### JAPAN

Until recently Japan was the only industrial country in the Orient. She has long had a well-developed iron and steel industry, but it has grown rapidly in post-war years, largely due to the expansion of her engineering and ship-building industries. Though possessing some coal, the industry is based almost entirely upon imported supplies of coke, iron ore, pig-iron, and scrap metal. Japan produces some 66 million tons of crude steel annually. The chief iron and steel centres lie around the Inland Sea and notably at its western and eastern ends: at Yawata, Moji, and Shimoneseki in the west and at Osaka and Kobe in the east.

### CHINA

Since the Communist Revolution, the Chinese People's Republic has made a determined effort to become an industrial country and, in fact, an industrial revolution has been telescoped into a period of some twenty years. China has expanded the iron and steel centres originally set up by the Japanese in Manchuria (at Anshan, Fushun, etc.), increased production at the traditional iron- and steel-producing centre of Wuhan, and established new plants at Shanghai, Tsingtao, Paotow, Taiyuan, Maanshan, Pingsiang, Chungking, and Canton. In addition, many new plants are being built up and down the country. Production of steel is already around 10 million tons annually and may be much more.

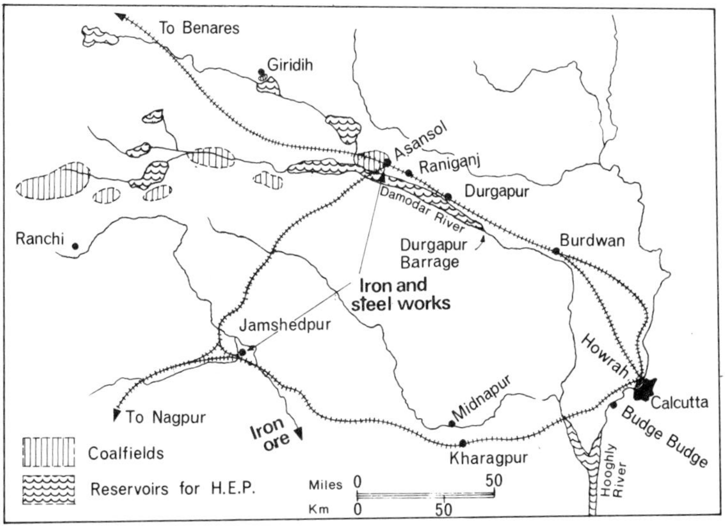


FIG. 117.—DAMODAR VALLEY, INDIA

The Damodar Valley in north-eastern India has been given "T.V.A. treatment." The Damodar River, a tributary of the Hooghly, has been harnessed to provide hydro-electric power, to store water for irrigation, and to provide a navigable waterway. The region has become the most important industrial area in India; indeed, one of the greatest in Asia.

The story of its development really goes back fifty years. A Parsee named Jamshedji Tata, an industrialist and public benefactor, began to exploit the iron-ore deposits which were discovered in Orissa, the area to the west of Calcutta. The presence of coal in the locality led Tata to set up an iron and steel works. This grew and a great iron and steel plant, the largest in the Commonwealth, arose at Jamshedpur, a mere jungle village fifty years ago.

The industry was fortunately placed for not only was there iron ore of good quality at hand but both coal and limestone, whilst the great city of Calcutta was a mere 100 miles (160 km) away. Later, manganese and chromite (used for special steel-alloys), together with mica and copper (both used in the electricity industry) and bauxite (for making aluminium) were also found in the locality.

In a sense the great Tata works became the symbol of the new India. It was decided to develop the Damodar area as a great integrated industrial region. During very recent years other great iron and steel plants have been constructed at Durgapur, Bilhai, and Rourkela. And there are numerous other industries—engineering, cement, chemicals, textiles—which are all expanding. So important among the manufacturing areas of India has the Damodar region become that India today ranks fifth amongst the industrial countries of the world.

Although nine-tenths of India's coal comes from the Damodar Valley and although this is India's chief industrial region, the scheme is also directed towards developing the agricultural possibilities of the region by means of irrigation (especially in the Burdwan area), soil improvement, and flood control.

**INDIA**

The Damodar Basin, immediately to the west of Calcutta, possesses several coalfields together with deposits of iron ore, and in this region an iron and steel industry was founded at Jamshedpur by Jamshedji Tata at the beginning of the present century. In a sense the great Tata iron and steel works, the largest in the Commonwealth, became the symbol of the new India for, like China, she aspires to become a great industrial country. During recent years three new iron and steel plants have been built at Rourkela, Durgapur, and Bhilai and these, together with expansion at Asansol and Bhadravati, combined with the output from Jamshedpur, raise India's iron and steel production to 6 million tons.

## *AUSTRALIA*

The possession of large supplies of iron ore (at Iron Knob and Yampi Sound) together with coal from the Newcastle field led to the establishment of the Australian iron and steel industry. Self-sufficient in the basic requirements of the industry, she is now able to produce all the steel she needs. The industry dates back to 1916, but it was the Second World War which stimulated and led to its great expansion. Iron and steel are produced in three chief centres: (*a*) around Newcastle and Port Kembla in New South Wales, using ore from Iron Knob; (*b*) at the small port of Whyalla, on Spencer Gulf, using coal from Newcastle; and (*c*) at Kwinana, south of Perth in Western Australia, where a new steel plant uses iron ore from Yampi Sound on the north-west coast and coal from Collie, about 100 miles (160 km) south of Perth.

## *AFRICA*

Africa appears to be rich in deposits of iron ore, which are widely scattered throughout the continent, but relatively poor in coal, most of which occurs in the southern part of the continent. Only in the Republic of South Africa has a large modern iron and steel industry been developed; Vereeniging, on the Transvaal coalfield, is the principal iron and steel centre, but steel is also produced at Johannesburg, Pretoria and Newcastle. Small quantities of steel are made in Rhodesia where there is a big plant at Que Que, using local supplies of iron ore and limestone, and coal from the Wankie coalfield. In the Atlas region iron ore is smelted at Oran and there are plans for the construction of a steel plant at Bone. Fedhala, in Morocco, is the site of that country's first iron and steel plant. Egypt is constructing a large iron and steel plant near the new Aswan High Dam. Here local iron ores will be electrically smelted by power derived from the hydro-electric station which is associated with the dam.

## *LATIN AMERICA*

Central and South America have large and valuable deposits of iron ore, notably in Cuba, Venezuela, Brazil, Peru, and Chile, but the region is sadly lacking in coal. Many countries, however, in part for purposes of national prestige but also to diversify their one-sided economies, are beginning to set up iron and steel industries, following the example of Brazil, which led the way. The chief plants are in Brazil (at Volta Redonda) in Argentine (at Zapla), in Chile (at Huachipato), in Peru (at Chimbote), in Colombia (at Paz del Rio), and in Venezuela (at Matanzas). Mexico's steel industry, three-quarters of which is located at or near Monterrey, has a capacity of about 3·5 million tons. It was planned to step this up to 10 million tons by 1968, but this target was not achieved. Output of steel rose from 2·3 million tons in 1964 to 3·25 million tons in 1968. In Brazil production in 1968 was 3·8 million tons, as against 3 million tons in 1964. In Latin America generally the fairly rapid expansion of the early 'sixties has not been maintained although the industry in some countries is progressing reasonably well. Even so, the total output of iron and steel in the whole of Latin America is still less than 10 million tons a year, less than half the amount produced by Britain alone.

### EXERCISES

1. Give a reasoned account of the iron and steel industry in *either* the U.S.A. *or* the U.S.S.R.

2. Select *three* important steel-making areas in Great Britain. (*a*) For each, outline the factors responsible for its original location and continued importance. (*b*) What transport problems are involved in the acquisition of raw materials and the disposal of the finished product?

3. Give reasons for the changing relative importance of the location of the U.S.A.'s iron and steel industry.

4. Discuss the geographical setting of the iron and steel industry in the European Common Market area.

5. Compare the iron and steel industries of France and West Germany, paying particular attention to the sources of iron ore and fuel and their transport to the centres of production.

6. In general the British iron-smelting and steel-making industries are found in three types of location—on coalfields, on ore fields, and at ports. Give examples of each of these and explain the reasons for the developments in such areas.

# *Chapter XXI*

# OTHER MAJOR INDUSTRIES

THE range and complexity of modern industry is so great that one could easily devote a book to it alone. Here, the best we can do is to consider very briefly a selection of these industries, and we have chosen the engineering (including ship-building which is a specialised branch of engineering) chemical, and textile industries which are the most important of the manufacturing industries.

## *ENGINEERING*

"The engineering industry" is an omnibus term for a wide variety of industries which fundamentally are based on the use of metals. Originally it was concerned with the making of engines and machines but to this has been added a host of other manufactures and engineering now forms a very complex branch of industry. A basic division of the industry into "heavy" and "light" aspects can be made; in the former are included ship-building and bridge-building, in the latter are included the manufacture of bicycles and radios. Important branches of the engineering industry are the mechanical, the electrical, and the motor car and aircraft engineering industries.

### THE UNITED KINGDOM

Coal, iron, and steel form the basis of those industries embracing the production of castings and mouldings, of machines and engines, of implements and tools, and of hardware and cutlery, which *in toto* are usually referred to as the engineering industries. Nowadays, a wide range of other metals, and other materials, too, are used in the industry. But it was the presence of coal and iron which gave rise to some of the leading engineering areas in the United Kingdom, for example, the Central Valley of Scotland and the Derbyshire–Nottinghamshire areas. The Manchester area, however, now one of the leading engineering areas, owes its development to the manufacture of machinery for the cotton textile industry. The engineering industries of the Birmingham district are the outcome of centuries of tradition. The presence of a huge market requiring a wide variety of engineering products has largely contributed to the growth of the engineering industry in the Greater London area. Local needs, especially of agricultural implements, have fostered the scattered engineering centres in the eastern counties.

*Machinery and tools.* The English Midlands are the principal area for the manufacture of machines and tools. Birmingham and the Black Country and the Derbyshire–Nottinghamshire area are the two chief centres. In northern England, the two industrial areas on either side of the Pennines, South Lancashire and the West Riding, have grown into major engineering areas. Manchester is now one of the world's greatest centres of engineering. The West Riding towns of Bradford, Huddersfield, and Leeds, and particularly Sheffield, are important centres. The engineering industries of Sheffield are highly specialised; marine engineering is especially important, and cutlery is a well-known speciality.

*Textile machinery.* British textile machinery has an international reputation; for example, the Oldham firm of Platt is known the world over. The making of textile machinery is one of the important branches of the engineering industry. The centres of manufacture of spinning and weaving machinery are located, as one might expect, in the great textile manufacturing areas. The most notable centres are Oldham, Wigan, Manchester, Keighley, Leeds, Nottingham, and Glasgow.

*Railway engineering.* In the early days, when British railways were run by private companies, railway engineering was largely the concern of the companies and they established their works at places, usually important railway junctions, on their own lines, in central positions, where land was cheap, rather than in the vicinity of the coal and iron fields, although some, it is true, were in the centres of the engineering areas near iron and coal. The chief centres are Ashford (Kent), Eastleigh, Swindon, Crewe, Birmingham, Derby, York, Manchester, and Glasgow. The Doncaster and Darlington centres are gradually being closed down.

*Electrical engineering.* It is the great engineering towns that have become centres of the heavy electrical industry, *i.e.* the construction of electrical machinery and power plant. Rugby is the only important exception. Electrical machinery and cable manufacture uses a variety of raw materials, especially non-ferrous metals which are mostly imported. A large part of the finished products is for export. These factors have led to a concentration of the industry in the Thames estuary and in the non-ferrous metal district of south Lancashire. Bradford is an important centre for the production of electrical transmission machinery. Glasgow is the chief centre of the industry in Scotland. The light electrical industry and the radio and television industries are spread over many parts of the country. Transport facilities are of minor importance to these industries since neither the raw materials nor the finished products are very bulky.

*Agricultural machinery.* Farm machinery is made in two distinct groups of towns: first, in towns in the traditional iron and steel areas, *e.g.* Derby, Sheffield, Chesterfield, Huddersfield (tractors), Leeds, Darlington, and Preston; and, second, in the market towns of east England which supply the needs of the local agricultural industries, *e.g.* Newark, Grantham, Peterborough, Bedford, Lincoln, Ipswich, and Norwich.

*Automobile engineering.* The presence of an already well-established engineering industry contributed to the rise and progress of motor car manufacture in the English Midlands. Recently the most powerful locating factor has been the Department of Trade and Industry which may control the siting of new works. For example, the Ford Motor Company opened their new plant on Merseyside in 1963 at Halewood, and Vauxhall commenced operations at Ellesmere Port in 1966. Both companies were granted expansion permits on condition that they built their plants in areas designated by the government. Almost all the motor car factories are in, or close to, the Midlands, *e.g.* Birmingham, Oxford (Cowley), Slough, Luton, Dagenham, Derby. The only outlying areas are at Linwood (Chryslers plant) and at Bathgate in Scotland (British Leyland's truck division). The materials necessary for automobile production (sheet steel, glass, leather, plastic, wood, rubber, paint, etc.) are now so diverse that the influence exerted by supplies of raw material has little influence upon the location of the industry.

*Aircraft engineering.* The aircraft industry, apart from wartime developments, is concentrated in the same areas as the automobile industry. Some manufacturers produce aeroplanes and aero engines as well as motor cars and car engines. London, Bristol, Gloucester, Derby, and Yeovil are important centres.

## THE UNITED STATES

The engineering industry with its many facets is by far the greatest of all the industries in the United States. The ramifications of the industry are legion and it is impossible in the space of a paragraph or two to do anything more than hint at the size, complexity, and importance of the industry. The engineering industry is very widely spread and large quantities of machinery and machine-tools are produced around all the major steel-making centres. But the New England and eastern Pennsylvania districts contain a large part of the industry.

The New England area, lacking fuel and minerals, has not developed heavy industry to any extent; rather has it concentrated upon the lighter aspects of engineering. A large variety of engineering products are made, including precision instruments, typewriters, sewing machines, clocks and watches, and electrical and radio apparatus, *i.e.* "Birmingham goods." A number of towns in the south-west, *i.e.* in western Massachusetts and Connecticut, are important engineering centres, *e.g.* Bridgeport, Springfield, New Haven, Hartford, Holyoke, and Waterbury. In the south-east, *i.e.* in Rhode Island, Manchester, and Worcester, machine tools are made, while Fall River has firearms and cycle industries. Altogether, the engineering industries of New England provide more employment than any other group of industries.

The region around Pittsburgh, which as we have seen is the premier steel region, is also a major engineering region, but throughout the Middle West there is scarcely a town without engineering industries of some kind. Around the basic industry of steel-making have grown up numerous engineering industries which are users of steel, *e.g.* locomotives, automobiles, agricultural machinery, etc.

The production of agricultural machinery is largely concentrated in the Middle West since this is the United States' greatest farming region. The towns of Chicago, Springfield, and Peoria are major centres of agricultural machine production and the state of Illinois, in fact, produces about half of all the country's farm implements. Other major centres are Cincinnati and Milwaukee. The agricultural engineering industry continues to expand as farm mechanisation is still advancing.

Automobile engineering, though widely scattered throughout the country, has its main concentration in the Detroit area, *e.g.* at Detroit, Lansing, Dearborn, Flint, etc. Chicago, Milwaukee, Toledo, Cleveland, and Cincinnati also make cars and lorries. The United States has the world's greatest motor car industry and some 9 million vehicles are produced annually. There is an enormous domestic market which absorbs some 95% of the output.

Locomotive engineering is localised at the major route foci of Denver, Chicago, Pittsburgh, Philadelphia, and New York. The United States is the world's largest producer of railway engines and rolling stock and accounts for some 60% of total world production.

The manufacture of aircraft has become most important in the west, especially in California, with Los Angeles as a major centre of production.

Electrical engineering is especially significant in the Hudson–Mohawk valley with New York, Schenectady, and Rochester as major centres. Philadelphia, Pittsburgh, and Milwaukee are other important centres. The manufacture of electronic apparatus is especially important in San Francisco.

## FRANCE

The metallurgical and associated engineering industries are one of the three major branches of French industry. The chief engineering areas are the north-east and Lorraine, and heavy engineering is naturally concentrated in these two main iron- and steel-manufacturing areas. A wide range of engineering products, embracing almost everything except automobiles, is produced in the north-east. The chief centres are Dunkirk, Lille (textile machinery), Douai, Denain (locomotives), and Valenciennes (textile machinery). Lorraine has little heavy engineering but the lighter side of the industry is well developed. The Lyons–St Etienne district is also an important engineering area specialising in machine-tools, cutlery, and bicycle manufacture. Paris and its industrial suburbs have important engineering works especially concerned with machine tools and electrical engineering. France's motor car industry is largely concentrated in the Paris area, with Lyons, Clermont-Ferrand, and Sochaux as subsidiary centres. The aircraft industry is widely dispersed, being found in Paris, Toulouse, Bordeaux, and Marseilles. Le Creusot and St Etienne are important centres of locomotive and railway rolling stock manufacture.

## BENELUX

Belgium, Holland, and Luxembourg all have iron and steel industries but only the first two have important engineering industries. In Belgium most of the engineering is carried on in the towns of the coalfield zone, *e.g.* in Liège and Charleroi, although both Brussels and Ghent share in the industry. Typical products of the Belgian engineering industry are mining equipment, furnaces, locomotives, drills, and pipelines. The Dutch engineering industry is especially concerned with hydraulic and drainage equipment and bridge-building, while the enormous Philips factory at Eindhoven is a major centre of manufacture for electrical machinery. Holland also has important non-ferrous smelting and refining industries at Arnhem and Ijmuiden.

## WEST GERMANY

The Federal Republic has a very well-developed and very scattered engineering industry. Traditionally she has been a great producer and exporter of engineering products. The heavy branch of the industry is concentrated in the Ruhr region (where it has always been) mainly in Essen, Duisburg, Bochum, Oberhausen, Gelsenkirchen and Dortmund. The principal products are girders, bridges, mining gear, pipes, boilers, furnaces. The other important engineering provinces are: the Bremen–Hamburg area; the Hanover–Brunswick area; the Mainz–Frankfurt–Darmstadt area; and the Heilsbronn–Stuttgart area. Nuremberg and Munich are important centres of electrical engineering. Solingen, Remscheid and Hagen in the Ruhr region are noteworthy as centres of cutlery and tools. Agricultural machinery is produced mainly in the towns of the rich loess farming belt, *e.g.* at Düsseldorf, Hanover, and Brunswick.

Locomotive and railway rolling stock are built at the major railway foci such as Kassel and Munich. The motor car industry, which has grown very rapidly during recent years and which has now a large export trade, is widely spread. Wolfsburg (near Brunswick) is the home of the well-known Volkswagen cars; Rüsselheim (near Frankfurt) has the Opel works; while there are automobile plants also at Bremen, Cologne, Düsseldorf, Stuttgart, and Nuremberg. Germany, with an annual production of 3 million vehicles, ranks third in the world in motor vehicle output.

[*Courtesy: Bullock & Turner.*

FIG. 118.—FIAT FACTORY, TURIN

The north Italian town of Turin is almost synonymous with the Fiat motor car company. The company has twenty factories and employs 127,000 people. This is a view of part of the Fiat automobile works. It illustrates well the enormous size of some modern industrial plants. Mechanisation and the assembly-line system, first introduced by Henry Ford in the United States, characterise such large-scale plants.

## SWEDEN

Mention should be made of Sweden for, though not a major engineering country, she has developed a highly specialised engineering industry. Lacking coal, Sweden could not compete with many other European countries in the production of either iron and steel or heavy engineering products. But she has specialised in engineering goods requiring great skill and using only relatively small quantities of raw material. Thus numerous towns in the Lake Belt of Central Sweden have engineering industries. Ball-bearings, springs, drills, precision instruments, turbines, telecommunications equipment are representative products. Eskilstuna is a great engineering centre and is noted for its cutlery. The capital, Stockholm, has important electrical industries. It is surprising how many Swedish engineering firms have international reputations, *e.g.* Electrolux, S.K.F., Ericsson, Aga, etc.

**SWITZERLAND**

Like Sweden, Switzerland specialises in the lighter side of the engineering industry. Developing out of the early demands for textile machinery, the engineering industry grew steadily until now the industry accounts for almost half of total Swiss exports. Diesel-electric locomotives are manufactured at Zürich and Winterthur, electrical machinery at Zürich, Basle, Berne and Geneva, and textile machinery at Neuchâtel. In addition, Switzerland produces machine tools, precision instruments, typewriters, calculating machines and, of course, watches and clocks.

**THE SOVIET UNION**

Only the briefest mention can be made of the vast engineering output of the U.S.S.R. The great industrial progress made by the Soviet Union since 1928 (the start of the First Five-Year Plan) has been based upon the iron and steel and heavy engineering industries. Industrialisation, at first mainly confined to European Russia, has gradually spread into Soviet Asia so that industrial centres (including engineering centres) are now fairly widely spread throughout the country.

The major centres of the engineering industry are: the Moscow district, which specialises in light engineering, *e.g.* machine tools, electrical equipment, and motor cars; the Donetz region which, with its sources of coal and iron, is primarily concerned with heavy engineering; the Ural region where Sverdlovsk and Chelyabinsk are the chief centres; and the Kuznetsk basin with Novosibirsk as the main focus.

Most of the heavy structural engineering takes place in the Donetz Basin. Marine engineering is carried on at Leningrad. The major locomotive and railway rolling stock centres are Briansk, Kolomna, and Voroshilovgrad; minor centres are Taghil, Krasnoyarsk, and Ulan Ude. Automobile engineering is carried on at Gorki (where there is one of the largest motor car plants in the world), Moscow, Yaroslavl, Chelyabinsk, and Novosibirsk. The Soviet Union produces 1 million motor vehicles annually.

## *SHIP-BUILDING*

The construction of ships is really a branch of heavy engineering but it is of sufficient importance to warrant separate treatment. It is an activity mainly of the chief iron- and steel-producing countries and of the trading and maritime nations. Ship-building requires great quantities of steel plates and girders; this fact, together with the growth in the size of ships, favoured the concentration of the industry in large estuaries or deep water coastal locations adjacent to steel-producing areas.

The ship-building industry was dominated by the United Kingdom until the turn of the present century. In 1870 the U.K. was responsible for some 70% of the tonnage launched in the world. In 1900 the U.K. was still producing about two-thirds. Thereafter, however, certain continental countries began to appear as competitors—notably Germany—and, more recently, Japan. During the 'thirties the U.K. was responsible for little more than a third of the total tonnage launched, although she continued to hold first place as a builder. Today, U.K. production is less than 10% of world output. Japan, which assumed the lead in the late 'fifties, now produces 8½ million tons annually.

The following table gives the tonnage launched in 1968 by the major ship-building countries, *i.e.* all those launching about half a million tons a year.

TABLE XXXVIII

*Tonnage Launched: 1968*

| *Country* | *Total tonnage* | *Tanker tonnage* |
| --- | --- | --- |
| Japan | 8,583,000 | 3,973,000 |
| West Germany | 1,352,000 | 508,000 |
| Sweden | 1,113,000 | 391,000 |
| United Kingdom | 898,000 | 81,000 |
| Italy | 506,000 | 102,000 |
| Spain | 506,000 | 239,000 |
| France | 490,000 | 350,000 |
| Norway | 490,000 | 182,000 |
| Denmark | 483,000 | 357,000 |
| United States | 441,000 | 111,000 |
| Poland | 424,000 | 43,000 |
| Netherlands | 303,000 | 156,000 |
| World total | 16,908,000 | 6,613,000 |

(*Note:* the U.S.S.R. is probably a major producer nowadays, perhaps launching 1 million tons annually, but accurate figures are hard to come by.)

**UNITED KINGDOM**

Not only was the United Kingdom for many decades the greatest producer of ships in the world, but she enjoyed a great reputation for the quality of her vessels. The reasons leading to this superiority were: (*a*) a long tradition and acquired skill in ship construction dating from the period of the wooden ship; (*b*) the pre-eminence of Britain as a mercantile and naval power demanding vessels of all types; (*c*) the presence of deep, sheltered tidal estuaries suitable for launching; and (*d*) the possession of iron and steel industries on coalfields lying close to the estuaries.

The principal ship-building areas are:

1. The estuary of the Clyde.
2. Tyne-, Wear-, and Tees-side.
3. Belfast Lough.
4. Birkenhead on the Mersey.
5. Barrow-in-Furness.

Minor ship-building centres are found at Aberdeen, Dundee, and Leith in Scotland; at Hull; and at Portsmouth, Devonport, and Chatham where there are Royal Naval Dockyards. Cardiff, which once had an appreciable shipbuilding industry, is no longer of any significance in this respect.

Normally, the two leading ship-building areas, the North-east Coast and Clydeside, build about 70% of the total tonnage launched. At present the industry is in great trouble and some of the shipyards on Clydeside have become bankrupt and closed down. The other ship-building centres, Belfast, Birkenhead, and Barrow, are much smaller both in size and in their output, averaging in a normal year approximately 8%, 6% and 4% respectively of the total production.

Nowadays there is less specialisation in types of ships produced though each region has a reputation for particular types of building. Most of the large passenger liners have come from the Clyde. The North-east Coast has tended to build smaller vessels and especially tankers. Belfast has tended to specialise in cargo-liners, Birkenhead in submarines, and Barrow in aircraft carriers and other naval vessels. The minor shipbuilding centres are mostly concerned with the construction of fishing vessels.

Closely linked with ship-building are the marine engineering and ship-repairing industries. Marine engines are built mainly near to the shipyards, though some marine engineering is carried on in Sheffield, Birmingham, and London. Repairing is not confined to the building centres; it is shared by most of the great ports of the United Kingdom.

## JAPAN

In 1957 Japan wrested the lead in ship-building from the United Kingdom. In 1960 she launched 1,731,656 tons as against the U.K.'s 1,331,491 tons and by 1962 she had doubled the U.K. output. In 1968 Japan produced $8\frac{1}{2}$ million tons of shipping, a third of which was tanker tonnage. In pre-war days the Japanese ship-building industry was stimulated by the expansion of her own naval and mercantile fleets; in post-war years she has built ships for export. The Japanese have developed many new techniques in ship construction. Ship-building is centred chiefly at Nagasaki, Osaka, Kobe, and Yokohama.

## UNITED STATES

Although possessing about half of the world's ship-building capacity, the U.S.A. has the greater part of this capacity lying idle and she produces, on the average, only about half a million tons a year. The main reasons for this are the high cost of the raw materials and the high labour costs. Unable, therefore, to compete in the world market, the U.S.A. has had to close down many of her yards; in fact, many of those still operating do so only because of government subsidies and naval contracts.

The mid-Atlantic coast is the most important ship-building area; steel is produced here in large quantities and there is convenient access to tide-water. Philadelphia, Newport News, Camden, and Chester are the chief centres of construction. Boston in New England is also a centre. The second major area is the Great Lakes region where several lakeside centres, notably Chicago, Detroit, and Buffalo, build lake craft. Though handi-capped by lack of steel supplies, ship-building has grown up during more recent times at Seattle, Portland, and Tacoma in the Puget Sound area, and at San Francisco. There is also some building in the Gulf Ports.

## WEST GERMANY

Germany was the second most important ship-building country in the world prior to the Second World War, producing about half a million tons annually. About 60% of the gross tonnage launched came from the shipyards of Hamburg and Bremen–Bremerhaven, the two main con-struction areas. During the past decade, West Germany has made a great comeback as a builder and in recent years has been turning out over 1 million tons of new construction each year. Normally, about half the tonnage launched comprises oil-tankers.

Hamburg and Bremen–Bremerhaven are still the chief centres, account-ing for about 60% of the output, roughly in equal proportions. Kiel is the third centre, with about 20% of the tonnage launched. Minor ship-building ports are Emden, Lübeck, Rendsburg, and Flensburg. Duisburg–Ruhrort on the Rhine is noteworthy as a centre for the building of Rhine barges.

## FRANCE

Though not one of the "big three," France ranks usually about fifth or sixth among the ship-building countries of the world. Her output remains consistently steady at around half a million tons. France's most important yards are in the Loire estuary at Nantes–St Nazaire, at La Ciotat (Mar-seilles), and Dunkirk. Brest, Lorient, and La Seyne (Toulon) all build merchant vessels as well as naval craft. There are smaller ship-building yards at Boulogne, Le Havre, Rouen, Rochefort, and Bordeaux, which build fishing vessels, coastal craft, and smaller type ships. Most of the ports carry out ship-repairing but Brest is of especial importance on this account largely because it is well-placed with respect to the North Atlantic traffic routes.

## HOLLAND

The Dutch have a long tradition as shipbuilders, but the growth of the Dutch ship construction industry since the war has been nothing short of spectacular. During recent years her output has been around a quarter of a million tons (566,993 in 1960; 303,000 in 1968). Holland has great building and repair yards at Rotterdam and Amsterdam. Much of the construction has been in tankers. The Dutch have gained a well-merited reputation for quick and efficient ship repairing and, due to increasing competition in the building side, they have begun, increasingly, to specialise in the repairing industry. For this, Holland, situated astride the outlet of the Rhine and alongside the main sea lanes of the North Sea, is ideally placed. In northern Holland there are many small boatbuilding yards turning out small coasting craft and fishing vessels, usually of under 600 tons gross.

## SWEDEN

Ship-building is a well-established industry in all the Scandinavian countries but is particularly important in Sweden. In 1967 Sweden ranked second in the world and was just ahead of the U.K. She has some of the largest shipyards in the world. The chief Swedish yards are found at Göteborg, Malmö, Uddevalla, and Oskarshamn. Three-quarters of the tonnage launched consists of oil-tankers, a large proportion of which are for the Norwegian mercantile marine. It is worth noting that Sweden, in proportion to its population, is incomparably the greatest ship-building country in the world.

## THE SOVIET UNION

Currently involved in the building up of a great navy and of a mercantile marine to challenge that of the U.K. and the U.S.A., the Soviet Union is expanding its ship-building capacity. But in spite of this, the output from her own yards is incapable of meeting her growing needs and she is having to place orders elsewhere. The chief Soviet ship-building centres are at Leningrad and other small Baltic ports, at Nicolayev and other places on the Black Sea coast (where great new shipyards are being developed), at Archangel on the White Sea, and at Vladivostok on the Pacific coast. The tonnage launched by the U.S.S.R. is not known with any certainty.

## *TEXTILE MANUFACTURING*

The manufacture of textiles is one of the world's leading industrial activities. Not only is it highly developed in most of the world's principal industrial countries, but it is frequently the major source of industrial employment among the less-industrialised countries. As an industry, textile manufacturing has advantages for a country embarking upon an industrial career: technically it is a relatively simple activity, equipment is easy to procure and install, countries can often produce their own raw materials, and the cottage textile industry is usually in existence, making the transition to factory industry a relatively easy one. Japan, for example, when first launching itself into modern industrialism, commenced with textiles. India, Brazil, and Egypt provide other examples.

Textiles are manufactured from natural fibres of animal or vegetable origin, *e.g.* wool, silk, cotton, flax; from man-made fibres, *e.g.* rayon, nylon, Terylene, glass fibre; or a mixture of both. Increasingly, the natural fibres are being "fortified" by artificial or synthetic fibres; mixtures of rayon and cotton, rayon and wool or wool and Terylene are now quite common. This mixing of fibres is in fact leading to the gradual break-down of the traditional distinction between the cotton textile industry on the one hand and the woollen industry on the other. In spite of the very rapid development of man-made fibres, the textile industry as a whole remains primarily dependent upon natural fibres, which account for about three-quarters of all the fibre used.

Of the natural fibres cotton is easily the most important. It is a versatile textile used for wearing apparel, bed "linen," household furnishings, and in industry. It has a great advantage in that it will bleach, dye, and print well. Of the other vegetable fibres, flax is the most important for textile production but ranks a long way behind cotton. Jute and the so-called hard fibres, abaca and sisal, are used almost solely for the manufacture of sacking and cordage. Wool is the principal animal fibre and is used in the manufacture of worsteds and woollens. The bulk of the wool comes from sheep, but very small quantities of alpaca and vicuña wool are also used, along with mohair from the Angora goat, and camel hair.

Rayon, the most important of the man-made fibres, is made from wood-pulp or cotton linters. Rayon falls into the category of regenerated fibres, that is, it is made from the chemical treatment of raw vegetable materials. On the other hand, the true synthetic fibres, such as nylon, Terylene, Acrilan, are synthesised from chemical or mineral substances, notably the by-products of coal and petroleum. These various man-made fibres are beginning to compete strongly with the traditional fibres, cotton, wool, flax, and silk, and the rapid decline of the last of these

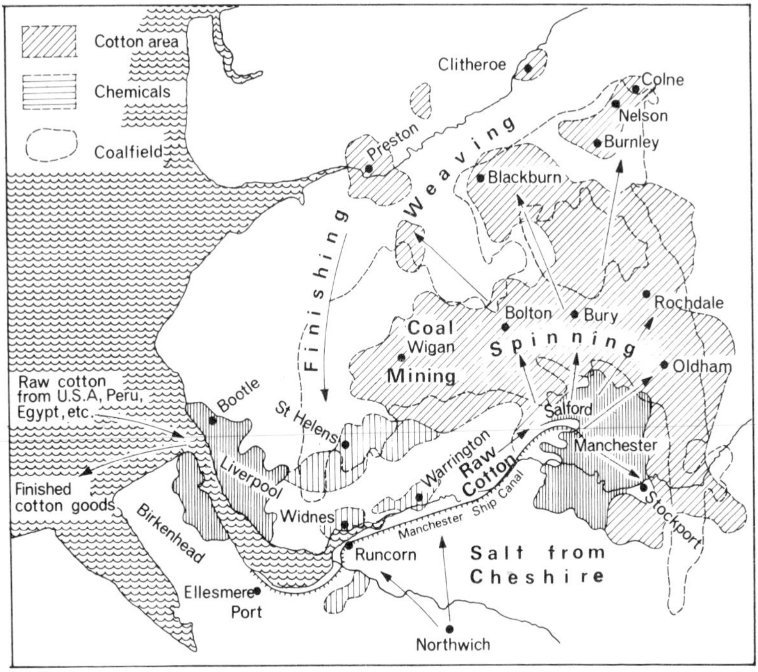


FIG. 119.—LANCASHIRE COTTON INDUSTRY

For a hundred years the manufacturing of cotton cloth was one of Britain's major industries and a major export item. The industry may be said to have grown up out of the "triangular trade" when ships returning home from the West Indies brought cotton to Liverpool and Glasgow. Lancashire, however, became *the* great centre of the cotton industry. A number of factors helped to stimulate it: (i) raw cotton could be easily imported via Liverpool and, later, via the Manchester Ship Canal; (ii) a domestic woollen and linen industry was already in existence in the area; (iii) soft water from the Pennine hills was available in quantity for the finishing processes; (iv) the damp atmosphere assisted the spinning process; (v) coal from the Lancashire Coalfield was available for power-machines; and (vi) the nearby chemical industry provided the necessary soap, chemicals, and dyes required by the finishing section of the industry.

The Lancashire cotton industry was very well organised. Specialisation led to a threefold organisation of the industry. Raw cotton was brought into the Manchester market, then distributed to the encircling spinning towns, next sent to the weaving centres in the Ribble valley, and then finally to the finishing towns which were close to Liverpool.

Today the industry has fallen on hard times because of competition from countries such as Japan, India, China, and Brazil which have advantages over Britain in such things as cheaper labour and home-grown supplies of cotton. Today fewer than 130,000 are employed in the Lancashire cotton industry, three-quarters of the finished products are sold in Britain, and the value of the export trade has been reduced to about £50 million.

is very closely linked with the rapid growth of the rayon and nylon industry.

## THE UNITED KINGDOM

The United Kingdom processes all the important textile fibres. Although the textile manufacturing industry has shown a relative decline, and in some sections of the industry an absolute decline, notably in cotton, the United Kingdom remains one of the world's leading textile producers and exporters. She still provides about 40% of the total world exports of woollen manufactures and about 10% of the cotton goods. The basic mechanical inventions and technological processes associated with large-scale, commercial textile manufacture had their origin in the United Kingdom during the Industrial Revolution; these not only gave the United Kingdom a lead in textile production, which led to her having almost a virtual monopoly of the world market for close on a century, but enabled her to develop a technological know-how which is still unsurpassed and which enables her to produce the finest fabrics in the world. Few countries, if any, can challenge the United Kingdom in the production of fine worsteds, high-grade cottons, or high quality linens. Here is a skill born of many generations of workers "in the trade."

Cotton has long been the chief textile industry but this has greatly declined during recent decades due to strong competition from India and Japan with their low production costs. The chief seat of the industry is South Lancashire where a number of favouring factors relating to the spinning process, the finishing process, water supplies, and power needs promoted and facilitated the establishment and expansion of the industry (*see* Fig. 119). Cotton is also manufactured in the lower Clyde area which had humid atmospheric conditions, nearby coal, and a port for import and export purposes. Glasgow and Paisley are the chief centres, the latter being the greatest producer of cotton thread in the world. Nottingham is primarily concerned with the manufacture of lace and hosiery. The growth of the industry here is largely due to the local invention of lace-making machinery.

The wool textile industry is more widely spread since it was the original domestic textile industry and has tended to linger on in some of the older centres. With the coming of the Industrial Revolution the woollen industry came to be specially developed in the valleys of the Aire, Calder, and Colne in the West Riding of Yorkshire. Here were found favouring conditions: (*a*) a domestic woollen industry already used to the production of coarse cloths; (*b*) supplies of raw wool from the Pennine sheep; (*c*) soft clean water from the gritstone uplands necessary for washing wool; (*d*) numerous streams providing power for mill water-wheels; (*e*) local coal from the Yorkshire coalfield when steam displaced water-power; and (*f*) facilities for transport by river and canal in the early days. Over 75% of all the woollen and worsted operatives in the United Kingdom are found in the West Riding.

It is interesting to note that the industry has always been a big employer of female labour; that after the Second World War large numbers of continental refugees (*e.g.* Poles, Latvians) found work in the wool mills and that more recently Pakistani and other coloured immigrants have secured employment in the industry. Bradford, Keighley, Shipley, Halifax, Huddersfield, Batley, Dewsbury, Morley, and Wakefield are the chief centres. A wide range of fabrics is produced—high-grade worsteds (Huddersfield), woollens, shoddy made from rags, clippings, and wool waste (Batley, Dewsbury), carpets (Halifax), and knitting wool. Bradford is the great marketing centre and possesses a Wool Exchange of international renown. Leeds, the largest city of the woollen district, plays little part in the actual manufacture of wool, but it is a great centre of ready-made clothes, and makes textile machinery.

Outside the West Riding the industry is found in the West of England centred on Bradford-on-Avon, Frome, Stroud, Trowbridge, where "broadcloth" is manufactured, and on Witney, noted for its blankets. The third main area is the Tweed basin of southern Scotland where the chief towns are Galashiels (high quality suitings), Hawick (hosiery), Peebles, Jedburgh, and Selkirk. The speciality of this region is high quality tweed cloths and knitwear. In the Islands of the Hebrides the famed Harris tweed is woven in homes and small workshops. Leicester and a number of small towns round about, *e.g.* Hinckley, Nuneaton, are noted for their knitwear goods, especially hosiery. Alloa in Scotland is a notable centre of knitting wool manufacture. The old flannel industry of mid-Wales has all but disappeared.

Northern Ireland (Ulster) is the home of an important linen industry, originally based on home-grown flax. The industry is now entirely dependent upon imported flax. The chief centres are Belfast (specialising in fine linens and damask fabrics), Newry, Lisburn, and Londonderry (shirts). Outside Northern Ireland the other manufacturing centres are Dunfermline (damasks), Dundee and Kirkcaldy (heavy linens), and Leeds.

Silk manufacture, not nearly so important as the other textile industries, is principally carried on at Macclesfield, Congleton, and Leek. The artificial silk or rayon industry has grown by leaps and bounds during the past fifty years and to it has now been added the synthetic fibre industry. Among the more important centres are Bradford, Manchester, Derby, Wolverhampton, Coventry, Flint, Lancaster, Spondon (near Nottingham), and Pontypool. Jute processing is mainly carried on at Dundee and Barnsley.

## FRANCE

France is a major producer of textiles and uses all kinds of fibres. The cotton industry is located in three main areas: in north-eastern France near the coalfield, around Rouen at the mouth of the Seine, and on both flanks of the Vosges. The chief towns in the north-east engaged in cotton manufacturing are Lille, Amiens, St Quentin, and Valenciennes (lace). In the Vosges area, the industry was originally focused in Alsace, at Colmar and Mulhouse in particular where it continues still. But, after the cession of Alsace to Germany, following the Franco–Prussian War (1870), many refugee workers settled on the western side of the Vosges and established the cotton industry in such towns as St Dié, Epinal, Remiremont, and Belfort.

The woollen industry is very widely spread throughout France. The chief centres, however, are in the north-east where originally local supplies of raw wool from Picardy and Champagne fed the industry. Much of the wool is now imported from Australia and Argentina through the ports of Dunkirk and Le Havre. Lille, Roubaix, and Tourcoing are important woollen towns producing the finer types of goods. In the same region are the more dispersed centres of Fourmies, Reims, Sedan, and Troyes (hosiery). Many of the small towns of the Central Plateau have woollen manufactures.

The linen industry is chiefly associated with the other textile industries of the north-east. Based originally on local-grown supplies of flax and now largely upon imported Belgian flax from the Lys valley, Lille and Cambrai are the chief centres. The term "cambric" derives from Cambrai.

The French silk industry, which is of great repute, is based upon Lyons. Lyons is the great organising centre and market for the industry, but most of the actual weaving is carried on in the villages and small towns around (*cf.* Manchester and the cotton industry). Little home-produced raw silk is now used; the bulk of it is imported from Italy and Japan. As elsewhere, the rayon industry has grown rapidly and much of the rayon production comes from Lyons, St Etienne, etc.

## WEST GERMANY

Like the United Kingdom and France, West Germany has a highly developed textiles industry. Although fairly well dispersed, the industry is concentrated in particular areas. The Westphalian region is the great textile area. Cotton manufacturing is carried on at Bocholt, Gronau,

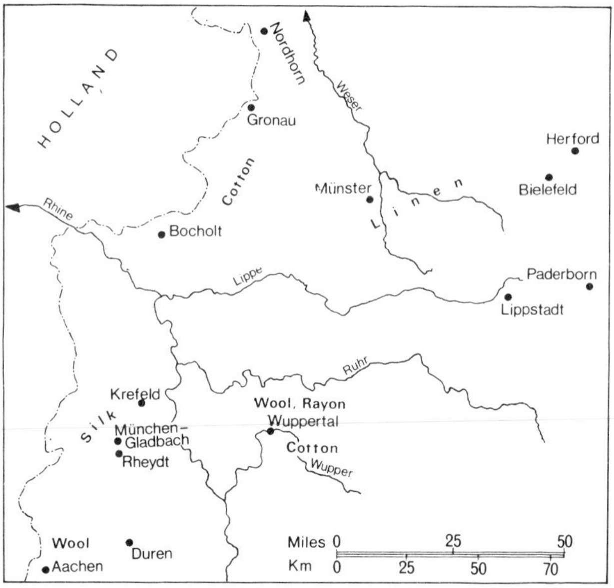


FIG. 120.—TEXTILES IN WEST GERMANY

The West German textile industry is fairly well scattered throughout the country although most of it is to be found in the northwest. Apart from the Wupper valley, there is a tendency towards regional specialisation: the towns near the Dutch frontier are mainly concerned with cotton, the towns in the Münster basin tend to carry on their traditional linen manufactures, the Krefeld area specialises in silks, rayons and the like, while Aachen and Duren are more particularly woollen centres.

and Nordhorn near the Dutch frontier. Münster, Bielefeld, Lippstadt, and Paderborn (an old wool market) in the Münster basin, now contain almost all the linen industry which formerly was more widely scattered. To the south of the Ruhr coalfield in the valley of the River Wupper lies the twin town of Barmen-Elberfeld (now known as Wuppertal) with cottons, woollens, and rayon. To the west of the Rhine are Krefeld, München-Gladbach, and Rheydt, noted for their silk manufactures. The first town has long been famous for its silks and velvets, while the second is an important cotton centre also. Aachen and Duren are old-established woollen centres. Geislingen in Swabia is a centre of knitwear production while the ancient town of Augsburg, originally mainly concerned with woollens, now manufactures cotton and rayon fabrics.

We may note that the post-war division of Germany greatly affected the textile industry. The worsted and stocking industries of Saxony and the woollen and linen industries of Lower Lusatia were lost to East Germany. On the other hand, the Federal Republic has gained large numbers of textile workers, refugees from East Germany. Following upon this disturbance of the textile industry, West Germany has made a great effort to become self-supporting in all branches of the textile industry and factories making specific textiles have been established which are not tied to the traditional textile areas. This post-war development, along with the increased use of synthetic fibres, *e.g.* rayon, Perlon (a German variety of nylon), etc., has helped to break up the old pattern of specialisation.

## ITALY

Italy has a fairly wide dispersal of both textile towns and textile industries but they occur mainly in northern Italy and especially north of the River Po and at the outlets of the Alpine valleys. However, a recent development has been the setting up of new textile factories in the relatively under-developed south of Italy. The cotton industry is the most important of the various textiles and has had a quick modern growth.

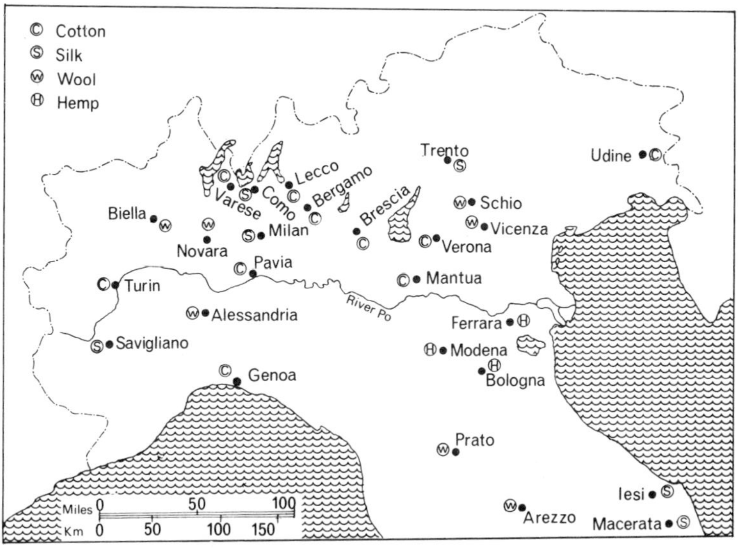


FIG. 121.—TEXTILES IN ITALY

Italy has a very wide range of textile industries: cotton, wool, silk, rayon, hemp, as well as many man-made synthetic fibres. The textile industry is mainly concentrated in northern Italy and especially the Plain of Lombardy, although some new factories have been set up recently in southern Italy. The silk centres lie mainly along the Alpine fringe and the hemp towns near the Po delta (this is the principal hemp-growing area).

Milan is the major centre, but around it are numerous small towns *e.g.* Legnano, Gallarate, Busto Arsizio, also engaged in cotton production. As against some 900 cotton factories employing some 215,000 people, the woollen industry, the oldest of the textile industries but now ranking second, employs some 125,000 workers in about 720 mills. The industry, which is mainly dependent upon imported wool, is localised chiefly in Piedmont, where it has been carried on for many centuries, at Biella, Novara, Pinerolo and Alessandria, and in Venetia at Asiago, Schio, and Vicenza. Some of the small towns of central Italy, such as Prato, have woollen manufactures.

Silk is one of the oldest of the textile industries, and until fairly recent times it was predominant. Italy still produces a small quantity of raw silk (861 metric tons in 1968). Como is the chief silk town; Milan, Trevigliano, Brescia, Bergamo, and Trento are other centres, while many villages in the foothills of the Alps also engage in silk weaving. Synthetic fibre manufacture, notably of rayon, has grown very rapidly; indeed, Italy now ranks fifth as a world producer of rayon. The industry is carried on in Aosta, Turin, Milan, Pavia, Cremona, and Padua. Linen, hemp, and jute are also manufactured. Hempen fabrics, made from home-grown hemp, are produced at Bologna, Ferrara, and Modena in the province of Emilia, while jute, imported from Pakistan, is processed in factories found in Turin, Pallanza, Lucca, Terni, and Naples.

### UNITED STATES

The cotton industry, the predominant industry, is found in two areas: the "old area" of New England and the "new area" in the Southern States. The New England industry, centred in Lawrence, Lowell, Providence, Fall River, Manchester, and New Bedford, owed its origin to the demand of the early settlers for cotton goods. Raw cotton was imported from the plantations of the South and the industry flourished because of: (*a*) its early start and the skill of the immigrant textile workers; (*b*) the abundance of water for processing and power and, later, hydro-electricity; (*c*) the rapid growth of a large local market.

The New England cotton industry has lost ground to the newer industry of the South and now accounts for less than a quarter of the total cotton goods production. The industry, however, continues to survive, in spite of high wages and transport costs, by concentrating on the production of higher grade cottons. The New England cotton industry provides a good example of the principle of geographical inertia.

The newer cotton area of the South, located principally around the southern edge of the Appalachian Mountains, dates from the end of last century. A string of Fall Line towns, Montgomery, Columbus, Macon,

Atlanta, Augusta, Columbia, and Charlotte, produce cotton manufactures and now account for approximately three-quarters of the national output. The Southern States have certain advantages over the traditional cotton textile area of New England: (*a*) proximity of supplies of raw cotton; (*b*) cheap hydro-electric power from the Appalachian streams and coal from the Birmingham coalfield; (*c*) lower costs in respect of land for

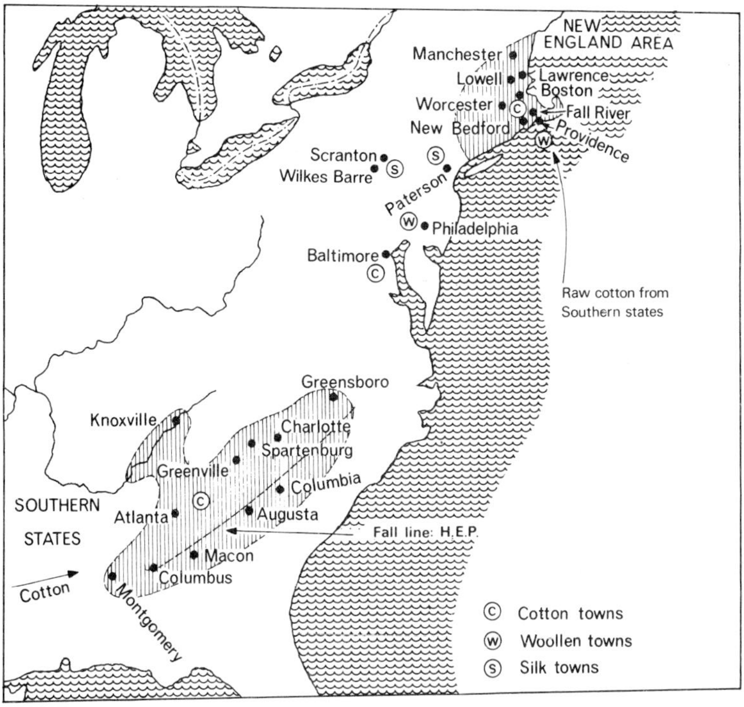


FIG. 122.—TEXTILES IN THE UNITED STATES

New England is the traditional textile area. The "South" is of more recent growth. Note how many of the cotton textile towns of the South lie on the Fall Line; here falling water provided a source of power for the factories.

factories; (*d*) relatively cheap negro labour, though this is becoming less so nowadays.

Woollen manufacturing is much more widespread and though much less important is gradually growing. The industry was first developed in the New England states and here are the main centres and the greatest expansion. Lawrence and Providence are the leading woollen towns of New England and they manufacture woollens, worsteds, and tweeds. The largest single woollen town, however, is Philadelphia which lies outside the New England region. It is a centre of carpet production. Boston is the chief wool market.

Just as the United States dominates the world in cotton goods production, so she leads in the manufacture of man-made fibres, *e.g.* rayon, nylon, Orlon, etc. Paterson, in New Jersey, and Scranton, in Pennsylvania, the traditional silk-manufacturing centres of the United States, have become important centres of the artificial fibre industry. The growth of the latter has, incidentally, much reduced the scale of the silk industry. Buffalo is also a centre of rayon manufacture. But important developments have occurred in the Southern States and Virginia is the leading state for rayon production, especially in the towns of Roanoke and Rock Hill. The manufacture of artificial fibre fabrics is widely spread, however, throughout the Piedmont zone.

## THE SOVIET UNION

The U.S.S.R. is one of the world's greatest producers of textiles and in the manufacture of cotton goods she ranks next to the United States. Textile production is centred mainly in the region around Moscow: in the capital itself and in a ring of towns encircling it. Cotton, woollen, linen, and rayon goods are all manufactured here. The long-established linen industry in the north still exists in Yaroslavl but now on a modern basis. Ivanovo and Kostromo in the Volga basin are now the centres of the U.S.S.R.'s greatest textile region; they are especially concerned with cotton and rayon manufacture. Serpukhov, south of Moscow, is also an important cotton town. Outside this central region are a number of outlying textile centres chief of which are Leningrad, with important cotton and woollen manufactures, Riga and Kaunas in the Baltic Republics, Tbilisi (Tiflis), and Leninakan in Transcaucasia, and Tashkent in Soviet Central Asia which has developed into a large cotton-manufacturing centre using locally-grown cotton cultivated by irrigation.

## JAPAN

Japan's modern industrial development was based to a very large extent upon textile manufacture—a branch of industry well-suited to countries aspiring to industrialisation. Importing textile machinery and borrowing ideas from the West, she quickly forged ahead as a producer of textiles, especially cotton goods, and soon became, in fact, a serious competitor in the international textile trade. Although cotton is the predominant textile, Japan has important rayon, silk, and woollen manufactures. A number of factors favoured the development of the Japanese textile industry: (*a*) raw materials were easily procured, *e.g.* home-produced silk and raw cotton from China; (*b*) coal and hydro-electric power were both available;

(*c*) the humid climatic conditions were suited to cotton spinning; (*d*) there was abundant cheap female labour; and (*e*) in the nearby countries of Eastern and South-east Asia there was a huge market for cheap cotton goods.

Cotton-spinning is mainly based in big, well-equipped factories, while the weaving tends to be done in smaller workshops. The industry is located primarily in the towns around the Inland Sea of Japan, especially in the great city of Osaka, dubbed "the Manchester of Japan," Kobe, and Nagoya. These towns are also centres of rayon manufacture. Japan is dependent upon imported supplies of raw cotton which come from the U.S.A., Mexico, India and Pakistan. Japan now produces just about twice as much cotton cloth as the United Kingdom and is the world's leading exporter, selling most of her cloth in markets which formerly were the preserve of Lancashire. The woollen industry, though much less important and again dependent upon imported wool, is growing. Nagoya and Osaka are the chief woollen towns. The silk industry is the traditional textile industry of Japan. The silk reeling, spinning, and weaving industries are more widespread though they tend to be concentrated in central Honshu; the chief centres are Kobe, Kanazawa, and Niigata.

**CHINA**

Not much is known about the progress of the textile industries of China, but it is clear that she must be a major producer. Much of the cloth produced is probably still manufactured on a domestic basis, although there are numerous large factories, some set up in pre-war years by both European and Japanese industrialists—mainly in coastal locations—others established by the Communist Government in more recent years. China ranks next to the U.S.A. as a grower of cotton and vies with her for first place as a producer of cotton yarn. The great coast ports of Shanghai, Tsingtao, and Tientsin were the original centres of the factory cotton industry, but newer factories have been established in inland centres such as Sian. Hankow, in the mid-Yangtze basin, has cotton- and silk-mills as, also, has its near neighbour Wuchang, which has a reputation for its silk manufactures. Canton has important textile industries processing cotton, silk, and jute.

**INDIA**

India, like China, is a major grower of cotton and a major manufacturer of cotton goods. Cotton has been spun and woven in India for several thousand years and at one time she possessed a great reputation for her cotton cloth. There has always been a vast production of cotton cloth from domestic looms, but there has always been a large population and a large internal market so that half a century ago she was compelled to import large quantities of cotton goods. Today, the tables have been turned and India produces enough cotton cloth to have a considerable surplus for export. Home-grown supplies of raw cotton, plentiful supplies of power (either coal or hydro-electricity), good transport facilities, and cheap labour have greatly assisted the industry's development. Indian cotton manufactures are, generally speaking, of low grade. Bombay is the chief centre of the cotton industry, but other important manufacturing towns are Ahmadabad, Nagpur, Kanpur (Cawnpore), and Madras. The production of woollen goods is much less important, although woollens are manufactured in a number of centres in northern India, notably in Kanpur. Jute manufactures, based on the jute grown in the Ganges delta, are of great importance and in the mills of Calcutta and Howrah and other centres along the Hooghly distributary large quantities of gunny cloth (used for making sacks) and hessian (used in upholstery) are produced.

## *THE CHEMICAL INDUSTRY*

This is a key industry, for chemicals are needed in a host of activities: in agriculture, metallurgy, textiles, leather, paper, glass, pottery, soap, and food preparation; in fact, in virtually every branch of industrial activity. In addition, however, the chemical industry is continually producing new chemical materials—synthetic fibres, plastics, and detergents are well-known modern examples of such new materials—hence the importance of the chemical industry will continue to grow. So vital and so important is the chemical industry in modern industrialism that the output of the chemical industry of a country may be said to be an index of its total industrial capacity.

The chemical industry is based upon: (*a*) natural raw materials: either minerals, chiefly salt, potash, sulphur, limestone, coal and petroleum, or vegetable products, such as wood-pulp, potatoes, vegetable oils, etc.; and (*b*) the by-products of various industries such as the iron and steel industry, the gas-making industry, the timber and wood-pulp industries.

"The range of the chemical industry's products is difficult to describe in a few words since, unlike most other industries, it produces a vast multitude of dissimilar and often unclassifiable products, instead of a few characteristic items such as iron and steel, leather, textiles and the like. In addition to a long list of typical chemicals with diverse applications, the industry supplies a number of products for specific purposes, including dyestuffs, pigments, solvents, refrigerants, rubber chemicals, synthetic perfumes, adhesives, plastics, synthetic fibres, emulsifying and foaming agents, fertilizers, insecticides, fungicides and weed killers."*

* *Oxford Economic Atlas*, second edition, 1959, p. 95.

The development of the chemical industry in any area is usually the result of five main factors:

1. The demands of manufacturing industries for chemical products, *e.g.* the textile industry which requires bleaching agents, dyes, etc.
2. Supplies of essential raw materials, *e.g.* salt and coal which in the early days of the industry formed the basic raw materials.
3. The presence of abundant supplies of clean, pure water, for the chemical industry is a greedy user of water.
4. Facilities for assembling and importing raw materials; since many of the materials are bulky, cheap water carriage is desirable.
5. The availability of a source of power, either coal or oil or thermal-electric or hydro-electric power.

The chemical industry has many branches, chief of which are: (*a*) the production of heavy chemicals, *e.g.* caustic soda, chlorine, hydrochloric and sulphuric acid; (*b*) the production of coal-tar chemicals, *e.g.* dyes, disinfectants, explosives, fibres, etc.; (*c*) the petro-chemical industry, *e.g.* petroleum distillates, sulphur, synthetic rubber; (*d*) the electro-chemical industry, *e.g.* wood-pulp, calcium carbide, ammonia, fertilisers; (*e*) the production of pharmaceuticals, *e.g.* drugs, medicines, vitamins, lotions, etc.; (*f*) the derived chemical manufactures, *e.g.* soap, paint, leather, paper, and glass.

As the world demand for chemicals grows apace, so the chemical industry is one of the most rapidly expanding, if not *the* most rapidly expanding, of all modern industries. It is most highly developed in the great industrialised areas, especially in the United States and the countries of north-western Europe. The United States and West Germany are the world's leading manufacturers of chemicals and vie with one another for first place in world trade in chemicals. The United Kingdom, France, Holland, the Soviet Union, and Japan possess important chemical industries.

## THE UNITED STATES

Rich in natural resources, the United States has a very large output of sulphur, potash, salt, and phosphates, as well as an abundance of coal, oil, and natural gas which greatly assist her chemical industry. Largely developing after 1880 and expanding rapidly after the First World War, the United States' chemical industry grew to be the world's largest, and she has retained this ascendancy up to the present day. And, until very recently, she was the chief world exporter of chemicals. Some indication of the size of the chemical industry is given by the simple fact that the United States produces some 15 million tons of sulphuric acid, which is about 40% of the total world output. The chief products of the United States' chemical industry are: sulphuric acid, used in the manufacture of fertilisers, petroleum-refining, and the making of other chemicals; soda ash (sodium carbonate), the leading alkali, used in the making of glass, caustic soda, and the production of pulp; explosives such as nitrocellulose, nitroglycerine, and dynamite; fertilisers; soap; plastics; synthetic fibres; and dyestuffs.

The chemical industry is fairly widely spread, although it is chiefly concentrated in the industrial north-east, because here is the principal internal market for chemical products. The main chemical centres are New York, Solvay near Syracuse, Pittsburgh, Buffalo, and Cincinnati. The Delaware River has long been the chief area for the manufacture of explosives. Outside the north-east, the most important area is the Gulf Coast area of Louisiana and Texas; here is a rapidly growing chemical industry based upon the local salt, sulphur, and petroleum resources. Altogether, the United States' chemical industry musters some 10,000 establishments and has a labour force of some 650,000. The industry is dominated by a few great companies, especially Du Pont, Allied Chemicals, and Union Carbide.

## WEST GERMANY

Germany has long had a great chemical industry. Many of the world's greatest chemists have been Germans. Prior to the Second World War, the German chemical industry was dominated by a great concern known as I. G. Farben. After the war the industry was controlled for a number of years for political and military reasons, but during the 'fifties it began to expand (as Allied military control was relaxed) and by 1960 the chemical industry of West Germany had become the giant among the chemical industries of Europe. In 1962 the industry's turnover reached the record figure of £2230 million, which accounted for nearly 12% of the total industrial production. The chemical industry now ranks alongside the machine industry as the country's second most important industry. It employs about half a million workers. The industry is organised in a number of major concerns, notably Farbenfabriken Bayer, Badische Anilin und Soda-Fabrik, and Farbwerke Hoechst, the present-day successors of I. G. Farben.

Originally based upon plentiful coal, salt, and potash deposits, the industry has come to make much use of other minerals and organic products such as wood and vegetable materials. Today, much of the industry is oil-based, and there has been a rapid development of the petro-chemical branch. The heavy side of the industry came to be located either in those areas where raw materials were found, *e.g.* at Strassfurt, or where the necessary materials could be easily and cheaply assembled, *e.g.* in the Rhine–Ruhr area. At present, the chemical industry is located primarily in or near the Rhine valley where there are three concentrations:

1. At Leverkusen and nearby Cologne and Düsseldorf (the great "4711" Eau de Cologne works is found in this district*).
2. At and around Frankfurt-am-Main and Darmstadt.
3. At Ludwigshafen–Mannheim and Heidelberg.

**THE UNITED KINGDOM**

The United Kingdom has a very highly developed chemical industry with almost all branches represented. Though fairly evenly distributed and present in all the important manufacturing districts, the two most

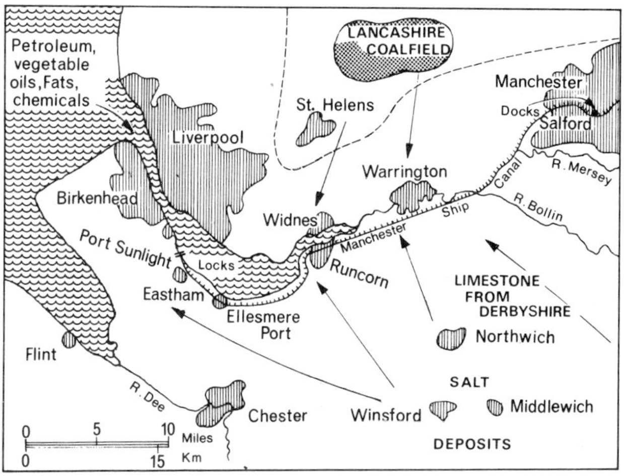


FIG. 123.—CHEMICALS IN BRITAIN

The heavy chemical industry in Britain is located in two main areas: Tees-side and South Lancashire. In the Merseyside region raw materials are easily assembled, pure water from the Pennines is available, while the textile, important paper-making industries of the region, soap, and glass manufactures have all stimulated chemical production.

important areas are the Tees-side and Merseyside localities which are the centres of heavy chemical manufacture. Plentiful supplies of salt and coal, the basic requirements for the heavy chemical industry, occur together in the Tees estuary area and in the South Lancashire–North Cheshire area.

* "4711"—so-called because the founder of the Eau de Cologne works had four sons, seven daughters, making eleven children; at least this is the story told by the local German people.

Both areas also have nearby limestone resources and easy water communications for the import and internal movement of chemicals. Imperial Chemical Industries Ltd. (I.C.I.), the predominant British chemical concern, has great plants at Billingham and Wilton on Tees-side producing heavy chemicals, plastics, synthetic fibre, and many other products, at Huddersfield, which is largely concerned with dyestuffs (for the demands of the local textile industry) and the making of paint, and at Northwich, where 50% of the working population is employed by I.C.I.

Along the Mersey estuary, using salt from the Weaver valley, and coal from the South Lancashire and North Wales coalfields, together with lime from Derbyshire and imported vegetable oils, there is the second major chemical area. Runcorn, Widnes, Warrington and Northwich, are the chief centres producing chemicals. There are also derived chemical industries in this area, *e.g.* glass at St Helens and soap at Port Sunlight.

Among the other more important chemical centres are Bristol, London, Birmingham, Manchester, Hull, Newcastle, Leeds, Glasgow, and Grangemouth. I.C.I. has built a great new factory on Severnside.

## FRANCE

The French chemical industry is concentrated in four main areas:

1. In the north-east on and near the Nord and Pas de Calais coalfields where the textile and metallurgical industries have led to a demand for chemicals.
2. In Lorraine where there are supplies of salt and nearby coal and potash and a demand from the cotton textile and metallurgical industries for bleaching agents, dyes, carbon, etc.
3. The Lyons–Alps–Marseilles area with its hydro-electric power, textile and leather processing requirements, local supplies of lime and olive oil, and easy import through Marseilles of tropical vegetable oils.
4. The Bordeaux–Haute Garonne area, where the industry is largely of recent development and is largely based on hydro-electric power and the recently discovered natural gas deposits.

## ITALY

During recent decades a great chemical industry has grown up which is still expanding. It is based mainly on home-produced raw materials, *e.g.* sulphur, borax, limestone, olive oil, methane gas, etc., and hydro-electric power. Italy produces a wide variety of chemical products and associated chemical products, *e.g.* sulphuric acid, boric acid, carbide, synthetic fibres, synthetic rubber, fertilisers, glass, and cement. A growing petro-chemical industry is located at the mouth of the Po based upon the

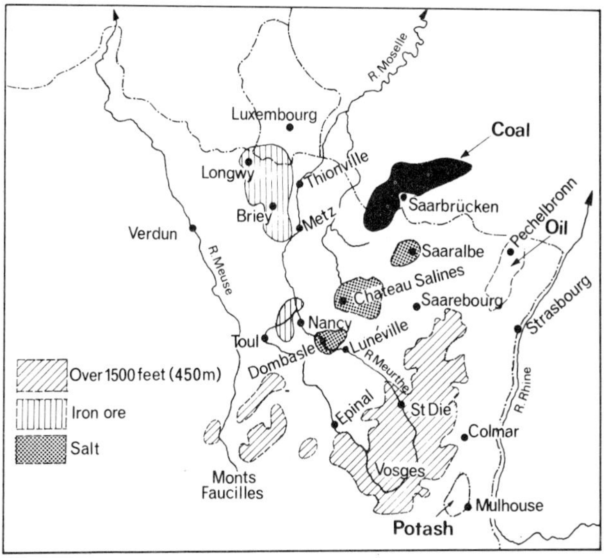


FIG. 124.—CHEMICALS IN FRANCE

Natural deposits of salt in Lorraine are the basis of the industry but coal, oil, and potash deposits together with slag, gases and other by-products from the iron and steel industry of the region have helped the growth of the industry. Over 2 million tons of salt are extracted annually. Large chemical industries have grown up in and around Nancy while the potash deposits of the Mulhouse district have given rise to fertiliser production.

rich natural gas resources of the Plain of Lombardy and imported petroleum. The chief centres of the chemical industry are Milan, Ferrara, Venice, Ravenna, Terni, and Naples. Montecatini is the chief Italian chemical concern.

## THE NETHERLANDS

The chemical industry has grown by leaps and bounds during recent years. This growth has been linked mainly with the growth in oil-refining and the Dutch have a very flourishing petro-chemical industry. The recent discovery of large reserves of natural gas in northern Holland is bound to stimulate the industry further. Farming in Holland, notably with respect to land reclamation, requires large quantities of fertilisers, and this has led to the development of the heavy chemical industry. There are chemical plants on the Limburg coalfield, at Ijmuiden, and at Rotterdam and its satellites, Pernis and Botlek.

## *THE PLASTICS INDUSTRY*

Special mention should be made of the plastics industry, a subsidiary of the chemical industry, which is the fastest-growing and probably the most important branch of the chemical industry today. The post-war development of the plastics industry has been truly phenomenal and world production of plastics has grown from 2·9 million tons in 1955 to 16·3 million tons in 1966; currently production is running at over 20 million tons. The tremendous growth of the plastics industry, especially during the past decade, has been due mainly to the very varied use of plastics and their wide applications in other manufacturing industries: for example, plastics are used for such purposes as floor coverings, furniture, car and boat bodies, disposable containers, and many other types of domestic consumer goods.

### UNITED KINGDOM

Britain has enjoyed a leading position in the development and production of plastics and at the present time is one of the world's major producers. The British plastics industry is expanding twice as fast as the chemical industry and four times as quickly as manufacturing generally. In 1968, for instance, production rose by 12% to give an annual total output of $1\frac{1}{4}$ million tons. At present, the most important uses are in the car and construction industries and it is anticipated that the demands of these industries will increase rapidly during the next decade. It is interesting to note that some authorities predict that the output of plastics will equal that of steel by the end of the century! As the current demand for plastics is constantly increasing, there are fears that the supply will not keep pace with demand. Clearly we must see that this situation does not arise, for Britain can ill afford to import a high proportion of its plastic raw material (currently running at some 300,000 tons); it seems likely, however, that we shall always have to import some plastics since no national industry could produce every single type of plastic in commercial quantities.

#### EXERCISES

1. Examine the factors which have influenced the growth and development of cotton manufacturing in *two* of the following areas: (*a*) Lancashire, (*b*) New England, (*c*) the Bombay region, (*d*) Japan.
2. Locate precisely the world's major ship-building centres and explain the development of the industry in these areas.
3. What is meant by the term "engineering industries"? Choose one branch of the engineering industry in Britain and give a geographical account of its location and growth.
4. Discuss the geographical factors influencing the development of the industries of Japan which produce manufactured goods for export.

5. Give an account of some of the raw materials and the natural resources of India that will help in her attempt to become an industrial power. What manufacturing industries already exist in India on a large scale and where are they situated? (*Chartered Institute of Secretaries.*)

6. Japan now produces considerably more cotton cloth than Britain and sells her cotton goods in markets which were formerly the preserve of Lancashire. Explain why Japan has come to have this predominance.

7. Write a brief account of the textile industries of the United States, emphasising especially the advantages she possesses for textile manufacturing.

8. Describe the location of the chemical industry in Great Britain, indicating the factors which have assisted its growth.

9. Account for the continued importance of *either* wool textiles in the West Riding *or* steel in the Don Valley.

10. Write a geographical account of the engineering industries of the Soviet Union.

11. "Italy has a well-developed and very diverse textile industry." Elaborate upon this statement.

12. Attempt to explain the recent growth and importance of the chemical industry in Holland.

## Part Four

# THE MECHANISM OF COMMERCE

## *Chapter XXII*

## TRADE AND COMMERCE

It will be obvious from what has already been said that man—especially when he lives in a highly-developed, industrialised country and enjoys a high standard of living—eats, drinks, and uses a great variety of commodities. In earlier times man had to be content for his food and drink and for most of the things that he needed upon the locality in which he lived. He lived an almost self-sufficient existence, as do many of the primitive societies we described in Chapter III. But even among peoples who are backward and isolated we find that certain commodities—such things as grain, salt, trinkets, weapons, poisons—are often procured by barter, or exchange, from neighbouring peoples with whom they come into contact, if only occasionally, or perhaps from farther afield. It becomes clear that even among the most primitive and isolated groups some trade takes place. Between highly-developed countries trade is of a very varied, complex, and large-scale character. The items which Britain, for instance, exports and imports each month forms a lengthy list, while the value of the exports and the imports runs to between £650–750 million per month.

There are many questions connected with trade, *e.g.* why and how did trading begin? what are the bases of trade? what kinds of trade are there? what is meant by the balance of trade? why are there restrictions sometimes on trade? how is foreign exchange transacted? In this chapter we shall try to provide an answer to these different points.

### *THE BEGINNINGS OF TRADE*

Exactly when man first began to trade is not known, but certainly by the fifth millenium B.C. a brisk trade had sprung up between Ancient Egypt and Mesopotamia, and the wall carvings in Egypt leave us in no doubt that a thriving trade existed several thousands of years ago. The famous inscription on the wall of the temple at Deir el-Bahari, which depicts the expedition to Punt, despatched by Queen Hatshepsut, gives us some idea of the commodities of trade in antiquity—gems, gold, spices, perfumes, unguents, leopard skins, fine feathers, peacocks, and strange animals. Trade in early times was mainly of a luxury nature.

Exchange of a more mundane character did, however, take place. The frequent Biblical references to "corn in Egypt" seem to imply that there was an abundance of grain, and there seems to be little doubt that occasionally this was shipped to Mesopotamia and elsewhere. Another important item of trade in early days was timber. Egypt, as a desert land, lacked timber, but wood was required for her ships; hence a considerable trade in cedar, fir, and oak timber took place between Egypt and the Phoenician exporting port of Byblos.

Even in antiquity there was a trickle of trade between Europe and Eastern Asia, and the Greeks, Romans and Arabs all played a part in this traffic later on. As early as the first century B.C. the Chinese were using their famous Silk Route, which ran from the end of the Great Wall of China westwards to south-west Asia. Along this route, caravans wended their way carrying precious silks, brocades, furs, iron, cinnamon, etc., from China and Central Asian oasis cities; along it also, from India and the East Indies, travelled pearls, rubies, ivory, perfumes, incense, pepper, cloves, and drugs. All these commodities were joyfully received by the richer Europeans through the agency of the merchants of Greece, and, later, Rome and the medieval Italian trading cities. Especially valuable, perhaps, were spices, used mainly for preserving meat and other foodstuffs and making them more palatable. Until the eighteenth century, it was customary to kill off the cattle and sheep because of the lack of winter feeding stuffs. Hence fresh meat was scarce, and by spring pickled and salted meat were turning and had an offensive smell. Only by spicing it well could it be stomached. Besides the landways linking west and east, there were seaways, of which the Red Sea and Persian Gulf formed important parts.

In return for the luxuries which the Europeans imported, Mediterranean merchants, with the assistance of others from the Baltic ports, were able to supply the Orientals with metals such as lead and copper and silver, with linen and woollen textiles, with amber, and with glass. It was never easy to maintain a large volume of trade between the eastern and western lands of Eurasia, and it is unlikely that more than a mere handful of Europeans actually visited China and the East Indies. Besides the lack of maps, and the great distances involved, the existence of the vast Old World Desert was a great obstacle. Reinforcing the desert barriers were the great mountain ranges of Central Asia. The physical conditions were likely to deter the most dauntless travellers. The wonder is not that Europeans did not travel as far as China, nor that Chinese or other East Asian peoples did not reach Europe, but that merchants did exist in both areas who were prepared to risk going to meet each other in the halfway markets of Western and Central Asia.

The invasion by the Turks, a Central Asian people, of the Near East, together with the fall of Byzantium, led to the eventual collapse of the Mediterranean trade and the corresponding decline of the Italian city republics such as Venice and Genoa. Trading activity in Europe now switched to the region north of the Alps where a rather loosely organised, but none the less powerful, trading league known as the Hansa had come into being. The Hanseatic League comprised certain ports and inland centres around the Baltic and North Seas and in the Low Countries, northern Germany, and the Scandinavian countries. Its influence extended even into Russia where the town of Novgorod was a Hanseatic centre, and to England where the Hansa had a number of depots, *e.g.* in London and Hull. Bruges, Hamburg, Lübeck, and Visby in the island of Gotland were the leading commercial centres of the League.

The great maritime discoveries towards the end of the fifteenth and at the beginning of the sixteenth centuries ushered in a new era of trade. The quest for gold and spices had been an important motivating factor in the great voyages of discovery but they opened up the world to a greatly enlarged trade. The leaders in this maritime discovery, Portugal and Spain, were the first to benefit and during the sixteenth century Lisbon took on the role hitherto held by Venice. Soon, however, all the countries on the Atlantic seaboard—France, Holland, Denmark, and England—began to participate in discovery, exploration, colonial activity, and commercial endeavour. Thus the centre of European trade and maritime activity shifted once again: to the Atlantic margins facing the "New World."

The Spaniards, at first, ransacked their American possessions for their gold and silver and later focused their attention upon the gold and silver mines. The Portuguese, who discovered Brazil in 1500, found no great stores of precious metals and so turned to develop the agricultural wealth of their colony. They began to grow sugar on the east coastlands of Brazil, and to export it to Europe. Finding the Brazilian Indians unsatisfactory plantation-workers, the Portuguese began to ship Negro slaves from their West African territories as a labour supply. Some of the Spaniards, who had commenced plantation farming in the West Indies, followed suit, but it was not until the British and the French had begun to govern many of the islands, *e.g.* British Jamaica and French Martinique in the seventeenth century, that the slave trade came to be fully organised in the Caribbean region. For nearly 200 years thereafter, ships periodically left the "slave ports" of London, Bristol, and Liverpool for West Africa, where they exchanged their cargoes of iron bars, brassware, and trinkets for Negroes, who were then shipped to America. To complete this "triangular trade," merchants transported sugar, tobacco, and later cotton, back to Western Europe.

For over two hundred years European trade with these newly-found and newly-settled lands was small and one-sided. Trade between the mother and daughter countries consisted, in the main, of shipping colonial products to Europe with little return trade. But gradually during the eighteenth and nineteenth centuries, as a result of the Industrial Revolution, the growth in European productivity, and the rapid increase in the population of the manufacturing countries, trade began greatly to expand. The demand for industrial raw materials which could not be produced in Europe or could not be produced in sufficient quantity to meet the growing needs of industry, such as raw cotton, wool, rubber, copper, and tin, together with the greatly increased demands for foodstuffs, especially wheat, meat, sugar, and beverages, led to a vastly expanded volume of trade which continued to increase with the development of modern means of rapid transport and such inventions as refrigeration, which made possible the carriage of perishable commodities. Thus at the present day, an enormous variety of goods, from oil to orchids and from tractors to tea, is moved by land, sea and air from country to country.

## *THE BASES OF TRADE*

The differences in the natural resources of countries, resulting from differences in climate, soil, relief, and geology, form the one enduring basis of trade between countries. They can never be overcome by man; it will never be possible to grow wheat in the Congo basin or rubber in the British Isles, except under artificial conditions in very small quantities and at a great and prohibitive cost. Nor can man exploit mineral wealth except in those areas which have been endowed by Nature. For example, coal is almost entirely absent from the Scandinavian countries and they must rely upon outside resources for their coal needs; similarly, although Britain produces a very small quantity of petroleum from a few oil wells, she is compelled to import 99% of her oil requirements. Trade arising from differences in commodities has become increasingly important and is undoubtedly the trade of the future. The latitudinal difference of climate gives rise to varying geographical zones with distinctive vegetable products; thus it would appear that there will always be interchange between temperate and tropical regions. Likewise, the uneven distribution of the world's mineral wealth implies the interchange of mineral resources. Such regional differentiation of vegetable and mineral products has always stimulated trade and was, most likely, its initial basis.

The second reason why peoples began to trade with each other was that some possessed plentiful supplies of commodities which others were unable to produce; they had more than they required for their own needs or, in other words, they had a marketable surplus. While some districts or countries lack, say, grain or timber or metals, others have an abundance; thus, obviously, there is a call for the exchange of goods. In normal times Britain has more coal than she needs, but not enough timber to meet her requirements; the Scandinavian countries, on the other hand, have an abundance of timber, but insufficient coal; thus an exchange to mutual benefit is arranged. The surplus wheat of Canada, the cotton of the United States, the meat of Argentina, the wool of Australia, the tea of India, the tin of Bolivia, the oil of Iran, etc., are exported and pay for the import needs of these several countries. It is sometimes said that a surplus is a pre-requisite for trade: while this may be said to be generally true, it is not strictly true. Countries sometimes export commodities even when there is no surplus: sometimes even when there is a shortage within that country; for example, Poland in pre-war days exported large quantities of meat even though there was an acute meat shortage in that country; and Britain, in the immediate post-war years, was exporting motor cars even though there was an insatiable demand for them internally. In such cases a more important economic factor than internal demand is at play.

When a country has a shortage of a commodity, there is an incentive for exchange. The shortage may be absolute, as in the case of rice in Britain, or relative, as in the case of iron ore. Although Britain mines a considerable proportion of her iron-ore requirements, this is insufficient for all her needs and she must import large quantities from such countries as Sweden and Spain. Likewise, although the United States produces some wood-pulp, the almost insatiable demand in that country for paper necessitates the large-scale import of pulp from Canada. Countries lacking certain commodities for which there is a great need must perforce enter into trade to get them. Britain must have such things as cotton, oil, iron ore, sulphur, etc., to keep the factories going on which she depends for her livelihood; also wheat, meat, sugar, and dairy produce to feed the population which cannot adequately sustain itself by its own efforts. In order to pay for these things, textiles, motor cars, engineering products, whisky, etc., must be exported. The needs of the community (essential things such as foodstuffs) or the products desired (luxuries such as French perfumes) form a basis of trade exchange between countries.

Differences in the culture of a people cause differences in production. The skill, initiative, and inventive capacity of a people may enable a country to make more rapid technical progress than its neighbours and so forge ahead of them in the production of manufactured goods. On the other hand, some peoples are backward or relatively backward in terms of industrialisation and technical development; for example, Iraq, Pakistan, Thailand, Kenya, Nigeria, Bolivia, and Cuba are in need of manufactured goods and exchange raw materials such as oil, rice, coffee, tin, sugar, and tobacco for them. The exchange of primary products for manufactured goods forms one of the chief bases of present-day trade. Peoples of different regions often produce distinctive and characteristic goods because of their culture; for example, the Chinese and Japanese produce silk goods and brocades, lacquer ware, and porcelain; the Iranians rugs and carpets; the Guatemalans basketry and colourful blankets, etc. While certain specialised products, such as Persian carpets, still command a world market, they are of decreasing importance as a basis of trade. Then, too, the skill and taste of the French have made them famous for artistic products and Paris has become one of the foremost centres of the world's trade in luxuries. However, this difference of production due to special skill and degree of culture is gradually losing its significance with the general advance of civilisation. Countries are becoming more akin, and in many cases the modern machine-made article is replacing the more artistic product of the craftsman.

The development of adequate transport facilities is another necessary factor for trade. Means and methods for the carriage of goods were developed early and land-borne and water-borne transport have grown in importance throughout the centuries. Recently air transport has taken its place alongside land and water transport. While trade is dependent upon transport, it is equally true that developments in transport and communication have stimulated trade. Again, the demands of trade have led to improvements in transport. Throughout history the two principles underlying the development of transport have been greater carrying capacity and greater speed of transit. In former times trade was largely confined to articles of small bulk, great value, and of a non-perishable nature, *e.g.*, gems, gold, spices, perfumes, silks. Today bulky goods such as timber and cereals, goods of small value like coal and fertiliser, and perishable commodities like fruit and meat are literally transported from one end of the world to the other.

A further necessary condition essential for the carrying on of trade and commerce is the normal functioning of social life and stable political conditions. War is a disorganiser of trade; not only does it interfere with and disrupt the normal pattern of trade during hostilities but it may affect trading relations for long periods thereafter. Peace is essential for the development of trade, for although war has always interrupted commerce to a greater or lesser degree, in these days war is likely completely to disrupt commerce on an international scale. Other world conditions beside war affect trade. State nationalism in the 'thirties greatly influenced trade; in the post-war years it was the dollar shortage. And governments may influence trade by their attitude towards the freedom or restriction of international commerce.

In sum, then, the main bases of trade are: (*a*) a differentiation of products; (*b*) a surplus of products; (*c*) a demand for commodities; (*d*) differences in culture; (*e*) adequate transport facilities; and (*f*) suitable world conditions.

## *TYPES OF TRADE*

The exchange of commodities—and this is what the term "trade" signifies—together with the ways and means by which this exchange is accomplished, forms the study of commerce. Here we are not concerned with the detailed technicalities of commercial transactions but it will be useful to note the different types of trade that take place and to see how countries manage their "housekeeping."

In a broad way, the trade of any country falls into two parts: (*a*) the Home Trade, and (*b*) the Foreign Trade. The older textbooks often distinguished a third type of trade, the Colonial Trade, which was distinct from Foreign Trade because it was intimately bound to that of the home country and because it usually enjoyed preferential treatment. For example, the system of "imperial preference" still operates, if in a rather limited way nowadays, between the U.K. and the members of the British Commonwealth.

The Home Trade is the internal trade of a country; it refers to the buying and selling of goods by the individuals of the community. The volume of the internal trade varies widely from country to country. Much depends upon the size of a country, the variety of its resources, the numbers of its population, the standard of living of the people, and the degree to which the internal communications and transport facilities are developed. In countries such as Burma, Egypt, and Peru there is little internal trade; in others, such as Britain, Western Germany, and the United States there is much. Historically, China has always had a very considerable internal trade; this has been due partly to the wide geographical variety found within the country and partly to the fact that China has always had a substantial population. But, in contrast to this well-developed internal trade, China's external foreign trade has always been very small for such a large country. The answer to this, of course, lies in the simple fact that China was very nearly self-sufficient in her requirements and there was little call for outside trade, except for a few luxury commodities. Geographers when discussing trade are apt to dwell upon the external trade and to forget, or at least to dismiss with a brief word, the internal trade of a country. But the internal trade should not be neglected nor underestimated; in the case of the United Kingdom, for instance, the home trade accounts for some 75% of the total trade.

The Foreign Trade is the external trade of a country; it connotes the exchange of commodities—foodstuffs, raw materials, manufactured goods—between one country and another. Again, the significance of the foreign trade varies greatly from country to country. China, as mentioned above, has always had a relatively small trade, whereas the United Kingdom's external trade, for many decades, has been of major proportions partly because it depends upon other countries for something like a half of its foodstuffs and for the greater part of its raw material needs, and partly because the country has lived upon the export of manufactured goods. The slogan "export or die" has a very real meaning for countries such as the United Kingdom and Japan.

Foreign Trade has two separate aspects: the import trade, which involves the bringing of goods from abroad into the home country, and the export trade, which involves the sending of goods to external countries, either neighbouring countries or overseas countries, for sale. A country such as West Germany, for example, carries on a large export trade with her immediate neighbours (especially her Common Market neighbours) but she also exports goods to distant countries such as the United States, Canada, Brazil, India, etc.

## *THE PATTERN OF TRADE*

The Industrial Revolution, as we have already mentioned, not only resulted in an enormous expansion of trade, but completely altered its character as well. Whereas commerce prior to about 1800 had been primarily concerned with precious and luxury commodities, thereafter it came increasingly to be concerned with necessities, necessary raw materials for industry, such as mineral ores, textile fibres, chemicals and timber, and necessary foodstuffs, such as grain, meat, sugar, and beverages. Note that sugar, tea, coffee, etc., which originally had been "luxury" commodities, had by the nineteenth century come to be regarded as "necessities." Here is a good illustration of the changed value of commodities: what may be luxuries to one generation may become necessities to another. Can you think of any more recent examples of this change in value?

As industrialism developed in Europe, some countries, especially those in the van of industrialisation, and particularly Britain which had been the leader, ceased to produce much of their food and came rather to concentrate upon manufacturing; they found it easier, and often more economic, to purchase their food requirements in exchange for manufactured goods or with the profits they made out of their industry. The classic example of this, of course, was Britain where agriculture came to be grossly neglected.

As a result of these developments, a radical change in the pattern of world trade emerged. For over a century now the general pattern of world trade has consisted of a flow of primary products, such as grain, vegetable oils, wool, rubber, minerals, etc., from the tropical regions and the "new" lands of the southern hemisphere to the industrialised countries of the northern temperate zone and, conversely, of a return flow of secondary products or manufactured goods, such as machinery, vehicles, textiles, chemical products, etc. One has only to compare the import and export trade of two representative countries such as the United Kingdom and Argentina to see the difference in the character of the trade. The United Kingdom's exports comprise overwhelmingly manufactured goods, her imports foodstuffs, raw materials, and mineral oil. Argentina's exports consist of cereals, linseed, meat, hides, and wool; her imports of machinery, vehicles, textiles, chemicals, together with some raw materials, such as coal, oil, and iron ore, for her own growing manufacturing industry.

This latter point leads us to note that the former clear-cut distinction between the economies of tropical and southern hemisphere countries, which were essentially primary producers, and those of the north temperate zone which engaged in industrial manufacture is gradually (indeed, in some cases rapidly) being modified. Industrialism is spreading to the underdeveloped countries of the world who have come to look upon industrial development largely as a status symbol and as a panacea to cure their ills. Hence the pattern of world trade accordingly has become modified although, in a broad way, it is still true to say that much international trade flows in a north–south direction. And, as we pointed out earlier in this chapter, latitudinal differences in climate giving rise to differences in vegetable products are likely to provide an enduring basis for north–south traffic.

The exception to the general north–south flow of trade is provided by the east–west traffic between Western Europe and North America. The historic, ethnic, and cultural links between these two regions have helped to forge intimate economic contacts and the greatest volume of trade flow in the world actually lies between them.

Finally, let us note the character of world trade. Over half, by value, of all international commerce is made up of manufactured and semi-manufactured goods, approximately a quarter of foodstuffs and beverages, and the remainder, roughly another quarter, of raw materials including fuels.

## *GREAT BRITAIN'S PLACE IN WORLD TRADE*

Six countries—the United States, the United Kingdom, West Germany, France, Canada, and Holland—account for approximately half of the world's total trade. The United Kingdom ranks next after the United States as the world's greatest trading country.

Both the volume of world trade and the participators in it are never static. The general pattern of world trade is constantly changing, though

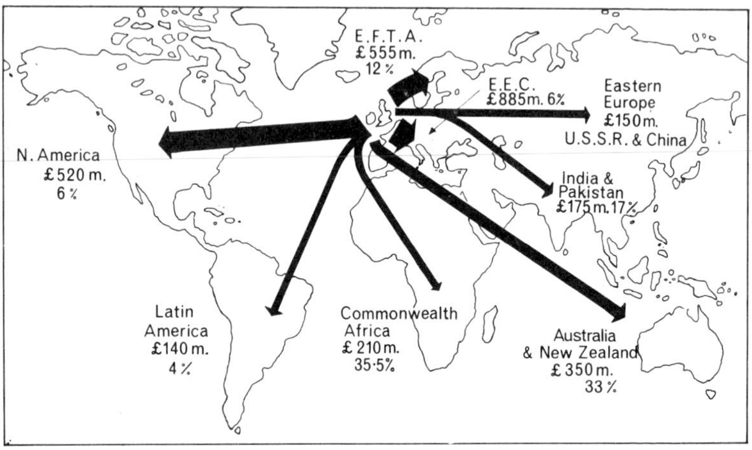


FIG. 125.—DIRECTION OF BRITAIN'S TRADE

This map, which illustrates the movement of Britain's trade, is a simple example of a "flow diagram," *i.e.* flow lines are drawn roughly proportionate in width to the "value" they represent—in this case to the monetary value of the goods exported. The percentage figures show Britain's share of the area's total imports. Fig. 86 is a similar flow diagram.

With respect to Britain's trade, note (*a*) the importance, in terms of value, of the trade with E.E.C., E.F.T.A., and the United States; and (*b*) the importance of the trade with the countries belonging to the British Commonwealth.

this change is usually slow. At different periods in world history different countries have taken the lead. Various factors, as we have already seen, have been responsible for changes in both the character and volume of world trade.

For some two hundred years Britain has been one of *the* leading commercial countries in the world. Napoleon disparagingly called the British "a nation of shopkeepers," although he subsequently came to realise the importance of Britain's commerce and attempted to cripple it, and destroy Britain, by preventing British goods being imported into the Continent.

But his "Continental System" was equally matched by Britain's blockade of Europe.

Already the leading commercial country in the world, Britain during the nineteenth century grew to be the greatest manufacturing nation. Since trade and industry went hand in glove with one another, Britain increased its lead over other countries as the century progressed. At the same time, it developed its maritime carrying trade for other countries so that its mercantile marine became greater than that of all other countries combined. And these things, in turn, led it to become the banking and financial centre of the world.

Long the leader in world trade, Britain retained the leadership, in spite of the competitive challenge of more recently developed countries, until the very eve of the First World War. When that war broke out in 1914 Britain was both the greatest exporting and the greatest importing country in the world. In the inter-war period Britain was compelled to yield pride of place as the world's number one exporter to the United States, although she continued to be the world's greatest importer. The Second World War dealt Britain a further economic blow and she fell behind the United States both as exporter and importer. The United Kingdom now ranks third after the United States and West Germany. The United Kingdom accounts for approximately 7%, by value, of the world's total export and import trade.

The United Kingdom is, to a very large extent, dependent upon trade: much more so than many countries which are better endowed geographically or have fewer people living at a lower standard. Both directly and indirectly the United Kingdom has a deep interest in the volume of world trade for, on the one hand, a recession in world trade affects adversely its export industries while, on the other hand, a falling off in trade reduces the demands which other countries make upon its shipping, banking, insurance services, etc.

Although the United Kingdom continues to increase the volume of her exports, she has a declining share of the total world export trade. A study of world trade figures in the past twenty years shows that world imports have more than doubled, but during the same period British exports have increased by only 50%. The more accelerated growth of world trade during the 1950–70 period has been due in part to the rapid post-war recovery by West Germany and Japan. At the same time it should be realised that many other countries, France, Italy, West Germany and Japan, for example, are increasing their exports at a faster rate than the United Kingdom.

## *THE BALANCE OF TRADE*

It is common practice these days to divide imports and exports into so-called *visible* and *invisible* items. Visible trade comprises goods—the manufactures, raw materials, and foodstuffs exported or imported. Invisible trade consists of services; these include among other things earnings from shipping, air transport charges, banking and insurance charges,

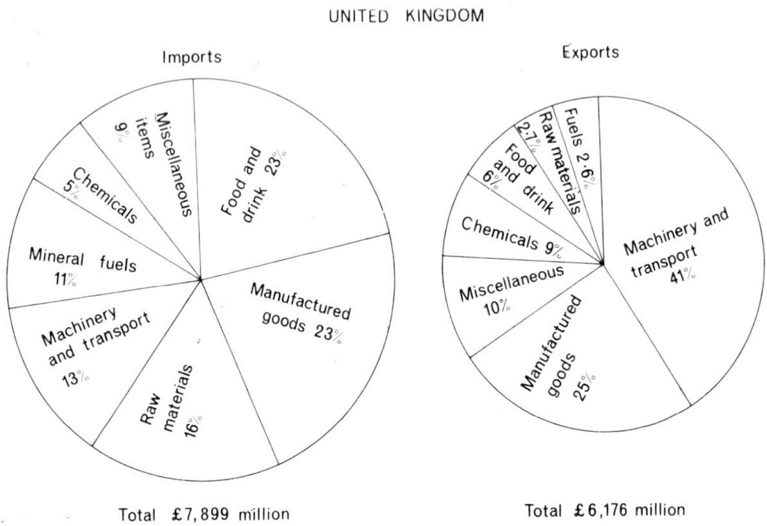


FIG. 126.—THE TRADE OF THE UNITED KINGDOM

The imports and exports of the United Kingdom, 1968. These pie-graphs, drawn to scale, show the main items in the import trade and the export trade. From these graphs it is obvious the United Kingdom buys more than she sells; how does she make up this deficit?

brokerage dues, royalties, interest on investments, tourist receipts, etc. These invisible items may be of considerable importance; in the case of the United Kingdom the invisible exports are of major significance.

A country is no different from an individual. An individual must live within his income; in other words, his expenditure must not exceed what he earns. If it does, then he will go bankrupt. Similarly, a country must earn as much as it spends or national bankruptcy will inevitably follow. Examination of the balance of trade figures of the United Kingdom shows that with respect to the visible trade the imports are always greater than the exports. Clearly, if this was the whole of the financial picture Britain would be a bankrupt country. However, every year there is an invisible income of several hundred million pounds, derived from the services mentioned above, and this "closes the gap" in Britain's trade; in other words, it raises the value of the exports up to the level of the imports and enables the country to pay its way.

In the past Britain's invisible payments have been an important source of income, not only helping to turn a deficit in the Balance of Trade figures (*i.e.* the value of the imports and exports of the visible kind) into a surplus but providing her with money for investment abroad. Britain's prosperity during the later part of the nineteenth century and the beginning of the present century enabled her to amass a vast amount of money and she possessed the greatest gold reserves in the world. This allowed her to lend money to other countries and to invest capital in overseas territories; for example, she developed, built and owned most of the Argentinian railway system. Interest was paid on such loans and profits accrued from overseas investments.

It is very likely that Britain would have continued to remain one of the world's richest nations had it not been for the two World Wars. Both exerted a heavy drain on her resources. For instance, during the Second World War Britain had to sell £4500 million of her overseas investments to help finance the war effort. Moreover, Britain's financial problem was aggravated by the fact that the export trade was drastically reduced at the very moment when she needed extra earnings to pay for the war. Thus at the end of six years, although Britain emerged as victor, she was significantly impoverished. When hostilities ceased, Britain's entire economy was geared to the war effort, international trade was practically at a standstill, and many countries lay ruined and helpless. Had it not been for the generosity of the United States and the help given by her under the Lend-Lease arrangements during the war and the Marshall Plan which provided for economic aid immediately after the war, Britain and many of the other countries of Western Europe would have been in dire straits indeed. The United States nursed Western Europe during the difficult immediate post-war years.

Britain, today, though having made a wonderful economic recovery, has fewer reserves to lean back upon than she had thirty years ago. Again, the economic growth rate in the Common Market countries has tended to exceed substantially that of the United Kingdom; the U.K.'s entry into the Common Market should provide the opportunities for the U.K. to catch up. Within the past two years there has been a decided improvement in the U.K.'s economic position; for example, in 1971 the visible trading account showed a surplus, for the first time for many years, of £295 million while the earnings from invisible exports were of the order of £600 million.

## *TRADE RESTRICTIONS AND TRADING BLOCS*

It would seem sensible and logical for each country to specialise in the production of those commodities, whether they be foodstuffs, industrial raw materials, or manufactured goods, for which it is best fitted, and it is true that usually a country will concentrate, or tend to concentrate, upon the production of those things for which it has the greatest comparative advantage simply because it is easier and more profitable to do so. For example, it is more economic for Britain to manufacture goods and buy most of her foodstuffs through the sale of such goods than to grow the foodstuffs that are needed. If such a condition was carried to its logical conclusion, each country would specialise in the production of the commodity or commodities which were most economic for it to produce, the world market would be perfect, as the economists say, and commodities would move freely between country and country.

However, we all know that man does not always do the sensible and rational thing and that the world is far from being a well-ordered place. Some countries, for prestige purposes, feel that they must possess an iron and steel industry, even if such an industry is not an economic proposition. Some countries grow crops or produce manufactured goods for security purposes. Others develop industries to give their economies a balance. And so on. Thus, instead of perfect freedom of trade in the world, we have a limited trade. Countries "protect" their agriculture or their industry by imposing tariffs or tax levies upon imported agricultural produce or manufactured goods. This state of affairs was carried to an extreme in the inter-war period when many countries, notably Germany and Italy, followed a policy of economic nationalism, *i.e.* an attempt to make their respective countries as self-contained and as self-sufficient as it was possible to do so: even to the extent of making *ersatz* (artificial and imitation) products.

After the Second World War a certain amount of mutual help became necessary, and some countries realised that there were benefits to be had by associating together and freeing trade between them. Holland, Belgium, and Luxembourg, for instance, formed themselves into a single customs union which came to be familiarly known as *Benelux*. This was followed, in 1952, by the setting up of the European Coal and Steel Community which co-ordinated the production of coal and iron and steel in six Continental countries.* The success of this co-operative effort led the participants to a desire to extend its scope so that all products would be included. As a result, by the Treaty of Rome (1957), the six countries established the European Economic Community, or the Common Market

* France, Belgium, Luxembourg, Holland, West Germany, and Italy.

as it is better known. Under the terms of the Treaty, "the Six" planned to reduce, in stages, all the restrictions between them by 1972. In fact, the abolition of internal tariffs and the acceptance of a common external tariff was achieved in July 1968, a year and a half ahead of the original target date. They now form a single common market in which goods pass freely between the member countries. With a total population of over 185

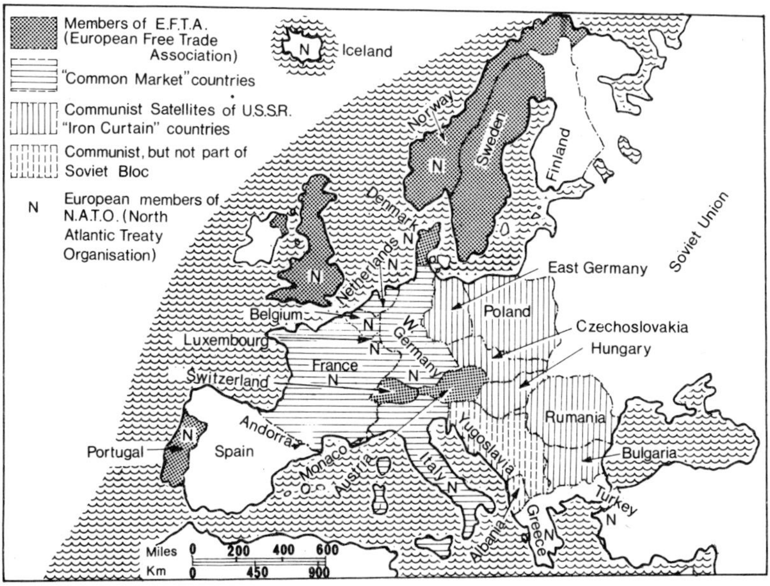


FIG. 127.—E.E.C. AND E.F.T.A.

The European Economic Community (E.E.C.), more popularly known as the Common Market Countries, has been a bold attempt by six West European countries to integrate their economies to promote large-scale production, economic efficiency, and European unity. The European Free Trade Association (E.F.T.A.), or the Outer Seven as it has been called, is a much less cohesive group. Its aim is to lower tariffs and relax restrictions between its members but, unlike E.E.C., E.F.T.A. does not aspire to a common market which will be protected by a tariff barrier. Note that Albania left Comecon in 1961 and is not part of the Soviet bloc although, of course, it is still a Communist state.

millions and a productive capacity approximately equal to that of the United States or the Soviet Union, the European Common Market will form one of the world's largest free-trade areas.

The United Kingdom is to become a member of the Common Market. When the association was first formed she stood apart and, later, in 1963, when she applied for membership, her membership was rejected. Great Britain's difficulty arose very largely from her association with the British Commonwealth with which she has certain commitments and whose members have preferential tariffs in one another's markets. Until these Commonwealth ties and trading problems were resolved *vis-à-vis* the Common Market countries, the United Kingdom had to remain outside the European Economic Community.

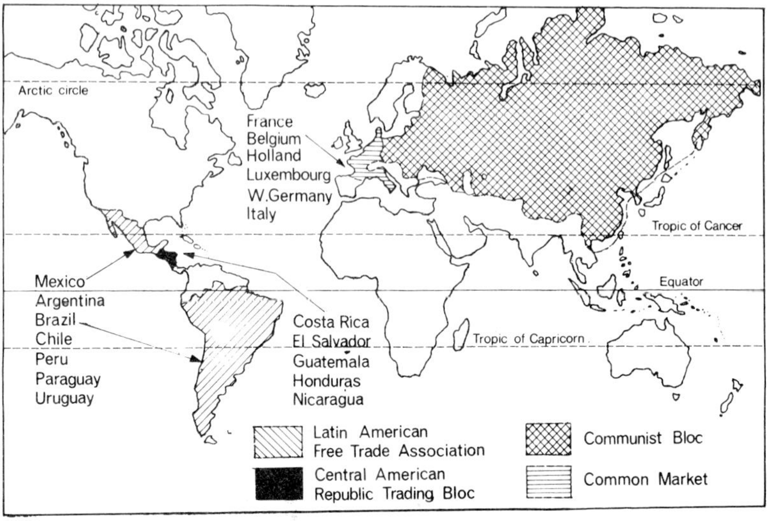


FIG. 128.—TRADING BLOCS OF THE WORLD

In the past, trade agreements were usually bilateral agreements, one country entering into some sort of commercial arrangement with another for their mutual benefit. Since the Second World War, however, there has been a distinct movement towards groups of countries, often affiliated through either a loose political confederation or a currency bloc, to form trading blocs, *e.g.* the Common Market, the Outer Seven, the Organisation for European Economic Co-operation (O.E.E.C.), the Communist Bloc, the Latin American Free Trade Association, etc. After the War an international trade confederation, The General Agreement on Tariffs and Trade (G.A.T.T.), came into being to which some three dozen countries belong and whose total aggregate trade accounts for over 80% of all international trade. The primary purpose of G.A.T.T. is to seek mutual agreements which will lead to a reduction of trading restrictions and foster more extensive and more unhampered exchange.

After "the Six" had joined together, the United Kingdom entered into a commercial agreement with six other European countries (Norway, Sweden, Denmark, Switzerland, Austria, and Portugal) to form the European Free Trade Association (E.F.T.A.) or, as it has become more popularly called, the "Outer Seven." Finland is also an associate member of E.F.T.A. But E.F.T.A. was much less comprehensive in its character than E.E.C. and could not be looked upon as a serious competitor to the latter.

The advantages accruing from some sort of commercial agreement have not been lost on other countries and the Latin American countries, for example, explored the possibilities of economic association. As early as 1952 the Central American Republics of Costa Rica, El Salvador, Guatemala, Honduras, and Nicaragua, held discussions to promote economic integration and the first pact was signed five years later. Likewise the Treaty of Montevideo (1960) linked the seven countries of Argentina, Brazil, Chile, Mexico, Peru, Paraguay, and Uruguay into a free-trade area. Since that date L.A.F.T.A. as the association is known for short, has come to include all the South American countries, except for the Guianas, along with Mexico. Here, then, is yet another attempt to set up a regional market and, although the progress of integration has been slow, some progress has been made. The latest development in this part of the world has been the formation, in 1969, of the Andean Common Market. The countries which have signed the agreement are Colombia, Ecuador, Peru, Bolivia and Chile. Trading between the member countries is as yet small and amounts to less than 5% of their total exports.

## *THE MECHANISM OF EXCHANGE*

Among primitive communities, commodities are exchanged by barter, a simple system of which the schoolboy "swop" is an instance. The value of one commodity is measured against another. But it was found, quite early on, that it was not always possible to exchange one commodity for another, and so things having a standard, acceptable value, such as a cow, a copper cauldron, etc., came to be used. Such things, however, were not easily portable and eventually metal pieces, stamped with a sign or seal, came to be used as a token of exchange. Croesus, King of Lydia, is supposed to have invented the metal coin. From time to time the metals used for making coin changed: from copper to silver and finally gold. Not all people, however, have used coins; some have used shark's teeth or cowrie shells. The important thing is to use something, no matter what it might be, whose value people would recognise and accept. Today, we use notes, but these in themselves are worthless, mere pieces of paper, but we accept them as mediums of exchange, as representing measures of value.

Coins are normally used for the payment of small sums, but if we wish to pay a large bill we use bank notes or sometimes a cheque. A cheque is really an order to one's banker authorising him to transfer part of the cash we have deposited in the bank to someone else from whom we have purchased goods. Cheques save us the trouble of carrying large sums of money around with us or of transporting it over long distances.

Just as individuals buy and sell things, so do countries, and international indebtedness arises when countries exchange goods and services. They could be paid for in gold (if there was no ban on the export of gold) but, more commonly, such goods and services are paid for by means of bills of exchange. These are similar to cheques. A bill of exchange is a document by which a debt is paid. For example, supposing a person or firm purchases goods from another party, but does not wish to pay for such goods immediately, he accepts a bill; this is done by signing a document. By this it is agreed to pay a given sum of money at the end of a certain time, or on demand. The party receiving the bill, if the money is wanted, asks a bank or a financial house to pay the money. This the bank or financial house will do, making the payment less a small amount charged as interest and for carrying out the transaction; this is known as discounting the bill of exchange. A bill of exchange may also be transferred, an operation called negotiation. Bills of exchange afford a ready means of adjusting commercial debts between interests in different countries. There are, of course, other means of remitting money abroad, but the details of foreign exchange do not concern us here. The bill of exchange has been mentioned because it is one of the oldest as well as most widely used methods of commercial exchange.

International transaction is a complicated business. Seldom do two countries engage in a straightforward exchange with each other, *i.e.* bilateral trade. Usually a large number of countries are involved in the exchange, debts and payments being transferred from one country to another so that the various payments and debts cancel one another out. Where several countries are involved in trading transactions, the trade is said to be multilateral. The cancellation of debts and credits between one country and another saves a great deal of time, trouble, and expense. In all essentials the organisation of foreign exchange is really an extension of the internal credit system used within individual countries to the wider field of international settlements.

## EXERCISES

1. Show how economic and geographical factors have led to the growth of Britain's trade.
2. "The differentiation of commodities between region and region is the fundamental basis of international trade." Discuss.
3. Attempt to explain why such a large country as China with its vast population has such a small external trade.
4. Which countries are Britain's major trading partners? Comment on the trade between Britain and these countries.
5. Indicate the sources of Britain's main imports and discuss the problems arising from this trade.
6. "Like Britain, Japan must export or die." Elaborate upon this statement.

7. Evaluate the resources available to the European Economic Community, indicating where the strength and weakness of the Community lies.

8. What is meant by the term "invisible exports"? Give examples of invisible exports and, in the case of Great Britain, indicate their importance.

9. "In the post-war years there has been a distinct tendency for large Trading Blocs to appear in the world economy." Amplify this statement.

10. How far has the geographical position of the British Isles assisted her development as an important trading country?

11. With the emphasis on products entering into world trade, discuss post-war developments in the processing, marketing, and transport of food.

12. The *per capita* foreign trade of New Zealand is approximately fifty times that of India. Give reasons for this enormous difference.

# *Chapter XXIII*

# THE TOURIST INDUSTRY

A CHAPTER devoted to the tourist industry may seem a strange topic to include in a geography book but it is by no means as out of place as may appear at first sight; and for two reasons: (*a*) in many countries, especially Western European countries, the tourist trade is now a major industry employing large numbers of people; and (*b*) the receipts from tourism form a very valuable invisible export, are especially important as a dollar earner, and contribute substantially to the balance of payments.

The aspects of tourism which concern us are: (*a*) why do people take holidays? (*b*) what factors predispose towards the development of tourist centres? (*c*) what kinds of tourist resorts are there? and (*d*) how valuable is tourism as a source of national income? Let us begin our enquiry by tracing briefly the history of tourism, holidaymaking, and the resort.

## *THE DEVELOPMENT OF TOURISM*

The annual holiday is an important feature of modern social life in the West. The migration of millions of people to the coast and country every summer is essentially a modern phenomenon, while the idea of "going abroad" is a very recent development. The practice of taking a holiday raises several interesting queries: how did the holiday arise? why does it take the form it does? and why do people go where they do? The institution of the annual holiday, which is very largely of English origin, has had important social repercussions: it has led to a greater mobility of the population, created a new industry, now of large proportions, and resulted in the growth of many large towns of distinctive function.

The term "holiday" derives, of course, from holy days, days associated with religious observances, though now it is generally used in a secular sense meaning a respite from the routine of workaday life and a time of leisure, amusement, and recreation. Holidays in one form or another are common to all civilisations, whether ancient or modern, though their character has differed from age to age and between one country and another. Public holidays were a feature of ancient Rome and among the most enjoyable of them was the Saturnalia—the feast of Saturn—in December, when all classes indulged in feasting and frolic, and even the slaves were permitted privileges. In Christian Europe certain days commemorating religious festivals and saints' days emerged as holy days on which there was cessation from work with fasting and prayer. An Act was passed in England in 1552 "for the keeping (of) holidays and fasting days," an Act which, incidentally, is still on the Statute Book. Subsequently, though public and semi-official offices in England frequently closed on certain saints' days, there were no general public holidays until the Industrial Revolution. The revolution in industry wrought drastic social, as well as economic, changes. One of these was the introduction in 1871 of the so-called Bank Holidays. Gradually the idea of a week's holiday emerged. In post-war years we have seen most people securing a second week's holiday; now three or four weeks' holiday is quite common.

The practice of "going away" for a holiday is of comparatively recent growth, but its history can be traced back to the seventeenth century. The medical profession was largely responsible for the growth of holidaying away from home. Doctors at this time began to recommend the medicinal properties and healing qualities of mineralised waters. They despatched their invalid patients to places such as Bath, Tunbridge Wells, etc., where mineralised waters were known to exist. As a result, places having mineral springs began to flourish as curative centres, *e.g.* Buxton, Matlock, Epsom, Leamington, Harrogate, and Scarborough. Of these watering places or spas—a term derived from the town of Spa in Belgium, a much renowned resort—Bath and Tunbridge Wells became the most fashionable and popular. But the discovery of a mineral spring in those days meant potential prosperity and places were quick to capitalise such good fortune. So fashionable did the habit of "taking the waters" at inland spas become that Horace Walpole, writing in 1790, commented sardonically, "One would think that the English were ducks; they are forever waddling to the waters."

The vogue for taking medicinal waters, which led to the growth of the spas, formed the first stage in the development of holiday resorts. The second stage centred on the movement to the coast which, in turn, led to the growth of the seaside resort. Doctors began to recommend sea bathing for those afflicted by gout and other ailments. A certain Dr Richard Russell, a Brighton physician, argued that sea water was as effective for medical use, internally and externally, as spa water. Russell not only urged sea bathing but insisted that sea water itself—a full pint of it—should be drunk! And so gradually the habit of sea bathing began, at first for its therapeutic value, but soon for pleasure. Royalty began to visit the seaside resorts and this greatly encouraged their development.

At the time when the Prince Regent was dipping in the briny very few people took holidays. Outside wealthy circles few ever travelled more than a few miles from their immediate neighbourhood and holidays away from home were the privilege of the well-to-do. Movement about the country was severely limited partly because of expense, but partly because of the difficulties of road transport. The coming of the railways had a profound effect upon the fortunes of the seaside resort since for the first time they allowed people to move relatively freely, quickly, and cheaply from one place to another. Largely because of the railway, a spectacular expansion of the habit of holidaymaking occurred in late Victorian times. Developments since then have been rapid and the past hundred years has seen the growth of scores of holiday resorts.

The habit of "going abroad" for one's holiday is very much a post-war feature. Prior to the Second World War, few, mainly the well-to-do, went abroad; today, hundreds of thousands leave Britain for a continental holiday. The continental holiday has its beginnings in the Grand Tour of the Continent which the sons of gentlemen undertook in the eighteenth century, but its recent staggering growth is the result of a combination of factors: the greater affluence of the working population, improved travel facilities, a wish to get some sunshine, and a desire to travel and see "foreign parts." Over 6 million people a year now take their holidays abroad.

### *THE WHY OF TOURISM*

We have noted the main reasons behind the development of holiday-making; let us now enquire: why do people these days engage in tourism? They do so for, perhaps, five main reasons:

1. For relaxation and refreshment of body and mind, which is becoming ever more necessary in modern life with its speed, strain, and stress.
2. For fun and excitement and the interest in "foreign parts," an appetite, whetted by enforced wartime travel, radio, television, and films.
3. For sporting activities such as skiing, mountaineering, boating, fishing, shooting, underwater swimming, etc.
4. For health purposes: to secure fresh air and sunshine and, sometimes, to take, and bathe in, medicated waters.
5. For interest, especially in places having important historical and cultural associations.

Clearly, several or even all of the above factors may operate at the same time but many are urged to take a holiday for a particular reason. Those who can afford the time and money frequently go to the winter sports for the specific purpose of skiing, tobogganing, and mountaineering. Many

Britishers are enticed to the Mediterranean lands by the attraction of nothing other than the promise of seven consecutive days of sunshine. And numerous Americans visit Europe just to take a look at the "old homeland" and at such historic cities as Oxford, Stratford-on-Avon, Bruges, Amsterdam, Heidelberg, Rome, and Venice.

## *THE FACTORS OF TOURISM*

A number of factors predispose towards the development of tourist areas or centres; these are:

(*a*) good weather; (*b*) amenities; (*c*) accommodation; (*d*) accessibility; (*e*) scenery; (*f*) beaches; (*g*) interest features.

Good weather is an important ingredient in holidaying and its importance, at least in Britain, is reflected in such advertising as "Come to *sunny* Sandsea." Most of the seaside resorts in England are on the warmer south, and sunnier east, coasts. The south Cornish coast has become the English Riviera because of the mildness of its winter, a mildness which permits sub-tropical vegetation to flourish. In the United States, Florida has become an important winter holiday area since it is one of the few parts of the Republic to enjoy warm winters.

Amenities, such as facilities for bathing, boating, recreation, dancing, and amusement, are an important feature of the seaside resort. Increasingly the holidaymaker has demanded entertainment in larger and larger measure and what has come to be known as "development" has preoccupied the resort managements. Piers, promenades and parks, ballrooms, bathing pools and putting greens, chair-lifts, funiculars, and illuminations are but some of the amenities provided.

Some countries, notably Switzerland, Austria, and Holland, have gained a worthy reputation for good food, comfort, and cleanliness. And the same applies, of course, to individual establishments. Many hotels pride themselves on their good cuisine and service. Good accommodation facilities will attract custom. The French Government, for example, paved the way for the tourist development of Corsica by launching a big hotel-building programme, a very necessary and wise move.

Accessibility is an important factor promoting tourist development. Physical isolation and inadequate transport facilities are handicaps to tourism. For example, although the North-western Highlands of Scotland possess much grander scenery than the Grampians, relatively few tourists in Scotland go beyond the Caledonian Canal. The lack of good roads, often the lack of any motorable highways, discourages many holidaymakers from touring Yugoslavia. Aircraft have revolutionised travel. Places which not so long ago were completely inaccessible to those having, say, only two weeks holiday are now within easy striking distance. Iceland, Majorca, and Crete are but a few hours flying distance from Britain. The popularity of the Belgian coast resorts for the English

[*Courtesy: Belgian National Tourist Office.*

FIG. 129.—BRUGES

Five hundred years ago Bruges was one of the most important towns in Europe. It was a great centre of woollen manufacture with flourishing trade in cloth and a major centre of commerce, for it was also a great port. Later it declined and now it is a town of charming old buildings and quiet canals, a Mecca for tourists.

holidaymaker is due in large part to the ease with which they can be reached by boat.

Scenic attractions are of great value to tourism. Dramatic mountain scenery and coast scenery exert a strange fascination and the tourist visiting the Alps or the Pyrenees or the Norwegian or Dalmatian coasts for the first time cannot but be thrilled by their physical impressiveness. When water is added to the scene, beauty is added to sheer physical splendour. What would the Trossachs, the English Lake District, or the Alps be without their lakes? Great natural wonders, such as the geysers of Iceland, the glaciers of Norway, the Giant's Causeway, the Grand Canyon, the Great Barrier Reef, exert an irresistible attraction to man and have led to the development of tourism.

Fine sandy beaches drenched in sunshine and offering good bathing conditions form a tourist attraction not lightly to be dismissed. The forty-odd miles of Belgium's coast, studded from end to end with a series of coast resorts, owes almost everything to the fine beaches created by the dune belt backing the shore. Some of the most famous resorts in the world, *e.g.* Palm Beach, Copacabana Beach (Rio de Janeiro), Montego Bay (Jamaica), owe much to their beautiful sands.

Features of interest, whether natural features or historical or cultural features, will give rise to a tourist trade. Think of the enormous attraction of Stratford-on-Avon because of its association with Shakespeare, or of Pisa because of its leaning tower, or of Bruges because of its old-world charm, or of Oberammagau because of its Passion Play, or of Petra with its magnificent ruins or, finally, the incomparable pyramids of Egypt.

These factors are all of significance in fostering a tourist industry, though their importance varies from place to place.

## *TYPES OF RESORTS*

It is not very easy to classify resorts. Though as a town type they are clearly distinguishable as a separate group, they tend to be like many ports, having varied functions. Ignoring capital cities with their cultural and historical attractions, such as London, Edinburgh, Paris, Rome, and Vienna, a broad fivefold grouping may be recognised.

*Seaside resorts.* These are coastal resorts relying basically on the attraction of their beaches and the sea. Most resorts of this type have undergone some sort of "development," *e.g.* promenades, piers, pools and gardens. Typically, they are close to industrialised urban areas with their large populations. Well-known examples are Brighton, Blackpool, Ostend, Punta del Este (near Montevideo in Uruguay), and Atlantic City (in New Jersey, U.S.A.).

*Scenic resorts.* Many resorts grew up originally upon the attractions of the local scenery, subsequently developing other features such as sporting activities. Some lakeside resorts have undergone such "development" that they now resemble the seaside resort, *e.g.* Lugano in southern Switzer-

[*Courtesy: Swiss National Tourist Office.*

FIG. 130.—UNTERWASSER

The Swiss resort of Unterwasser in north-eastern Switzerland lies in the middle of meadows and forests overlooked by majestic mountains. It is a popular centre for both summer and winter holidays.

land. Examples of this type are Bowness–Windermere, Keswick, Pitlochry, Llangollen, Interlaken, Banff (in Alberta), and Bariloche (in the lakeland of southern Argentina).

*Sports centres.* These fall, perhaps, into two sub-types: the winter-sports centres such as Zermatt, St Moritz, Klosters and Gstaad, all in Switzerland, Chamonix in the French Alps, Banff in Alberta, and recently even places like Aviemore in the Scottish Highlands, where skiing, skating, and climbing are the attractions; and other sports centres, such as St Andrews for its golf, Cowes for its yachting, and Deeside in Scotland for fishing.

*Watering places.* Many resorts have grown to importance through their possession of springs yielding mineralised water. These medicinal waters have been capitalised and in central Europe in particular spas are both common and popular. They have usually undergone some "development" so that gardens, music salons, recreative facilities, etc., have added to the attraction of their curative functions. Vichy, Baden-Baden, Karlsbad, Loeche on the Continent, and Buxton and Cheltenham in England are examples.

*Historico-cultural centres.* Many places having renowned historical or cultural associations, or possessing important works of art have become centres of tourist attraction and thus have developed into tourist centres. Notable examples are: York, Edinburgh, Oxford, Cambridge, and Stratford-on-Avon in Britain, and abroad Bruges, Chartres, Heidelburg, Florence, Jerusalem, Athens, etc.

## *THE ECONOMIC IMPORTANCE OF TOURISM*

The tourist industry in many countries has become a major industry employing large numbers of people and earning large amounts of foreign currency. Money earned by tourism falls into the category of so-called "invisible exports" similar to the income derived from such services as shipping, insurance, banking, etc. The receipts from tourism help to pay for imports of foodstuffs, raw materials, and manufactured goods, and in countries such as Switzerland and Greece they are of great importance. In post-war Europe when there was a "dollar shortage," the earnings from American tourists were invaluable.

Let us take a few examples to illustrate the importance of tourism. Outstanding in Europe is Spain whose total number of annual visitors has jumped from 4 million in 1960 to around 20 million in 1970—a quite phenomenal increase. Tourism is now the country's largest single source of revenue. A decade ago, tourism was largely confined to the Costa Brava and Majorca but today practically the entire Mediterranean coast of Spain and all the Balearic Islands have been developed. Hotels, holiday villas and all the other accoutrements of the tourist industry are mushrooming along these sun-drenched coasts of Spain.

The tourist business is France's second largest export industry: over 10 million tourists visit the country annually, thereby earning France much foreign exchange. To maintain this valuable source of income—even to extend it if possible—is the aim of the French Government. During more recent years France has begun to lose some of its tourist trade; many people who once went to France are now going to Spain, Italy, and elsewhere and there is an increasing tendency for foreign visitors to stay only one or two days in France, whereas previously they remained for one or two weeks. The number of foreign visitors has shown very little increase and the percentage increase is considerably lower than that of most of her neighbours. In order to claim a larger share of the tourist traffic, the French Government is to help the industry to enlarge and improve its existing facilities. There are plans—already under way—to develop new tourist areas, especially in Languedoc as well as in Brittany and along the Biscayan coast. The creation of some 50 new resorts is envisaged as well as many new medium-priced "two-star" hotels. During the past decade considerable development has, in fact, occurred in Corsica whose tourist potentialities are considerable.

It will be clear that tourism could be a valuable source of income to the under-developed countries who are desperately short of capital for their development. Some countries, *e.g.* India, Morocco, Thailand, appreciating the importance of tourism as a source of income, and especially of foreign exchange, are developing their tourist potential as quickly as they can.

## *THE TOURIST INDUSTRY IN BRITAIN*

London has long been a Mecca for foreign tourists but it is only in comparatively recent years that tourism in Britain has become a major industry. A record number of visitors—nearly 6 million—came to Britain in 1969; this was an increase of 21% on the previous year. Their expenditure here at £359 million was 27% greater than the figure for 1968, which itself had been 20% greater than that for 1967. All the indications are that the tourist trade will continue to increase and it is estimated that by the mid 'seventies the number of overseas tourists will be around 10 million. As an invisible export, tourism is Britain's biggest dollar earner and has become a very important factor in Britain's balance of payments. In spite of the fact that the number of British residents going abroad is now around 8 million, their expenditure is around £324 million. On balance, therefore, there is a surplus on the travel account amounting to around £35 million. Table XXXIX gives a breakdown of the number of visitors and their expenditure for the year 1969.

The British Isles have much to offer the tourist: a variety of landscape in miniature as beautiful and as interesting as one can get almost anywhere in the world; a wealth of ancient monuments, such as Stonehenge, the

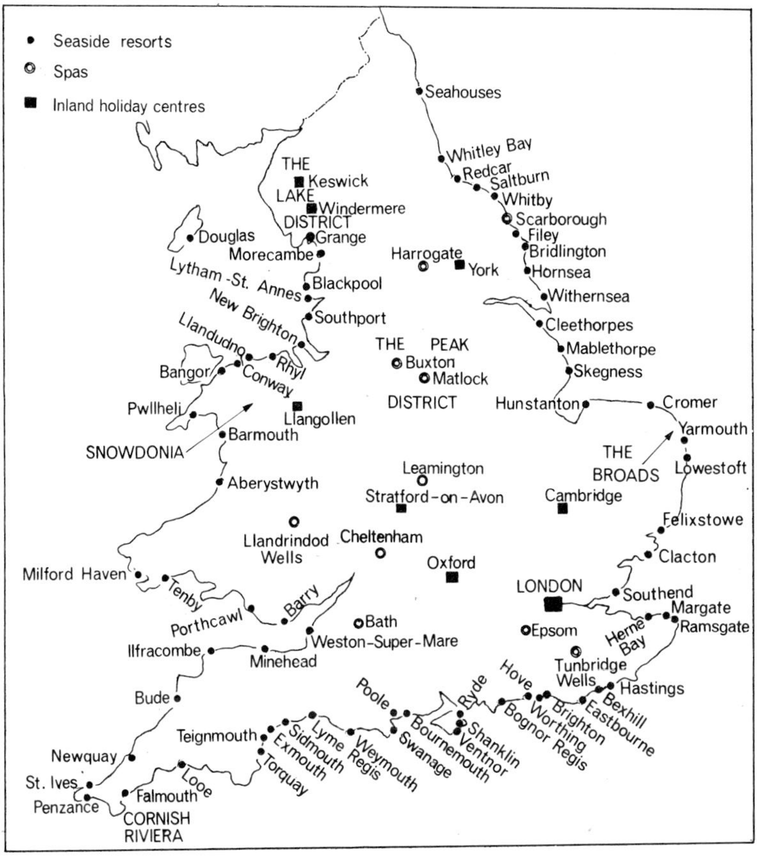


FIG. 131.—ENGLISH RESORTS

Not every resort is marked on the map: there are hundreds of small coast resorts in creeks and coves throughout the length of England's varied coastline and many inland centres, though not classed as resorts, have large numbers of visitors every year.

Devil's Arrows near Boroughbridge, and the Rudston monolith not far from Bridlington, as well as the famous Roman Wall; abbeys, monasteries, cathedrals, castles, and great country houses in infinite variety; many towns of great historical interest and attractiveness such as York,

Chester, Shrewsbury, Ludlow, and Norwich; the ancient seats of learning of Oxford, Cambridge, and St Andrews; Wimbledon, the Highland Games, the Edinburgh Festival, etc.; while London itself, with its mixture of ancient and modern, its theatres and galleries, its cosmopolitanism and pulsating life, is one of the most captivating and interesting of capitals.

From the point of view of the British economy, it is important that tourism be encouraged. The British Travel Authority (formerly the British Travel and Holidays Association), which must be given much of the credit for putting Britain on the tourist map, is trying hard to stimulate the tourism business. This will not be easy to achieve for there is a growing competition in the tourist business; for example, the United

TABLE XXXIX

*Tourism in Britain*

*Number of visits and expenditure of overseas visitors in 1969*

| *Country of residence* | *No. of visits (in thousands)* | *£ million* |
|---|---|---|
| United States | 1295 | 98·5 |
| Canada | 361 | 28·0 |
| North America | 1656 | 126·5 |
| France | 543 | 18·9 |
| West Germany | 570 | 20·7 |
| Italy | 180 | 12·2 |
| Belgium and Luxembourg | 257 | 7·8 |
| Netherlands | 357 | 10·3 |
| Austria | 36 | 1·9 |
| Norway, Sweden and Denmark | 287 | 14·0 |
| Switzerland | 156 | 10·3 |
| Spain | 72 | 6·2 |
| Other Western European countries | 99 | 9·3 |
| Western Europe | 2557 | 111·6 |
| Other non-sterling areas | 360 | 33·2 |
| Irish Republic | 764 | 32·0 |
| Australia and New Zealand | 126 | 16·2 |
| South Africa | 81 | 8·9 |
| Other sterling areas | 277 | 30·9 |
| Overseas sterling area | 1248 | 88·0 |
| All countries | 5821 | 359·3 |

(*Source*: British Travel Authority)

States has launched a big campaign to attract tourists, the continental tourist areas are proffering increased tourist allurements, while many "new" countries, *e.g.* India, Thailand, Japan, the North African and

Middle Eastern countries, have embarked upon a programme of tourist attraction. The greatest weaknesses in the British tourist industry lie on the servicing side: more good provincial hotels are needed, many more motels are required, catering needs improving, and more personal service is called for.

### EXERCISES

1. Give an account of the main routes from London to the following holiday areas: Switzerland, the Mediterranean coast of Spain, Vienna, and Stockholm.

2. An American visiting Britain for three weeks in June wishes to stay in three contrasting holiday areas outside London. Plan an itinerary, assuming travel by rail or air. Give an account of the types of scenery he will encounter, the routes he should follow, and the places of historic and other interest he should visit.

4. Give a full account of the tourist attractions to be found in any *one* major Mediterranean tourist area, paying particular attention to climate, scenery, and historical association. By what routes might the area be reached from London?

5. What factors of climate and scenery would attract tourists to *three* of the following areas: the Rhine Gorge district, the Ardennes, Brittany, the south-west coast of Norway, Finland, Slovakia?

6. What tourist attractions would a yacht cruise up the Rhine to Basle provide?

7. Indicate the importance of the tourist industry to (*a*) France, and (*b*) Britain.

8. Give an account of *two* of the following: (*a*) the growth of the North Wales holiday resorts, (*b*) the Norfolk Broads as a holiday area, (*c*) the potentialities of the Scottish Highlands for "winter sports," and (*d*) the scenic attractions of south-western Ireland.

9. Write an essay on the geographical factors affecting the development of the tourist industry.

10. Give a full account of the tourist attractions of any *one* major Mediterranean island. By what routes might the island be reached from London? What climatic or other disadvantages might the island have for the average British tourist?

11. Verify the statement that "tourism is an important part of international trade."

12. What are the main tourist attractions of *either* Finland *or* Portugal? For the country you select, compare the routes to the capital from London in terms of travel time and cost.

# *Chapter XXIV*

# LANGUAGE AND COMMUNICATION

## *COMMUNICATION*

In geography the terms transport and communication are usually used loosely: railways, for example, are spoken of as "means of communication" and in many textbooks one can find chapters headed "Transport and Communications" when their contents deal solely with methods and forms of transport. What is the difference between communication and transport? Communication may be defined simply as the transmission of thought by voice and message. Transportation may be reserved as an expression for the function of carrying. Often in the past, and sometimes even at the present time, the means of communication and the means of transport were the same; the carrying of messages by runners is a case in point. But, except among primitive communities, this is seldom so today.

The communication of thought has been of tremendous importance and significance since the very earliest days of human existence; today it is more important than ever before. Not even the simplest human societies can function unless men can communicate with one another. Imagine the limitations placed upon the individual (and upon social groups) if there was no such thing as the human voice. Because primitive man wished to express his thoughts and feelings he used his vocal chords to produce a series of grunts and squeals which, in due course, developed into speech or language. Thus from the very earliest times man has used his voice to communicate messages.

Wonderful as the voice is, it has its limitations. The human voice cannot carry very far and the spoken voice was, until very recent times when recordings could be made, transitory. As a result of these two facts early man developed signs and signals and noises which could be interpreted at a distance, *e.g.* the drums of African peoples, the smoke signals of the American Indians, and semaphore devised by Europeans. Somewhere about five thousand years ago, man invented writing so that he could perpetuate his thoughts. Indeed, the history of civilisation is very much concerned with the progress which man has made in expressing his thoughts more accurately and in disseminating those thoughts more widely and more speedily.

## *COMMUNICATION DEVICES*

Throughout human history man has invented a series of devices to facilitate and speed up communication. Drums, fire, and smoke signals were early devices. Runners, as in ancient Greece, were much used in ancient times. Then, horses and other animals were used to carry messengers, and relay teams made it possible to carry messages reasonably quickly over fairly long distances. But, until just over a century ago, communication was dependent primarily upon transportation. Messages, as with goods, came to be carried farther and faster as the means and methods of transportation improved.

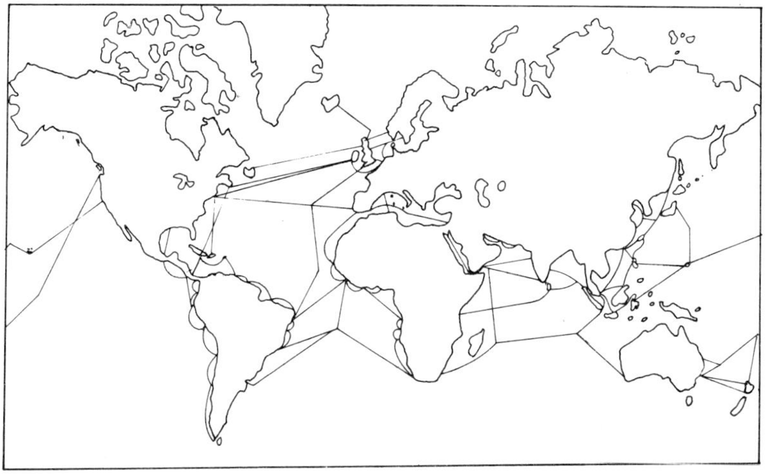

FIG. 132.—WORLD CABLES

Developments in systems of communication were slow until about the middle of the nineteenth century, when developments in electrical technology wrought a revolution. Quite suddenly communications were speeded up as never before. The invention, in 1844, of the electric telegraph by Samuel Morse ushered in a new era in communications. This new invention had two important results: first, communications ceased to be completely dependent upon transportation; and, second, it led to the establishment of widespread communications networks. "Within a few decades the more important cities of every Western country were linked by wire; news collecting and transmitting agencies were organised; and newspapers began to take on a national, as distinct from local, character. The installation of underseas cables soon followed, and before the end of the century the continents of the world were thus brought into as close and intimate communication via words as neighbouring towns had formerly been."* In 1851 England and France were linked by a telegraph cable. The first transatlantic cable was laid in 1866; and with it the time required to communicate between the two continents of Europe and North America was reduced from weeks to minutes.

In 1875 the telephone was invented by Alexander Graham Bell and electric land telephones came into use. Before long the land sprouted telegraph poles and came to be festooned with overhead wires. Now, for the first time, it became a practical possibility for the human voice to be heard from one end of the earth to the other. Another useful development was the marriage of the typewriter with the telegraph to produce the teletype machine; today, the teleprinter system is widely used and the British Weather Forecasting Bureau of the Air Ministry relies upon it to a great extent. Recently a teleprinter link-up was installed between Washington and Moscow.

The greatest development of all in communications was the coming of radio due mainly to the work of Marconi. Radio meant that messages could be heard instantaneously the world over. The influence of the radio, for good or ill, is enormous. The speed, reliability, and accuracy of radio is matched by no other communicating medium.

Visual communication came with the invention of the film, but the motion picture served entertainment rather than business ends. Television has now joined the other means of communication but, like the film, its importance is still mainly in the field of entertainment.

Insofar as communications are concerned, the conquest of space and time would seem to be complete, but there is no telling what new devices electro-technology will bring in the future. All that can be said is that modern science and technology which have produced radio, television, radar, and communication satellites are certain to yield further refinements in communications.

## *THE IMPORTANCE AND VALUE OF COMMUNICATIONS*

Modern communication devices, such as the telegraph, cable, telephone, and radio, carry messages and ideas with astonishing speed to almost every part of the earth. At the present time, there is, for all intents and purposes, world-wide coverage.

The facility with which messages can be despatched nowadays is significant geographically in many ways, *e.g.* in forecasting weather, in giving warnings of potential hazards, in securing information on market prices, in conducting business matters, in advertising for commercial ends, in

* LaPiere, R. T., *Sociology*, McGraw-Hill Ltd., 1946, p. 233.

international political relations, etc. Let us elaborate some of these points.

The ability to transmit information about weather conditions is of tremendous value to a host of people, to farmers, sailors, aircraft pilots, etc. Frost warnings, gale warnings, fog warnings and the like may save lives as well as money. To quote but one example: in summer ice is a frequent danger in northern seas, especially the North Atlantic, when ice from the Greenland glaciers breaks off and drifts southwards, sometimes into the shipping lanes which shift polewards during the summer months to take advantage of great circle sailing. The International Ice Patrol Service, whose job it is to locate icebergs (and to try to destroy them), warns ships of their whereabouts by radio. This valuable service was set up after the *Titanic* struck an iceberg on her maiden voyage, in 1912, and sank with the loss of 1513 lives.

There are many potential hazards—flood, fire, locust infestation, etc.—which can be guarded against if adequate warning is given. Speedy messages can help man to take due precautions. During recent years there have been many threats of floods in southern England but the fact that due warning has been given has helped man to mitigate the disasters of such natural calamities. Forewarning of some natural disasters such as earthquakes cannot be given but the news of such disasters is quickly made known by modern communications and assistance can be sent to devastated areas almost immediately to alleviate the distress.

The speed with which messages can be sent from place to place and particularly over long distances has been a great boon. Information on the supply, demand, and prices of commodities can be secured by businessmen which helps them to buy and sell and to plan for their markets. A good illustration of the value of communications is provided in the case of the tramp steamer. In former times tramps sailed the seas calling from port to port in the hope of picking up a cargo; often time and money were lost in seeking cargo. Nowadays, the owners can cable or radio their ships and direct them to ports where a consignment awaits them. This, clearly, is a much more efficient way of conducting business.

Radio and television have provided important media for advertising, to some extent taking over the functions of the newspaper. The value of radio, and especially television, for advertisement purposes needs no emphasis. Advertising is an important adjunct of business these days. Communication media such as radio and television which began as entertainment media have now become useful tools of the business world.

Again, rapid communications are of tremendous consequence these days in the sphere of international relations and politics. Leaders can speak directly to their peoples—think of the Queen's Christmas Message to the

Commonwealth—and communications can help to link together peoples, often widely dispersed. International crises may be solved by statesmen taking prompt action, itself only made possible as a result of speedily communicated knowledge.

## *THE LANGUAGES OF COMMERCE*

The development of speech and language is one of the fundamental developments of man's cultural evolution. Only by speech and gesture can man communicate with his fellow men. Business, it is clear, makes great demands upon communication, but language is especially important. Unfortunately, man has developed a multiplicity of tongues and these are

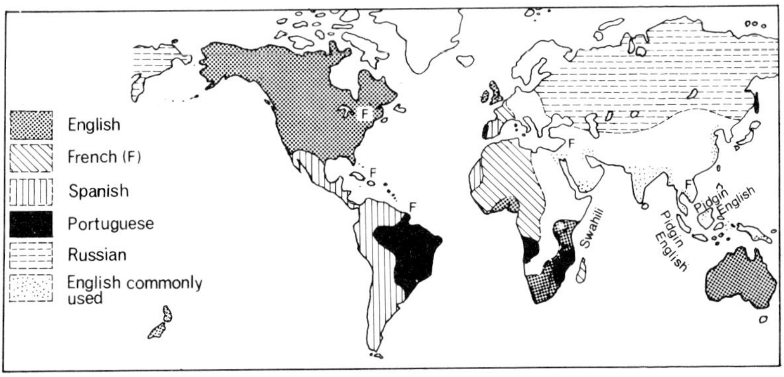


FIG. 133.—LANGUAGES OF COMMERCE

a great hindrance to commercial intercourse between peoples. Largely for this reason, a few languages, mostly European languages, have come to be used as "the languages of commerce" and most international transactions are carried out in these commercial languages. Man prefers to do business in his own language or in one with which he is familiar.

Very largely because European peoples spread throughout the world conquering or colonising various parts, the European languages have become predominant in international usage. The fact, too, that many of the new developments in communications were invented in Europe has tended to emphasise the importance of European languages; increasingly, with the greater use of telegraphy, telephony, etc., the languages of commerce have tended to become restricted to a few of the European tongues.

English is the most widespread of all the languages; not only is it the mother tongue of the British peoples in the United Kingdom, Australia, New Zealand, Canada, and East Africa, but it is the language spoken in the United States, in South Africa, and in Britain's remaining colonial territories. In Eastern and South-eastern Asia a mongrel kind of English, known as "pidgin" English is spoken; the term "pidgin" itself is a Chinese corruption of "business". Pidgin English is really a jargon used in dealings between peoples who cannot speak the Chinese or Malay languages on the one hand and the Oriental peoples who cannot speak English on the other; it is composed of English, Portuguese, and Chinese words which are arranged in accordance with Chinese syntax. It is worth noting that the Chinese themselves have spread widely throughout south-eastern Asia and the East Indian archipelago.

French, which formerly was the language of culture and diplomacy in Europe, and which was widely understood in North and Central Africa and South-east Asia, has lost ground as France has declined as a great power and as it has lost its colonial empire. We should note the presence of a large French-speaking minority in Canada.

The Iberian languages have currency throughout practically the whole of Latin America. Spanish and Portuguese are, of course, importations from Europe. With the major exception of Brazil, where Portuguese is the official language, Spanish predominates generally throughout Latin America. Thus Spanish is not only the commercial language but the prevailing tongue of Central and South America, except in Haiti (which speaks French), the Guianas, and Brazil.

German, though widely used in Central Europe and regarded as the "scientist's language," has not developed into an important commercial language except in Europe. Russian is growing in importance in every way. Russian has been systematically spread not only through the European territories of the Soviet Union but throughout the Asiatic territories where among the non-Russian peoples it is being compulsorily taught as the second language. Once strictly confined to the homeland and little known outside Russia, Russian is now becoming, largely because of the enhanced importance of the Soviet Union politically, much more widely learned.

"The languages of commerce, when carried on between peoples speaking different tongues, is generally of a very mongrel character. In the days when Italian trade was predominant in the Levant, there arose in all the coasts of that region a trade language, the basis of which was a corrupt Italian, but which borrowed numerous words from the local dialects in different places. This language is known as the *lingua franca*, and is still spoken in many of the eastern Mediterranean towns. The dominant languages of commerce at the present day have all begotten corrupt forms of speech of a similar nature."* Pidgin English has already been referred

* *Chisholm's Handbook of Commercial Geography*, ed. L. D. Stamp and S. C. Gilmour, Longmans, 14th edition, 1954, p. 116.

to; Swahili is another case in point. Swahili is a language derived from Bantu mixed with Arabic, English, Portuguese, and Hindustani. It is widely spoken throughout Kenya and Tanzania, as well as in Pemba and Zanzibar, and even among some of the Congo tribes. The Hausa language, again, serves as a sort of *lingua franca* over much of Africa north of the equator and west of the Nile. The development of these *lingua francas* illustrates the need for some mutually intelligible mode of communication between peoples speaking different tongues. Language facilitates the economic, as well as the cultural, progress of man.

It may not be inappropriate to refer briefly to Esperanto. This was invented by Dr Ludwig Zamenhof, a Pole, in 1887 to serve as an international auxiliary language. It consists of a few thousand root words, has a simple grammar and syntax, and a phonetic pronunciation. It can be easily acquired and makes possible ready communication between persons of very different native speech. At first, it achieved quick success but later interest waned and, although there are many Esperantists in the world, it has never really caught on as an international language. Lack of a truly international language is to be regretted, for linguistic differences do form a formidable barrier to communication and understanding.

## EXERCISES

1. Distinguish between the terms "transport" and "communication." What are the main communications devices?
2. Discuss the importance and value of modern systems of communications.
3. Which languages are the languages of commerce? Why are there so few languages of commerce?
4. "Not even the simplest human societies can function unless men can communicate with one another." Illustrate the truth of this statement.

*Chapter XXV*

# THE WORLD'S HIGHWAYS

MODERN business, industry, and trade are dependent upon transport and communications facilities. The carriage and movement of people and goods from place to place are vital aspects of world economic activity. More than ever before, increasing numbers of people engage in travel, and for a variety of reasons—for holidays, for political reasons, for scientific purposes, as well as for business reasons. More than ever before, the world is dependent upon the interchange of commodities, and there is a greater flow of goods around the world than at any time in the past. This movement of people and goods is based fundamentally and necessarily upon transport.

Travel by people, the spread of trade, and the increased volume of traffic have been helped enormously by the development of new, more efficient, and speedier means of transportation and by improved communications facilities. In times gone by transport was slow and costly and the quantity of goods carried small because man was dependent upon animal transport, wind-driven ships, etc. For example, it took Columbus' *Santa Maria* 79 days to cross the Atlantic while the *Queen Mary* did it in 4½ days and a jet aircraft makes the crossing in a few hours. Just over a century ago the French novelist Jules Verne wrote a book, *Round the World in 80 Days*; this was the story of a breakneck attempt to make a circuit of the earth within what was then thought to be a virtually impossible time limit; today B.O.A.C. run a 48-hour round-the-world service. Not so very long ago petroleum was carried in drums on cargo vessels, then specialised bulk carriers or tankers appeared, now much oil is transported by pipeline. These are only a few of the great developments that have taken place.

Improvements in transport have been closely connected with scientific and technological progress. During the past two hundred years important developments and rapid changes have taken place: in the eighteenth century canals and roads were constructed; in the nineteenth century railways and steamships appeared; and in the present century motor transport and air transport have been developed. The outstanding factor in transport development during the past two hundred or so years has been its mechanisation, largely owing to the invention of the steam engine (applied to the locomotive and the steamship) and the internal combustion engine (applied to the motor car and the aeroplane), which the progress in science made possible, transport has been revolutionised.

This revolution is still going on: in fact we are, in a sense, in another important phase of the transport revolution since during the last two decades we have seen very important developments in modern transport systems. These may be enumerated briefly as follows:

1. The use of bulk-cargo vessels.
2. The development of containerisation.
3. The development of motorway systems.
4. The introduction of high-speed trains.
5. The use of new novel craft, *e.g.* hovercraft.
6. Improvements in aircraft, *e.g.*, Jumbo Jet, Concorde.

These newer developments will be referred to again shortly.

Linked with the improvements in transport have been the equally important developments in communications. The inventions of the telegraph, wireless, radar, automatic pilots, and the like have all contributed to the astonishing speed-up in world communications. Anything that happens anywhere in the world can now be known the world over within the space of a few minutes. Clearly, all these developments in transport and communication have, in effect, made the earth a much smaller place.

## *MODES OF TRANSPORT*

The methods of transport used in the world at the present day fall into nine different categories:

1. Human porterage and human traction.
2. Animals used (*a*) as beasts of burden, (*b*) for draught purposes.
3. Motor road transport.
4. Railways, including light railways.
5. Ropeways and cableways.
6. Pipelines.
7. Inland water transport—rivers and canals.
8. Ocean transport.
9. Air transport.

A survey of the methods of transport used throughout the world is a fascinating study. Here, however, we cannot go into the matter in any detail and we shall have to be content with a very brief survey. But before we describe the chief modes of transport four points should be emphasised:

1. Transport systems vary widely from country to country and from region to region as a result of the influence of geographical, economic, social and historical factors.

2. Mechanisation in one form or another is responsible for the greater part of the movement and carriage of both people and goods at the present day.

3. Man has adapted himself to travel in three different media, by land, by water, and by air; each medium, moreover, has its own specialised type of vehicle.

4. A distinction should be drawn between the agency undertaking the transport, *i.e.* whether by man, animals, or machines, and the path used and followed by the agent.

Land transport is undertaken by man, pack animal, animal-drawn vehicle, motor car, railway, and by more specialised media such as cable-ways and pipelines. Water transport is effected by rafts and by boats, the latter being driven by oar and paddle, by wind, or by motor. Air transport is undertaken by the aeroplane alone for the airship or zeppelin has not proved to be very satisfactory.

Man is a very slow transport agent with a very limited carrying capacity: he can carry approximately 70–80 lb (31–36 kg) over a distance of about 10–15 miles (16–24 km) a day. Human porterage is still to be found in some regions as, for instance, in the tropical forests of central Africa and in China. Human porters are used where other forms of transport do not exist and in areas where human labour is abundant and cheap.

Animals are quicker and are usually capable of carrying heavier loads than man; if they draw vehicles, they can carry heavier and larger loads still. The wheel (which was one of man's great inventions) is a considerable economiser of labour. Where snow lies on the ground the sledge, which carries heavy weights economically, is often used. The horse is the most efficient of all animal-carriers. The carrying capacity, speed of movement, and cost of upkeep vary with the animals.

Motors and railways are the quickest forms of land transport as well as the most efficient; they are, however, expensive in outlay and upkeep. The railway can carry much greater loads than motor transport and for long-haul transport has a decided advantage. The motor car is a more flexible vehicle and better suited to short-distance transport. The motor vehicle is almost ubiquitous in the world today, though in some regions roads are either poor or non-existent and here the older forms of transport still prevail.

The latest development in overland transport is the pipeline. Pipelines are used for carrying liquids, *e.g.* water, petroleum, milk, and natural gas. There can be no doubt that increasing use will be made of pipelines in the future, especially by countries rich in oil and natural gas.

Water transport is generally cheaper than land transport. Though slower than most other kinds of transport, it is especially useful for bulk transport. Man, animal, or motor engine can haul much greater weights through water than on land by wheel or sledge. If the commodities carried are of the bulky, non-perishable kind and the time factor is of little importance, water carriage is, without question, the cheapest of all forms of transport.

### *SPEED AND COST*

The most important considerations in transport are: (*a*) the cost of carriage; (*b*) the speed of carriage; and (*c*) the quantity or load factor.

The cost of carriage is obviously a very important consideration in transport. Trade almost always follows the cheapest routes. The cheapest route is not necessarily the shortest route; neither is the quickest route always the shortest. Natural obstacles along the line of direct route may have to be avoided and a more circuitous route taken, although journeys round such hindrances may in fact take a shorter time than those through or over them. Sometimes a short route, though quicker than a longer one, is more expensive because of the need to use special equipment or vehicles whose running costs are heavy. In mountainous country, for example, rack railways, more powerful engines, bridging and tunnelling, etc., mean higher transport costs.

Where the time factor is involved, such as in getting perishable commodities like fish or fresh fruit or flowers to market, the quickest routes are commonly chosen, even though these on the face of it appear to be the most expensive; such routes are, in actual fact, the cheapest since perishable goods quickly lose their value once they begin to deteriorate. If time is of little or no consequence, as in the case of most non-perishable commodities, such as coal, ores, timber, etc., more leisurely routes can be followed. Similarly with regard to passenger traffic: if time is no object, the traveller from Britain to the United States, for instance, is likely to go by sea; but if time is important he is likely to travel by air. Frequently to business men "time is money;" when this applies, quicker travel, though perhaps more costly, is, in the final reckoning, cheaper.

The use of routes is primarily determined by cost of carriage and this is dependent upon four factors:

1. The distance over which goods have to be transported.
2. The means of transport used for carriage.
3. The obstacles and handicaps to free movement.
4. The nature of the goods carried.

Cheapness in transport is secured by: (*a*) uninterrupted carriage over long distances; (*b*) increasing the size of the carrying agent; (*c*) using level and direct routes as far as possible; (*d*) having full cargoes in both directions.

Speed in transport can be achieved in a variety of ways: (*a*) by using new and more powerful methods of propulsion; (*b*) by improving trackways to facilitate speedier movement; (*c*) by guaranteeing full cargoes, which enables transport to be organised to schedule; (*d*) by improving terminal facilities, which allows a quicker turn round; (*e*) by inventions such as radar, automatic signalling, radio communications, etc., which facilitate movement.

## *CONTAINERISATION*

In the long history of transport every now and again some introduction comes along which has revolutionary effects: containerisation may be said to be one of these. The use of containers for goods is, of course, nothing new, but containerisation is a new form of packing on a large scale. And this new form of transport technology is likely to have, indeed is already having, profound geographical repercussions.

There has developed in recent years a large and growing use of containers. These, simply, are unit loads in, usually, metal packing cases of standard sizes which fit motor vehicles, rail cars and ships' holds, and even aircraft (although the tubular shape of an aircraft fuselage makes the use of containers less easy) and so can be moved easily from one kind of transporting agent to another.

For certain types of commodities, *e.g.* minerals, wheat, sugar, the bulk carrier was devised which proved to be very satisfactory since it permitted large quantities of cargo to be carried which could be loaded and unloaded quickly by mechanical means. Non-bulk goods provided a problem. In an attempt to overcome this, *unitisation* was first introduced. This consists of the build-up of units of goods on pallets, which are essentially trays; these are useful for shipping small quantities. The concept of the pallet was, then, the first to gain acceptance and was widely used. But it had its shortcomings and these led to the development of containerisation —the idea of putting general cargo in large boxes.

Containers, as noted above, have long been used, but *standard* containers are new. In 1967 an international unanimous agreement was reached with respect to standard containers. The standard for large containers is based on an end section of 8 feet by 8 feet (2·4 m by 2·4 m) with lengths of 10, 20, 30 and 40 feet (3, 6, 9 and 12 m), or in other words on a length module of 10 feet (3 m). These standard boxes will fit any vehicle which complies with the International Standards Organisation recommendations.

A number of advantages accrue from the use of containers. Containers can be mechanically handled and this speeds up loading and unloading, thereby giving greater efficiency; quicker turn round reduces the time a ship spends in port. The container, being durable, can be re-used and gives a much greater degree of protection to its contents than does the pallet system. Vehicles can be purpose-built to carry the maximum number of containers. The traditional trans-shipment of cargo at the dock side is made unnecessary for containers can be packaged at inland collecting centres. In a word, containerisation reduces the cost, the delay, and the pilfering of goods in transit. Containerisation, of course, presupposes that all goods can be packed into containers.

The speed at which containerisation has been adopted has come as a big surprise, even to those intimately concerned with the industry. Containerisation is, however, very costly and, for example, it has been estimated that the capital cost of going over to containers on Britain's main deep-sea routes will be in the region of £1000 million. This does not include the short-sea routes, where investment will be much less and is already well advanced, *e.g.* in the case of Felixstowe which has services to the Netherlands and West Germany.

Containerisation has had a tremendous impact upon the transport industry and, in truth, has become a symbol of the current transport revolution.

## *ROADS*

In early times pack horses and camel caravans undertook a certain amount of carriage on land, but it is likely that the bulk of transport was by water. During the days of the Roman Empire wheeled vehicles were used on the magnificent system of stone roads that the Romans built. Unfortunately these early roads in Europe fell into disuse and disrepair and during the medieval era most land carriage was by pack animal. The routes followed by pack animals were trails rather than roads and until comparatively recent times such highways as existed were dirt trackways, dusty and rutted in summer and quagmires of mud in winter. It was not until the eighteenth century, when industrial progress demanded improved means of communication, that made-up roads (at least in Britain) began to be constructed. Telford, Macadam, and Metcalfe were the pioneers of the metalled road. It was the invention of the motor car which led to the great extension and improvement of road transport.

Today motor cars, motor lorries, and motor coaches play an important and ever-increasing role in the transportation of passengers and goods. In order to keep pace with their rapidly increasing numbers and use and also to combat a growing congestion and death-rate on the roads, existing

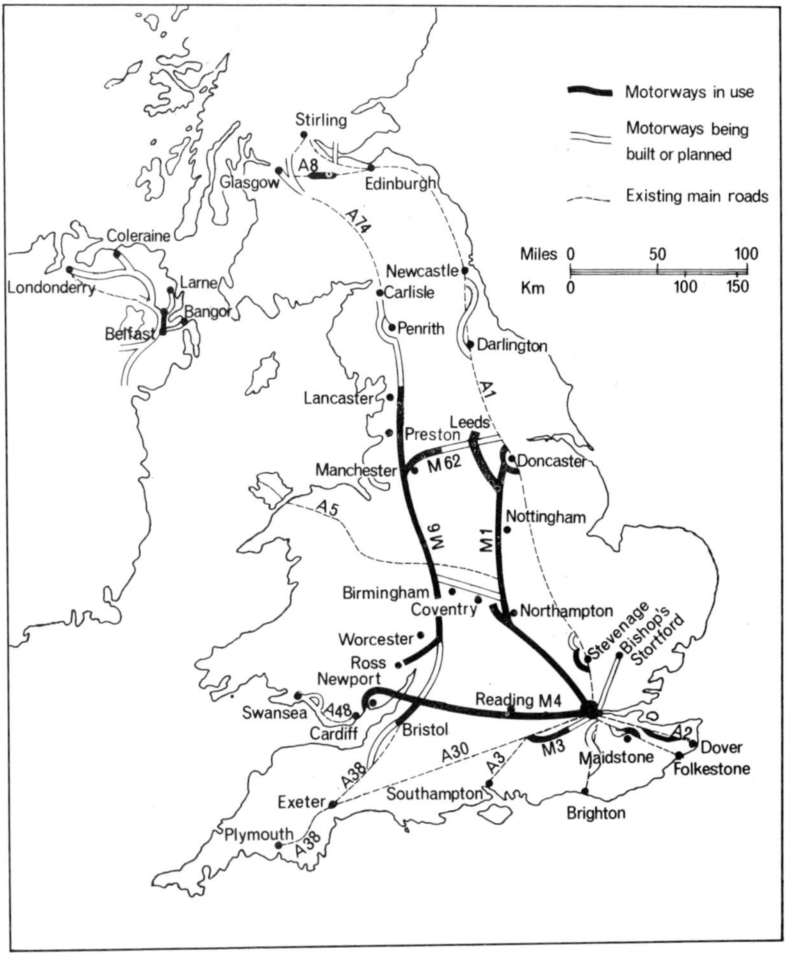


FIG. 134.—THE ROADS OF BRITAIN

Between about 1840 and 1914 the railways were the dominant medium of transport; and Britain had the densest and most intensively used network in the world. However, the coming of the motor vehicle and the development of the lorry and long-distance road transport, brought increasing competition to the railways. Since the Second World War the railways have run, increasingly, at a loss, and British Rail have been forced to discontinue uneconomic lines and to streamline and improve the railways.

Meanwhile road traffic continues to increase. But Britain's road system is far from adequate to meet modern needs. After a long period of patchy development and piecemeal improvement, Britain's road highway system is now beginning to take the shape of a national road network which is capable of carrying the vastly increased volume of motor traffic at high speeds.

highways are being straightened, widened, and cambered while special arterial roads, or motorways as they are called in Britain (*autobahnen* in Germany, *autostrade* in Italy), are having to be built. So vital has motor transport become that many bus stations and coach terminals are as large and as busy as the bigger railway stations.

During the past fifty years the automobile in its various forms has revolutionised methods of internal transport. Road transport has the advantages of flexibility of service, greater speed over short distances, and directness of communication over other forms of land transport. For long-haul traffic and for heavy bulk loads, however, railways are still required and, usually, to be preferred, although we should note the growth of long-distance motor traffic by heavy lorry and trailer. Most countries seek to integrate their road and rail services and the desirable end is to use the roads as "feeders" to main line railways. In undeveloped regions awaiting opening-up the prime essential is usually the laying down of a railway; all-weather roads to act as "feeders" to the railway come second.

Road transport is very unequally developed throughout the world. Western Europe and North America have the greatest number of vehicles, the closest road networks, and the most developed services. All the cities and ports of Western Europe are linked by inter-communicating road systems and most of them have linking coach services. Similarly, in North America coaches and lorries provide coast-to-coast services and link Canada with Mexico. In Canada a great new highway, the Trans-Canadian Roadway, runs from St John's in Newfoundland to Vancouver in British Columbia. Another major Canadian highway is the Alcan or Alaskan Highway, originally built as a military road during the Second World War, which links Edmonton to Alaska.

Outside Europe and North America good roads and well-developed road systems are scarce. Beyond the immediate environs of towns, made roads often peter out and become mere dirt tracks; these during rainy seasons frequently become impassable. Even so, motor transport has helped to solve many transport problems in regions where railways are lacking and where their presence is prohibited by the cost of railroad building. In parts of Africa, for instance, motor transport has been a great boon, providing useful links and overcoming many transport difficulties; for example, scheduled services now cross the Sahara from the Atlas region to the Gulf of Guinea. Similarly in Australia, where distances are great and settlements widely spaced, motor transport is much used. One of Australia's major road links is the trans-continental Stuart Highway which connects Birdum, the railhead in the Northern Territory to Oodnadatta in South Australia via Alice Springs and Tennant Creek. In

[*Courtesy: Italian Embassy.*

FIG. 135.—ITALIAN AUTOSTRADA

Modern Italians have the same skill in road-making as their ancestors, the Romans. This *autostrada* (national highway) cuts through the Apennines from Florence to Bologna. Note that in the whole of the stretch shown there is no way on to the *autostrada* from the surrounding country, which depends upon tortuous minor roads for local communication in this difficult limestone region.

China, which not so very long ago was very poorly served with inter-regional transport links, it is now possible to travel to almost any corner of the country by bus, though good roads are still few and far between. Roads form important links between ports and their hinterlands in South America, and some countries, such as Brazil, Bolivia, and Peru, are building great roads to assist national cohesion as well as economic development. But the major road project in South America is the Pan-American Highway, already largely completed except for one relatively short break (between Panama and Colombia), which links up all the South American Republics and joins together South America, Central America, Mexico, and the United States.

## *RAILWAYS*

The railway was in a very real sense a product of the Industrial Revolution, and for almost a century before the First World War the railway was the predominant agent in inland transport. Just as the use of motor transport revolutionised methods of inland transport in the years following the First World War, so transport was revolutionised by the introduction of railways in the middle decades of the nineteenth century. The advent of the steam-driven locomotive, running on fixed rails, solved two important problems:

1. It made possible the economic carriage by land of (*a*) materials in bulk, and (*b*) bulky goods.
2. It made possible the relatively rapid movement of people, mail, commodities, etc.

[*Courtesy: Canadian Pacific Railway.*

FIG. 136.—DIESEL-ELECTRIC TRAIN

A photograph of the new diesel-electric "Canadian" passenger train threading the Rockies. Note the great length of the train; the single track main line; also the glass-domed observation cars. The Canadian Pacific Railway (privately-owned) and the Canadian National Railway (State-owned) run from coast to coast and help to link the widely separated provinces of Canada together. As in the United States, the Rockies and other ranges of the Western Cordillera form a great physical obstacle separating the interior plains from the Pacific coastlands.

In addition to its carrying function the railway had two important effects:

1. It helped to accelerate the process of industrialisation, and the most highly industrialised areas have the densest railway networks.

2. It served as the advance guard of civilisation in the undeveloped "new lands" of the world, making possible their agricultural development and settlement.

The railway development and mileage in any area is largely the result of four factors:

1. The degree of progress in civilisation.
2. The density of population.
3. The degree of development of the natural resources.
4. The relief of the land surface.

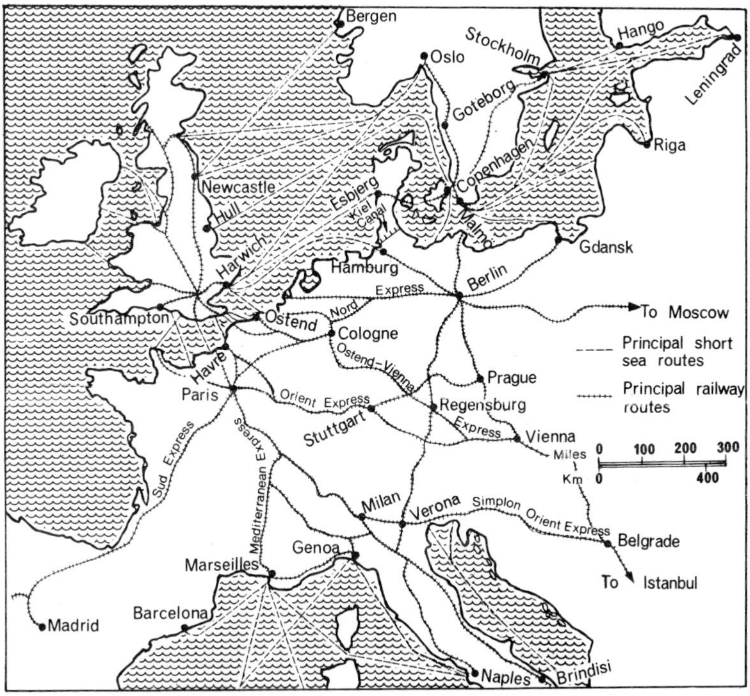


FIG. 137.—THE RAILWAYS OF EUROPE

Europe has a very close network of railways, and train services in most countries are fairly frequent and fast. The map shows only the main continental through lines: note their names. Note, also, the importance of London, Paris, Cologne, Berlin, Regensburg, and Milan as great railway foci. Most of the European railways are on the standard gauge of 4 feet $8\frac{1}{2}$ in. so that through services can be run across national frontiers.

In countries which are highly industrialised and urbanised and where commercial activity is highly-developed, as in England or Belgium, there is a very dense network of railways. Thickly populated areas do not necessarily have well-developed railway networks: think of the Great Plain of Northern China, or Java; but clearly dense populations are likely to create a demand for rail communications. The exploitation of mineral wealth, quite apart from the development of the coalfields, has influenced much railway construction, *e.g.* many of the railways in Africa, Australia, and South America were built to tap mineral resources. Railways must pay due regard to physical conditions, especially the relief of the land surface. Much more so than roads, railway building and alignment are determined by topographical factors, notably gradient.

Broadly speaking, the stage of development, social as well as economic, of a region may be judged from the adequacy of its railway network. In fact it has been said that "railroad mileage is an index of civilisation." Western and Central Europe and the eastern half of the United States have well-developed and, in places, very close networks; areas such as the Argentinian Pampas, the Coffee Region of Brazil, South Africa, and the Indian sub-continent have a reasonably close net; but some areas have few lines and nothing in the nature of a net, *e.g.* China, Queensland, Nigeria, while a few areas are practically devoid of a single line.

One of the biggest handicaps to the very desirable link-up and integration of railways is the existence of different gauges. Much inconvenience has resulted from the construction of national, and even state or provincial, railways on different gauges. Australia is a case in point: here the railways were built to three different widths, the narrow (3 feet 6 in.), standard (4 feet 8½ in.), and broad (5 feet 3 in.) gauges. Such variety greatly hindered travel over long distances and at one time a journey from Brisbane to Perth entailed thirteen (now reduced to five) changes. Some of the countries of South America, Brazil in particular, are bedevilled by a multiplicity of railway gauges.

Railways may be conveniently classified into three types:

1. Local networks such as serve industrial regions or densely populated areas, *e.g.* the London Basin, the Ruhr region, the humid pampas of Argentina.
2. Lines of penetration such as the isolated lines running inland from a port to tap a mineral deposit or a productive agricultural area.
3. Trans-continental lines such as the Canadian Pacific Railway, the Buenos Aires–Valparaiso Railway, the Trans-Siberian Railway, and the Australian Transcontinental Railway.

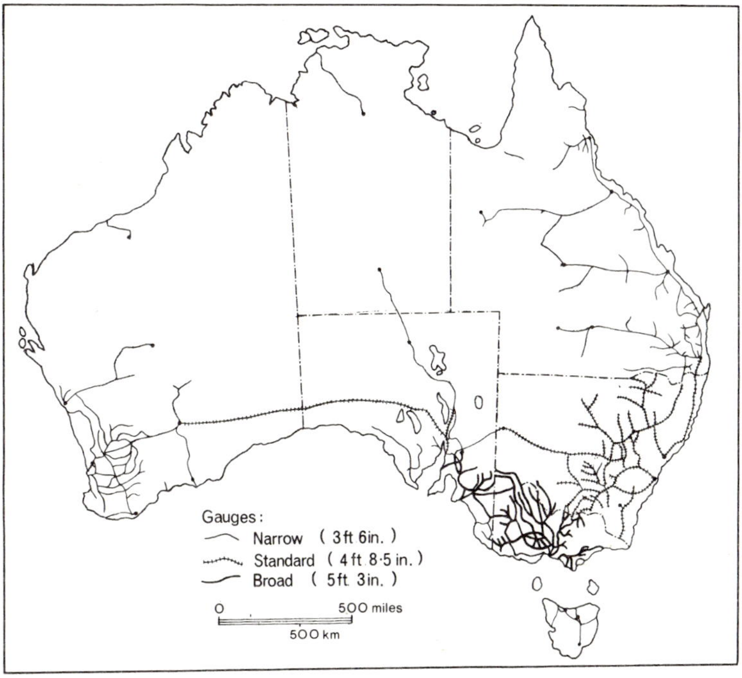


FIG. 138.—THE RAILWAYS OF AUSTRALIA

Australia's railway system illustrates a problem common to many countries (*e.g.* Brazil, India). The system has two distinctive features: first, it is primarily a coastal system, serving the capital cities and sending disconnected lines inland from the ports to mining or farming districts; secondly, it has three different gauges, the narrow, standard, and broad gauges, which serve different areas and which create difficulties of inter-communication. The improvement of the railway system constitutes one of Australia's most pressing economic problems. It involves: (*a*) linking up the inland "feeders" to provide a better and more integrated railway pattern, and (*b*) the reconstruction of the railway system to a uniform gauge.

The narrow gauge is mainly confined to Western Australia and Queensland; the standard gauge to New South Wales; and the broad gauge to Victoria. The reason for this variation is that the lines were laid down by different authorities at different times.

Clearly the losses resulting from the breaks of gauge are great. At each state border goods and passengers must be transferred thereby causing delays and also increasing the costs of transport. Goods in transit between Sydney and Perth take fourteen days or longer but on the distance involved they should be carried in about five days. The vexed subject of railway unification has been discussed for a long time but with little decision. However, in 1956, the Rail Standardisation Committee made specific proposals for the linkage of all the state capitals on the standard gauge and in 1961 the decision to proceed with the first link—Kalgoorlie to Perth—was agreed upon. It was completed in 1968.

Especially important are the great trans-continental railways. These were constructed for both political and economic reasons. The Canadian Pacific Railway, for example, was built partly for political reasons: British Columbia refused to join the confederation of Canadian provinces until a rail link with the east was guaranteed. Again, the fear that Western Australia might secede from the union of Australian states was partly responsible for the construction of the Australian Transcontinental Railway. The building of the Trans-Siberian Railway, which runs from Moscow to Vladivostok, was also undertaken primarily for political and strategic purposes—to provide quicker communication between Moscow, the centre of government, and the remote and isolated far eastern colonial territories of Russia. It is worth noting that the trans-continental lines of the Northern Hemisphere form an integral part of the great girdle of east–west routes.

## *ROPEWAYS AND CABLEWAYS*

We are all familiar with chair-lifts and cable-cars such as are used in the Alpine part of Switzerland, but it is seldom realised to what extent ropeways and cableways are used as a medium of transport. The main principle involved is that loads are suspended from overhead cables and carried clear of the ground. The great advantage of this method of transport is its ability to cross terrain, such as forest country, deep valleys and wide rivers, where no other form of transport is physically practicable or economically possible.

The use of aerial ropeways is not new; their use goes back for over a hundred years. British firms were pioneers in this mode of transport. "By 1914 British engineers designed, supplied and erected ropeways all over the world in climates varying from arctic to equatorial, and often in fantastically difficult terrain. One ropeway built by the oldest British company still in the field was begun in 1913 and is still in regular use carrying goods 47 miles over the Andes in Colombia. It terminates at 17,000 ft and crosses ravines 4000 ft deep."* After the First World War many important technical advances in ropeway transport were made, especially in tonnage capacity. "Between 1920 and 1939 many notable ropeways were exported [by Britain]. They included installations for Reynolds Jamaica Bauxite Mines, in Jamaica (160 tons per hour); Algoma Ore Properties, in Canada (500 tons per hour); Cement, in Eire (215 tons per hour); Dalen Portland Cementfabrik, in Norway (220 tons per hour); and British Aluminium, in West Africa (300 tons per hour)."*

* Quotations from "Transport by Rope," *The Times Review of Industry and Technology*, Vol. 1, No. 10, December 1963, p. 35.

Until 1939, however, ropeways were, in general, of comparatively modest capacity and length, but further important technical advances were made after the Second World War. In the first place, lengths of up to 50 miles (8 km) with capacities of up to 500 tons an hour were achieved, and, secondly, many automatic devices were introduced into ropeway systems to save manpower. At the present day, the tendency is towards much larger and more costly ropeway installations. "Because they run at higher speeds they have to be fitted with complex electrical control, automatic loading and unloading and devices to reduce servicing. New materials have also been developed. Nylon wheel treads are used to give kinder treatment to ropes and, in some instances, alloys have shown advantages over steel."*

A cableway is really a specialised version of a ropeway; they lift as well as carry and may be thought of as very long span overhead cranes. They are more especially used for dam construction.

Ropeways are used by collieries for the disposal of their waste material, by lumbermen who use high-level cables to "skid" the logs over the tree-tops or other obstacles, by quarrying and mining concerns to carry ore, etc., to a railhead, and by mountain countries which have scenic and winter sports attractions for tourists.

### *PIPELINES*

The recent increase in the use of pipelines for the transport of commodities "represents one of the most notable revolutions in the history of transport, and especially in the transport of energy."† Water, of course, has long been moved by pipe, but the use of pipelines to carry other things is relatively new. "The pipeline is as much a pioneer of transport facilities in the underdeveloped countries as are roads, or railways, or aircraft."†

For many decades now the overland transport of oil has been effected by pipelines. The United States has a great network of oil pipelines carrying petroleum from the producing to the consuming areas; most famous of these pipelines is the "Big Inch" which connects the Gulf Coast oilfields with the north–east. The United States has over 300,000 miles of oil pipeline. In the Middle East oil pipelines are much used to carry the crude petroleum from the great oil basin of Mesopotamia and the Persian Gulf to the Mediterranean Sea. The largest of these pipelines is "Tapline," a 30-in. (762-mm) diameter pipe traversing the Arabian–Syrian Desert for a distance of 100 miles (1600 km) and linking the Arabian American Oil

* Quotation from "Transport by Rope," *The Times Review of Industry and Technology*, Vol. 1, No. 10, December 1963, p. 35.

† MANNERS, G., "The Pipeline Revolution," *Geography*, Vol. XLVII, April 1962.

Company's concessions near the Persian Gulf with Sidon on the Levantine coast. One of the longest pipelines is the great 3000-mile (4800-km) trans-continental Comecon or Friendship pipeline (completed 1963) built by the Soviet Union which carries crude oil from the Ural-Volga oilfields to the Communist satellite countries of East Europe.

Natural gas, a natural commodity of rapidly growing importance, is also transported by pipeline. The United States has long made use of its natural gas resources and the mileage of gas pipeline is now greater than that of the railways. The discovery of natural gas deposits in Europe—in Italy, France, Holland, and Germany—has led to the laying of a gas grid system.

Here we might note that overland transport of oil and gas, especially over long distances, is most economically effected by pipeline. But carriage by water (*i.e.* by tanker) is considerably cheaper than by pipeline. The chief disadvantages of pipelines are that they are costly to construct, they are not flexible, and to be economic they must be used to their maximum capacity.

Pipelines are being used to transport an increasing range of commodities. For example, a coal pipeline 105 miles (168 km) long connects Pittsburgh with Cleveland in the United States—the coal being transported in the form of "slurry," crushed coal and water. Britain is now operating a pilot scheme. A 5-in. (127-mm) steel pipe runs from Walton Colliery to Wakefield power station, in Yorkshire—a little over one mile (1·6 km)—and delivers 40 tons of coal (in the form of a sludge) an hour. Chemicals are also being pumped from Fawley oil refinery to the new I.C.I. chemical plant at Severnside. Yet another recent development has been the introduction of milk pipelines in the Alpine parts of France, Switzerland, and Austria; milk is piped from the alps to the dairies in the valleys below.

## *INLAND WATERWAYS*

Inland movement by water is undertaken (*a*) by river, (*b*) by canal; such movement, however, is limited by the depth, width, and direction of the waterway. In early historical times much, perhaps most, of the inland carriage of commodities was by water. This was possible when vessels were small, the volume of traffic limited, and the time factor was not particularly pressing. But during the eighteenth century ships began to grow in size, trade began greatly to expand, and speed of carriage came to be of greater importance. In order to overcome the limitations of many rivers and, also, to provide many inland towns with water communications, canals began to be built. In England, a pioneer in canal construction, the building of these new waterways became almost a mania. Canal building on the Continent came somewhat later. The nineteenth-century development of the railway, and more recently, of roads, led to a decline in water transport in England, but in Europe, where conditions were more favourable, it tended to hold its own. It is of special interest to note that within very recent years water transport has tended to make a comeback; this is closely linked with the cheapness and capacity for bulk carriage of

[*Courtesy: Eric Kay.*

FIG. 139.—BARGES ON THE RHINE SYSTEM

Barges descending and ascending through locks on the river Main at Greisheim, Germany. The barges in the foreground are empty; when full they lie low in the water and sometimes almost appear to be awash. On the Rhine one can often see trains of such barges. The works in the background is a chemical plant.

water transport—witness the recent and current developments of the Albert Canal, the new canal between Zeebrugge and Ghent, the improvement of the River Moselle, and the Rhône Valley scheme.

Inland waterways have both advantages and disadvantages. The chief advantages are:

1. There is no track to lay or maintain, although dredging may be necessary, in the case of natural waterways.

2. They may form the only practicable routes, *e.g.* in very difficult, mountainous country.

3. Waterways, under favourable conditions, provide cheap transport for heavy, bulky, imperishable commodities such as coal, ore, timber, cement.

The main disadvantages are:

1. Rivers may involve devious journeys and may flow in the wrong direction from the point of view of trade.

2. Canal construction involves heavy capital outlay and the canals require constant maintenance and sometimes dredging.

3. Otherwise navigable rivers may be interrupted by falls or rapids while canals often require locks if there are differences in level.

4. Transport by water is slow in comparison with other forms of overland transport.

5. River levels may change seasonally and freezing may cause stoppages.

Although water transport is carried on to a greater or lesser degree the world over, there are only six major navigable systems of inland waterways:

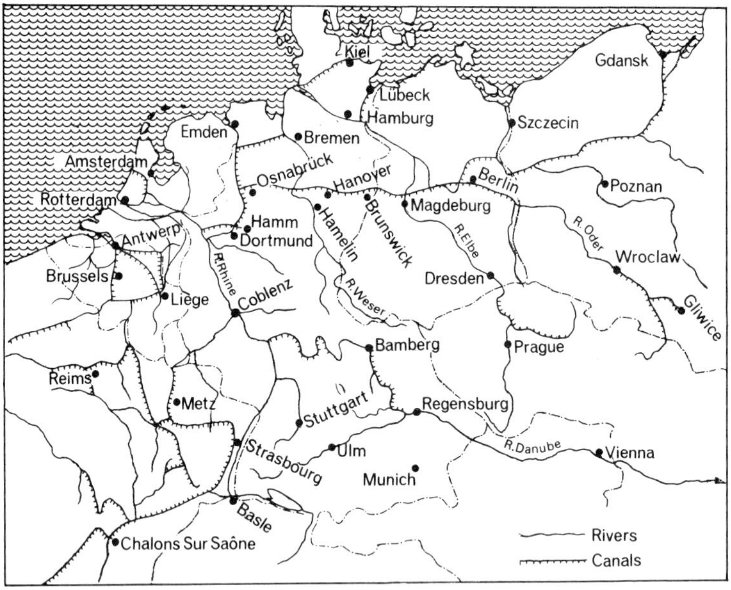


FIG. 140.—THE CANALS OF CENTRAL EUROPE

Central Europe possesses an extensive system of natural and artificial waterways. Water transport, by river and canal, is helped by (i) the levelness of the North European Plain, which seldom necessitates locks; (ii) the presence of great east–west valleys, left by the Ice Age, which facilitates lateral links; (iii) the soft rocks of the plain which make canal excavation a relatively easy matter; (iv) the heavy, bulky nature of many of the commodities requiring transport: water carriage, remember, has the advantage of cheapness; and (v) most of the rivers, which were amenable to improvement, flowed in the right direction, *i.e.* northwards towards the English Channel and the North Sea.

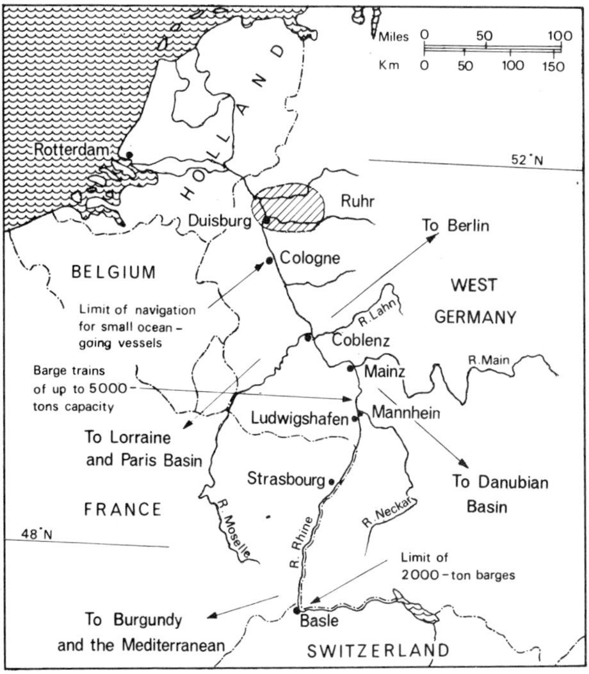


FIG. 141.—THE RHINE WATERWAY

The Rhine, though small in comparison with such rivers as the Mississippi and Yangtse, may claim to be the world's most important waterway; certainly it is the most important in Europe. The Rhine valley affords a route into the heart of Europe and, by links with the Rhône–Saône valley via the Belfort Gap and with the Danube across Bavaria, offers passages across the great mountain barriers of the Alpine system. Thus, by means of the Rhine, the North Sea has connections with the Mediterranean and Black Seas.

The Rhine has few natural obstacles, such as rapids and waterfalls, in its lower and middle courses and there is uninterrupted navigation as far as Basle in Switzerland. Regularisation of the river, especially in the Rift Valley section, has improved navigation. The Rhine is free from winter freeze-up. It has, also, numerous lateral water connections.

The Rhine serves the Ruhr, the greatest industrial region of the continent, and passes through areas of varied industry and agriculture. Oil, iron ore, cotton, wool, and foodstuffs move upstream (coal also goes via the canalised Moselle to Lorraine), while timber, chemicals, coal, and manufactured goods move downstream. The Rhine leads to Switzerland and functions as its life-line; hence the volume of traffic is swelled by Switzerland's imports (iron, cotton, grain, cacao) and its exports (machinery, watches, textiles, food products).

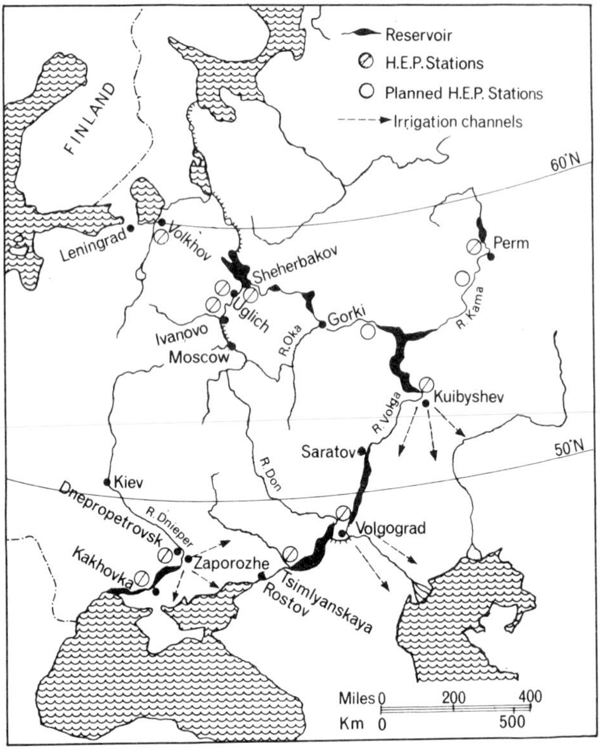


FIG. 142.—THE VOLGA SYSTEM

The rivers of the Soviet Union, in both its European and Asiatic parts, are long, broad, and slow-flowing. Historically, they have always formed important highways. Broadly speaking, the northern part of the country suffers from an excess of water, the south from too little. This offered a challenge to Soviet scientists and they have begun to harness their great sluggish streams.

The most important of all the Soviet schemes is the Volga–Don project; this serves a greater area than any other river control project, and its main features are: (i) the diversion of the upper Volga by means of the Moskva–Volga Canal to provide Moscow with additional water supplies; (ii) the diversion and distribution of water from the lower Volga to irrigate the arid lands around the Caspian Sea; (iii) the construction of dams at a number of points along the river to provide hydro-electric power; and (iv) the construction of the Volga–Don Canal (opened in 1952) gives access to the Black Sea from the Volga.

Formerly full use of the Volga was handicapped by the seasonal variations in its level. This problem has been surmounted by the building of the great dam near Rybinsk which has led to the creation of the "Rybinsk Sea." The water stored in this great reservoir is released during the summer months and helps to raise the level of the Volga.

*Western and Central Europe.* The rivers flowing into the North Sea from the Seine to the Elbe, together with their numerous inter-connecting canals, afford the best-developed waterway system in the world. Water transport is particularly important in France, Belgium, Holland, and West Germany. The River Rhine is *the* great artery of this European system and is of premier importance to Holland, France, Switzerland, and especially to Western Germany (*see* Fig. 141). Mention here might be made of the River Danube, Europe's second most important waterway, which is navigable for 1500 miles (2400 km). Unfortunately, the Danube flows in the wrong direction—away from the important part of Europe and into a semi-enclosed and rather isolated sea, the Black Sea.

*The Volga System.* Compared with the Rhine, the Volga, the principal artery of European Russia, is less important as a waterway; nevertheless, it is one of the world's great rivers, giving access to some 70,000 miles (112,000 km) of navigable waterway. The Volga's potential importance is rather restricted since it flows into the Caspian which is an inland sea. This handicap has been mitigated by the canals linking it to the River Don. The value of Russia's rivers is further limited due to prolonged winter freeze-up. The great rivers of Soviet Asia, the Ob, Yenesei, and Lena, are of very little value, partly due to the length of time for which they are frozen and partly to the direction of their flow which leads to the ice-closed and commercially unimportant Arctic Ocean.

*The Chinese Waterways.* From early historical times the Chinese rivers Hwang-ho, Yangtse, and Si, have been important highways. The Yangtse-kiang has been, as it continues to be, of especial value as a highway largely because of the absence of alternative means of communication. It is the great artery of central China, linking the delta region with the populous, productive, but isolated, Basin of Szechuan (the Red Basin). The Yangtse is navigable by ocean-going vessels as far as Hankow but smaller boats can get through the Ichang–Wanhsien gorges to Chungking and junks can ply upstream as far as Ipin. The Grand Canal, which links Hangchow with Tientsin, and which is hundreds of years old, has recently been brought into commission again. The Chinese lowlands form a veritable maze of canals which are much used for local carrying trade.

*The North American Rivers.* There are really two great systems in the North American continent: the Great Lakes–St Lawrence waterway and the Mississippi. The former, which has been vastly improved by the construction of the Seaway (*see* Fig. 147), allow ships to penetrate 2350 miles (3760 km) inland and has turned the lakeside ports into ocean ports. The Great Lakes form one of the world's busiest waterways. The Mississippi with its tributary the Ohio, though relatively less important than formerly, still carries over 1 million tons of freight annually. The river is navigable for steamers of considerable size as far upstream as Minneapolis. The Mississippi and the Ohio are linked by canals to the Great Lakes. The continent's inland waterways are especially important for the large-scale carriage of bulky commodities such as coal, iron ore, timber, wheat, and cotton.

[*Courtesy: National Film Board of Canada.*

FIG. 143.—IROQUOIS LOCK

The Iroquois lock on the St Lawrence Seaway is one—the most westerly—of the seven new Seaway locks. This lock can handle vessels up to 715 feet (234 m) in length. In the background of the photograph can be seen the Iroquois Control Dam.

*The Amazon.* The Amazon and its tributaries offer during the flood season some 36,000 miles (57,600 km) of navigable waterways. Ocean-going vessels can reach Manaos, 1000 miles (1600 km) upstream, while river steamers can ascend the river as far as Iquitos, in Peru, which is 2300 miles (3680 km) from the Atlantic. During the high-water season many of the tributaries are passable, but at other times of the year river craft must be man-hauled around the rapids. Traffic on the Amazon is light, however, because of the sparseness of the population and the limited production of the basin. The Amazon provides a good example of a waterway system which, though suitable for navigation by large vessels, is relatively little used.

*The Paraná–Paraguay.* This system, which drains into the Río de la Plata, is the most important river system in the whole of South America from the point of view of its economic value. It forms a separate system from the Amazon since the two, unlike the Great Lakes–Mississippi system, are unconnected; though with canal linkages, short-circuits, and dredging, the great rivers of South America could be connected up to form a magnificent system of interior waterways. The highly productive lands surrounding the la Plata estuary are readily accessible by water highways. The Paraná is navigable for ocean freighters as far as Santa Fé, some 150 miles (240 km) upstream, and for river steamers up to Asuncion, in Paraguay, 850 miles (1360 km) upstream, while the Río Uruguay is navigable to Salto where rapids obstruct further passage.

## *OCEAN TRANSPORT*

Ocean transport has certain advantages over land carriage. Water provides a ready-made carriage-way for ships which, unlike roadway or railway, requires no maintenance. Terminals (*i.e.* ports) at the ends of the sea "lanes" have, it is true, to be made, but between them there is need of neither track nor surfacing. Again, except for fog and floating ice, and occasionally stormy weather which may hinder progress, ocean-going vessels have no physical obstacles to surmount such as often handicap overland transport. Furthermore, large vessels are capable of carrying far greater loads and far greater weights than can be handled even by the longest railway train, the most powerful lorry and trailer, or the largest aircraft. Ocean transport is, therefore, the cheapest of all forms of transport.

### BRITISH SHIPPING

The fact that a country is a great trading nation does not necessarily mean that the country is also a great shipping nation. The United Kingdom ranks after the United States and West Germany as a major world trading country, but it is also one of the world's three great shipping countries. The United Kingdom's merchant fleet comprises 2,042 ships with a gross tonnage of 22,274,000 tons: this gives her 11% of the world tonnage. The United Kingdom's merchant fleet is, at the present time, the largest, and the most efficient, and the most modern ever. A third of the vessels are under five years old. The fleet contains almost every type of ship afloat—passenger liners, cargo liners, bulk carriers, container ships, methane carriers, tramp ships, coasting vessels, ferry ships, etc. The fleet comprises the following broad groups:

42·8%—tankers.
28·6%—cargo liners.
22·0%—tramps.
6·6%—passenger liners.

Table XL

## MERCHANT SHIP TYPES

- OCEAN-GOING
  - Passenger and Mail
    - Large Trans-Atlantic Liners
    - Cape and other Routes
  - Passenger Cargo
    - General
    - Refrigerator
      - Meat
      - Fruit
  - Cargo
    - Tramp
    - Bulk (ores, timber etc.)
    - Tankers
      - Wine
      - Oil
        - Refined
        - Crude
- SEA-GOING
  - Cross Channel and Train Ferries
  - Short Sea Traders
  - Coasters
- FISHING CRAFT
  - Trawlers
  - Seine Netters, etc.
  - Drifters
  - Whalers
- AUXILIARY
  - Tugs
    - Sea-going
    - River
  - Dredgers
  - Ice Breakers
  - Pilot Cutters

The tanker tonnage is growing at a faster rate than that of the rest of the fleet because it reflects the increasing world demand for oil. The fleet is manned by some 98,000 officers and ratings.

Gross earnings from the fleet's world-wide services totalled £822 million in 1969. The United Kingdom's ships contributed around £300 million net to the country's balance of payments and in fact were responsible for more than one-fifth of all the "invisible exports" earned by industry and commerce. The full value of the fleet to the British economy is even greater than these figures suggest, for, if the United Kingdom did not possess its own ships, something of the order of £250 million a year would have to be paid to foreign shippers to do the job done by British vessels.

**TYPES OF VESSEL**

From the very earliest times ships have been specialised in function: the Romans had longships or galleys for military purposes and round-ships for trade; and so had the Vikings. Even the Eskimoes had their kayaks and umiaks which served different purposes. During more recent times ships have shown an even greater specialisation and several distinct types of commercial vessel gradually evolved. Today, five main types of merchant ship are recognised.

*Liners*, which until very recently were the largest ships afloat (the *Queen Elizabeth* of 83,673 tons was the greatest), are primarily passenger vessels operating on the main shipping routes. But even liners carry a certain amount of cargo. The passenger liner has had to face strong competition from the aeroplane and the day of the mammoth liner has in all probability passed.

*Cargo-liners* are vessels combining the function of cargo and passenger transport. These vessels carry commodities such as grain, meat, fruit, and wool, but are also able to accommodate a limited number of passengers. Many of the vessels on the Australia–New Zealand run or the South American run are of this type.

*Tramps* are small cargo vessels (usually about 6000 tons) which do not normally carry passengers, though some these days do provide limited accommodation for a few passengers. Tramps pick up cargo as they go, following no set route, and they may be away from their home port for months on end. They are becoming less important as ocean carriers, though they continue to transport many bulky commodities such as coal, ore, lumber, and fertilisers.

*Tankers* are a relatively new type of ship; they are the product of the oil age. Recent years have seen a tremendous growth in the size of oil-tankers and some of them are the largest vessels afloat; a number exceed

500,000 tons dead weight. Today, more than half the world's merchant shipping tonnage comprises tankers, a fact indicative of the great importance of oil in the modern world.

*Coastal craft* are mainly used for shipping heavy, bulky commodities such as coal and ore from point to point along the coast. Some countries, *e.g.* Britain, Norway, Japan, the United States, make much use of coasters. Commonly they are small vessels of shallow draught to enable them to penetrate rivers and coastal creeks.

## SHIPPING ROUTES

Although ships have freedom of movement and are capable of going virtually anywhere they tend to keep to certain "lanes." They do so because of: (*a*) physical conditions, and (*b*) economic considerations. Clearly,

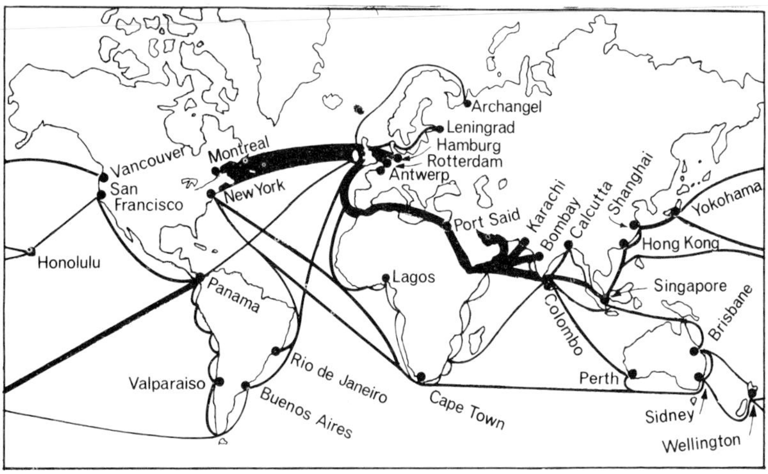


FIG. 144.—MAIN SEA ROUTES

The figure shows the main sea lanes and their importance prior to the closure of the Suez Canal in 1967.

ships will go only where there are goods or people to be carried and the most important shipping routes are those linking the most productive and most populous regions. Certain physical conditions also help to determine the routes followed by ships, *e.g.* the availability of harbours and ports on coasts, weather conditions such as fog and storm occurrence, and oceanographical factors such as sea ice and icebergs, submarine banks, and shallow waters.

The chief shipping lanes are:

*The North Atlantic route* between western Europe and eastern Canada and the United States, which is the busiest and most important of all the world's shipping routes and the most important passenger route.

*The Cape route*, via West Africa, to the Republic of South Africa and thence eastwards to Australia and New Zealand; this route carries passengers to the southern Dominions and fruit, wine, wool, tobacco, and meat come from these countries to Britain.

*The South Atlantic route* to Brazil, Argentina, and Uruguay; coal, machinery, and manufactured goods provide the chief items of traffic on the outward journey, raw materials and foodstuffs (meat, coffee, fruit) the chief items on the return journey.

*The Panama route* via the West Indies to the Pacific whence the route branches to Hawaii, Australia, or the western coast of North America or the western coast of South America.

*The North Pacific route* between the western coast of North America and the Far East; only a small volume of trade is carried over this route.

Until the closing of the Suez Canal in 1967, the Mediterranean–Red Sea–Indian Ocean route, which led to Australia and the Far East, was the second most important world route. Much of the traffic has had to be re-routed round the Cape.

## *SHIP CANALS*

To facilitate ocean shipping and to save travelling time a number of ocean ship canals have been constructed. A complete list of these canals is given in Table XLI. These vary greatly both in their commercial importance and their physical characteristics. Here we have space to deal briefly with three only: the Suez, Panama, and Great Lakes–St Lawrence Canals.

### THE SUEZ CANAL

The construction and opening of the Suez Canal almost one hundred years ago completely revolutionised the shipping routes of the Old World. Until the Canal was cut, ships had to round the Cape on their journeys to India, the Far East, and Australia–New Zealand. The linking of the Mediterranean with the Red Sea greatly shortened certain sea routes, *e.g.* between Britain and India the distance was reduced by some 4000 miles (6400 km), between Britain and the Far East by some 3000 miles (4800 km), and between Britain and Australia by 1200 miles (1920 km). The opening of the Suez Canal has led to the Mediterranean and Red Seas becoming a major route between Europe and the Indian and Pacific Oceans; it has also given a new lease of life to many Mediterranean ports, *e.g.* Genoa, Venice, and Marseilles, which, until 1869, were in decline.

For almost a hundred years the Suez Canal functioned as a major shipping route, providing a welcome short-cut for ocean shipping, but in 1967, as a result of the Arab–Israeli War, the Canal was closed to traffic and it still remains closed. This closure has created difficulties for the trading countries who regularly used the Canal but it has also brought great loss of revenue to Egypt who, of course, is no longer collecting tolls from the passage of ships. The refusal of Egypt to re-open the Canal is very short-

[*Courtesy: Hulton Press Ltd.*

FIG. 145.—THE SUEZ CANAL

A ship steaming through the Suez Canal which links the Mediterranean (at Port Said) with the Red Sea (at Suez). The Canal runs for 100 miles (160 km) without locks. Note the flat monotonous desert and the trees which have been planted to protect the canal banks from the wash of the ships. A speed limit is imposed to help prevent the sandy banks from being washed away.

sighted policy for many countries have re-orientated their sea traffic and much of this will have been lost to Egypt for good. This is especially true of the tanker traffic since the oil companies have turned to using more economic large-size tankers which, in any case, are incapable of negotiating the Canal with its limited draught.

Table XLI

*The World's Ship Canals*

| Name | Year opened | Recon-struction | Length in miles | Depth in feet | Bottom width in feet | Number of locks | |
|---|---|---|---|---|---|---|---|
| Göta Canal (Sweden) | 1832 | 1916 | 115 | 10 | 47 | 70 | Links Göteborg with Soderköping; of rather limited usefulness. |
| Saulte Ste Marie (Canada, U.S.A.) | 1855 and 1895 | | 1·6 | 22 | 100 | | A group of canals joining Lakes Superior and Huron; the levels of which differ by 21 feet. |
| Suez Canal | 1869 | | 100·6 | 39·3 | | nil | Eliminates long voyage around the Cape; deepening in progress.* |
| North Sea or Amsterdam Canal | 1876 | | 16·5 | 23 | 88 | | Links Amsterdam with North Sea; has maintained Amsterdam as an ocean port. |
| Welland (Canada) | 1887 | 1930 | 26·75 | 25 | 200 | 7 | Circumvents Niagara Falls and Niagara river rapids. |
| Corinth (Greece) | 1893 | | 4 | 26·25 | 72 | | Links Gulfs of Corinth and Aiyina. |
| Manchester Ship Canal | 1894 | | 39·7 | 28 | | 4 | Has made Manchester, 54 miles from the open sea, into one of Britain's major ports. |
| Kiel Canal | 1895 | 1914 | 60·9 | 45 | | 2 | Originally constructed for naval purposes, it shortens the North Sea–Baltic passage. |
| Panama Canal | 1914 | | 50·7 | 41 | | 6 | Eliminates the need for rounding the Horn. |
| Brussels or Rupel Sea Canal | 1922 | | 19·8 | 21 | | 4 | Has converted Brussels into an inland port. |
| Alphonse XIII Canal (Spain) | 1926 | | 4 | 25 | | 13 | Provides access to Seville for ocean-going vessels; length of waterway is 53 miles but actual canalised section is only 4 miles long. |
| White Sea–Baltic Canal | 1933 | | 140 | 16·5 | | 19 | Makes use of lakes, canalised river, and 32 miles of actual canal. |
| Princess Juliana Canal (Holland) | 1935 | | 20 | 16 | 52 | | |
| Volga–Moscow | 1937 | | 79 | | | | Links Moscow with the upper Volga. |
| Houston Canal (Texas) | 1940 | | 56·7 | 34 | | nil | Has made Houston, over 50 miles from the open sea, into a major port. |
| Albert Canal (Belgium) | 1940 | | 80 | | | | The canal gives Antwerp a water link with Liège and the industrial coal trough region of central Belgium. |
| Volga–Don Canal | 1952 | | | | | 13 | Links Don with Volga at nearest approach interconnects Black Sea with Caspian. |

* Closed since 1967.

## THE PANAMA CANAL

This, the second of the two major man-made inter-ocean passage-ways in the world, is cut where a fortuitous gap exists in the mountains at the narrowest part of the isthmus of Panama. The Canal, which includes three groups of locks, is one of the world's great engineering feats. It runs through a ribbon of Panama territory (the Canal Zone) leased to the United States.

The opening of the Panama Canal completely re-orientated the trade routes of the western hemisphere and focused many new shipping lanes into the Caribbean Sea. It has greatly improved communications between the eastern and western coasts of the Americas and between Europe and Western America and New Zealand. For example, the journey between Liverpool and San Francisco is shortened by 5000 miles (8000 km) (approximately two weeks' steaming time) compared with that round Cape Horn.

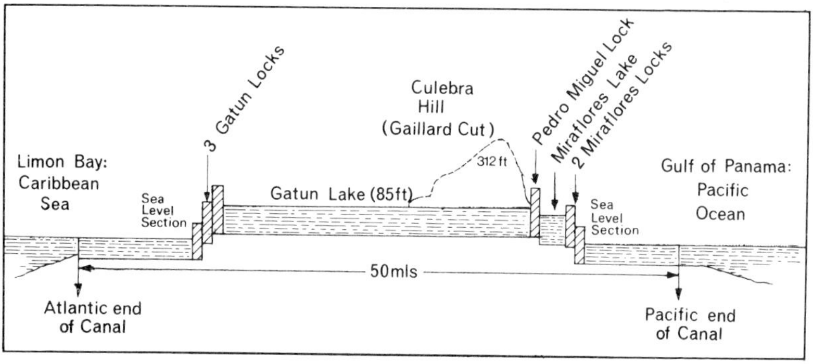


FIG. 146.—SECTION ALONG THE PANAMA CANAL

The canal provides a broad, deep waterway between the Atlantic and the Pacific. Unlike Suez, the Panama Canal has to make use of locks for along the lowest possible line of route the Culebra Hill rises nearly 400 feet (130 m) high. The diagram gives a cross-section of the Canal: note the three locks at either end. What are the Canal terminals?

The Canal, moreover, has made an important contribution to the development of the countries of western South America, which were previously remote and inaccessible.

The Panama Canal did not compare with the Suez Canal from the point of view of traffic or freight: slightly more than 15,000 vessels use the Canal annually, while the cargo amounts to some 60 million tons. The west to east traffic (chiefly Canadian timber and salmon for Britain, Californian fruit for Britain, Chilean nitrates and copper and Bolivian tin for the United States and Europe) is more than twice as great as that flowing westwards (chiefly refined petroleum products and manufactured goods).

## THE ST LAWRENCE-GREAT LAKES CANALS

The Great Lakes of North America with their connecting canals—the Sault Ste Marie, or "Soo," canals between Lakes Superior and Huron and the Welland Canal between Lakes Erie and Ontario—provide a continuous waterway for vessels drawing up to 25 feet (8·3 m) of water. These canals, though closed by ice for a third of the year, carry a greater tonnage of cargo than either the Suez or the Panama Canals. The traffic chiefly comprises grain, iron ore, and coal. Between Lake Ontario and Montreal there existed a 180-mile (288-km) stretch of waterway which were beset by rapids. Several canals, among which were the Lachine and Soulanges Canals, had been constructed to circumvent the rapids, but these were shallow canals which limited navigation to smaller ships. Thus Montreal was, in effect, the head of large, ocean-going vessels.

The construction of the St Lawrence Seaway, which involved the making of two new deep-water canals, with locks, on the south bank of the river, has provided a through passage-way for large ocean-going craft. Liners can now penetrate inland to the head of the Great Lakes, a distance of over 2000 miles (3200 km). Thus former lakeside ports have been turned into seaports. A notable feature of the Seaway is that it also provides for the generation of 2 million hp (1·4 million kW) of electricity. It was thought that the cutting of the Seaway would exert a profound effect upon the future economic development of both Canada and the United States which jointly undertook the project, but events so far have shown that the Seaway has not proved to be as successful as was hoped.

## *AIRWAYS*

The use of aircraft and the development of air communications belong to the twentieth century. Within the relatively short space of some 70 years truly amazing developments have occurred. Much of the accelerated development in the design, size, propulsion, and use of the aeroplane was due to the two World Wars, but even without the stimulus of war there can be little doubt that aviation would have made great strides. Although speed is the only respect in which air transport has an advantage over land or sea transport, this factor alone is of sufficient consequence in the modern world to have stimulated the development of aviation.

Air transport differs from all other forms of transport in that it is three-dimensional; as a result it is "the only one to enjoy the advantages and suffer the disadvantages of the third dimension."* Since aircraft are not

* SEALY, K. R. "London's Airport and the Geography of Airport Location," *Geography*, Vol. XL, 1955, pp. 255–64.

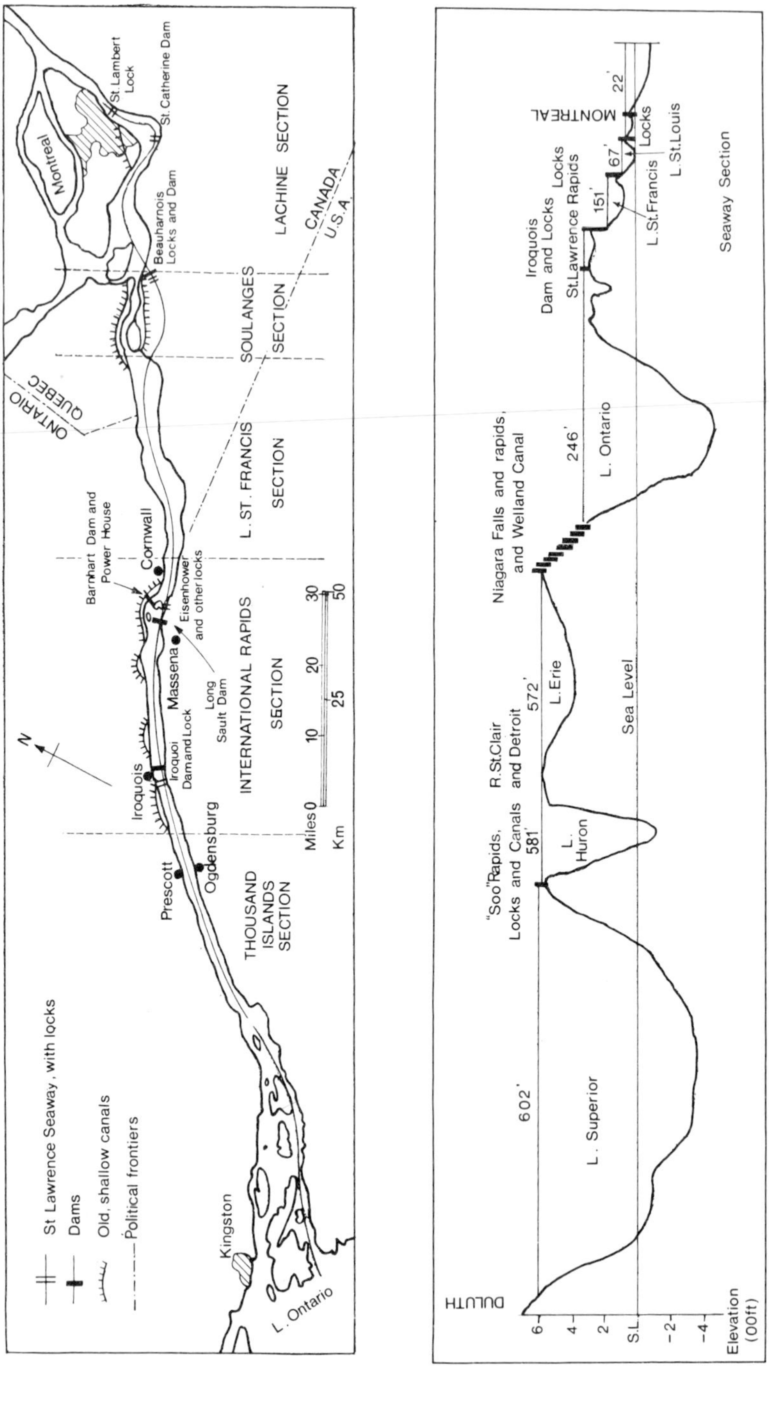
St Lawrence Seaway, with locks
Dams
Old, shallow canals
Political frontiers
Kingston
L. Ontario
THOUSAND ISLANDS SECTION
Prescott
Ogdensburg
Iroquois
Iroquoi Dam and Lock
INTERNATIONAL RAPIDS SECTION
Massena
Long Sault Dam
Eisenhower and other locks
Barnhart Dam and Power House
Cornwall
L. ST. FRANCIS SECTION
ONTARIO
QUEBEC
SOULANGES SECTION
Beauharnois Locks and Dam
LACHINE SECTION
CANADA
U.S.A.
Montreal
St. Lambert Lock
St. Catherine Dam
N
Miles 0 10 20 30
Km 25 50
DULUTH
L. Superior
602'
"Soo" Rapids, Locks and Canals
581'
L. Huron
R. St. Clair and Detroit
572'
L. Erie
Sea Level
Niagara Falls and rapids, and Welland Canal
246'
L. Ontario
Iroquois Dam and Locks
St. Lawrence Rapids
151'
L. St. Francis
Locks
67'
Locks
L. St. Louis
MONTREAL
22'
Seaway Section
6
4
2
S.L.
-2
-4
Elevation (00ft)

FIG. 147 (*facing page*).—THE ST LAWRENCE SEAWAY

The St Lawrence river and the Great Lakes provide an extensive natural waterway into the heart of the North American continent. In earlier times rapids, waterfalls, and differences in some of the lake levels, which handicapped easy, through navigation, had been partially overcome by building canals and locks. But navigation for sea-going vessels was prohibited by insufficient depth of water.

For nearly half a century many far-sighted people in both Canada and the United States agitated for the development of a deep waterway which would enable sea-going vessels to penetrate right into the interior of the continent. But there was also opposition to such a proposal from the port authorities on the Atlantic coast who believed a new waterway would steal their trade. Vested interests succeeded in delaying the project.

After the Second World War a scheme was proposed to provide deep-water navigation and, at the same time, to generate hydro-electric power. The project, which has come to be called the St Lawrence Seaway, was opened in 1959. The scheme was jointly undertaken by the United States and Canada since "the Seaway" covers that part of the river along which the international frontier runs.

Here, in this section, until 1959, there were several canals with a 14-feet (4·7 m) draught which allowed passage of vessels up to 3000 tons. Now, with the construction of new canals having a depth of 27 feet (9 m), sea-going ships of up to 9000 tons can negotiate the section and pass from the lower St Lawrence into Lake Ontario. Sea-going vessels are now able to reach such Lake ports as Duluth, Fort William, Port Arthur, and Chicago, some 2000 miles (3200 km) from the ocean, without having to trans-ship their cargoes at Montreal on to smaller craft. The Seaway also allows big "Lakers" to travel all the way from lake-side ports to Montreal instead of transferring their goods on to small "Canallers" in the International Rapids section of the river.

The Seaway project also included the building of several great dams to provide hydro-electric power, *e.g.* the Canadian-American Power Dam and the Long Sault Dam have ponded up the river creating a new 28-mile (45-km) long head-pond, called Lake St Lawrence. The growing industries of the riverside and near-by towns were needing more power and here was one means of supplying it.

A further reason for the building of the Seaway was to enable the iron ore from the recently discovered and developed ironfields in Labrador and Quebec to be transported in quantity to the industrial cities on the Great Lakes. Bulk carriage of iron ore, without trans-shipment, would give a valuable saving in transport costs.

The construction of the Seaway meant that a considerable portion of the river valley had to be flooded. In fact, several small towns and villages on both river banks had to be moved, or rather rebuilt, since their original sites were flooded. Nearly 10,000 people had to be found new homes. Some people, however, managed to take their own houses with them, for an American engineer, J. W. Hartshorne, "designed a giant house-moving machine which was able to pick up a full-sized house, furniture and fittings included, and wheel it from its old, to-be-flooded site to its new one, perhaps as far as 10 miles."

tied to the surface and have the ability to fly above terrestrial obstacles, it is commonly said that the aeroplane has the freedom of the air and a freedom of action denied to any other form of transport. Such freedom, however, is largely illusory. In the first place movement, for commercial purposes, is normally channelled along carefully prepared routes; secondly, movement may be restricted by political considerations, some countries banning flights across their territory for security or other reasons; thirdly, aircraft need airports just as surely as ships need harbours and port installations and aircraft are tied more closely to ground facilities than is usually imagined; and, fourthly, weather conditions impose a greater control upon air transport than upon any other form of transport.

In the early days of air transport, when aircraft had a relatively restricted range of flight due to their size, speed, and limited fuel-carrying capacity, air routes were so designed as to avoid extensive ocean crossings, high mountains, uninhabited deserts, and vast forest areas. Hence the very barriers which impeded human travel on the earth's surface also restricted man's movement by air. However, due to the greatly improved performance of aircraft and their greater reliability, physical features in themselves now place no limit upon the choice of routes.

In early days, too, weather conditions severely limited flight. Although weather conditions still cannot be ignored, the control exercised by the weather is less restrictive than formerly. Until about a quarter of a century ago flying was confined to the lower levels of the atmosphere; since then, as a result of technical developments, aircraft have begun to use the stratosphere where they can fly "above the weather." Radar, de-icing techniques, beacons for night flying, and a host of other inventions and aids have helped to combat the weather. But, notwithstanding all these things, it must be emphasised that low cloud, ground mist, fog, icing, dust-storms and other weather phenomena, which either interfere with visibility or affect the performance of the aircraft, still hinder flying and create hazards.

Air routes today are primarily determined by (*a*) adequate ground facilities for operation, and (*b*) availability of traffic for economic working. Because of the constantly increasing size and weight of aircraft, especially of the modern commercial airliners, larger and larger aerodromes are being required: for modern commercial aircraft there must be a" runway 3000 to 5000 yards long at sea-level in temperate latitudes."* Finding suitably large areas in, or near, areas of dense population—where clearly it is desirable to have airports—is becoming an increasingly difficult problem. Furthermore, the operation of modern aircraft requires much in the way of assistance from ground services, *e.g.* air traffic control, radio

* *Idem.*

and meteorological services, passenger- and goods-exchange facilities, and maintenance facilities.

## AIR TRANSPORT

Recent years have seen new developments in air transport. Giant-size craft, such as the Jumbo Jet and Concorde, are being introduced and these, if successful, are bound to have very important repercussions upon both the aircraft industry and air transport. The use of helicopters (which have been much used in the Vietnam War in South-east Asia) may revolutionise inter-city transport, but heliports, so far, have not caught on. The most recent development in air transport is the plan to resurrect the airship. The first airship was built by Count Zeppelin in 1900 and though a number were constructed ill-luck dogged them. However, it may well be that the 1970s will see the re-introduction of commercial airships.

The availability of traffic, whether passenger or goods traffic, obviously will be the over-riding factor in air communications which are commercially operated. Air transport is still, in general, costly, and this limits its use. It is best suited for the carriage of commodities which are low in bulk but high in value, such as mail, travellers' samples, diamonds, bullion, or commodities of a perishable nature which require speedy transport and careful handling, such as cut blooms and high-priced fruit and salad crops, or passengers wishing to economise on time.

## AIR SERVICES

The total amount of freight moved by air is still very small—less than 2% of the total world figure. However, there are strong indications, if we can believe recent trends, that we are on the threshold of a substantial increase in air freight traffic. During the two decades 1950 to 1970 the international movement of air cargo increased tenfold, while on the North Atlantic route it expanded twenty-five times! These trends lead us to believe that there will be a great upsurge in air carriage in the years immediately ahead and, indeed, the International Air Transport Association has forecast that by 1975 air freight traffic will increase seven or eight times over the 1965 level and that the growth in passenger travel will also increase greatly, although the Association also predicts that the growth rate in cargo transport will be double that in passenger transport.

At the present time most of the air freight is actually carried in the holds of passenger aircraft, but the use of all-freight carriers is gradually growing. We can point to three main factors which have stimulated the increasing use of aircraft for moving cargo: (*a*) freight rates, formerly prohibitively high, have begun to drop; (*b*) the carrying capacity of aircraft has been greatly increased; and (*c*) the use of standard type containers which can be handled as easily by aircraft as any other form of transport.

It is interesting to note that Heathrow Airport now ranks, in terms of value of goods passing through, as the third port in Britain, ranking after London and Liverpool. It handles more than a quarter of a million tons of cargo a year.

Air services are of two main kinds: (*a*) short-distance services such as those operating between important centres in a country or making short-sea crossings, and (*b*) long-distance services such as the trans-continental

FIG. 148.—WORLD AIR ROUTES

Great Circle routes are the shortest routes and aircraft, alone of all transport media, can follow such routes (although ships may follow such routes for part of their voyaging). But such short routes are not always taken for a variety of reasons: (i) the need for serving large centres of population; (ii) the limited range of aircraft, though this factor is being constantly reduced; and (iii) political barriers: some states refuse freedom of access to their territory.

and trans-oceanic flights. There is another distinction in services, too: although most aircraft operate to fixed schedules along regular routes, some aircraft, owned by private companies, are operated for charter, that is they can be hired for special purposes or special journeys. Charter aircraft are much used in Britain during the summer holiday season.

A world map showing the pattern of air routes brings out certain features: (*a*) an east–west girdle of routes in the northern hemisphere; (*b*) the great focusing of routes in three regions: Western Europe, eastern U.S.A. and the Caribbean, and South-east Asia; (*c*) several important foci from which numerous routes radiate, *e.g.* London, Moscow, Cairo,

Karachi, Singapore, Manila, New York, Rio de Janeiro; (*d*) the network of air routes shows a world-wide coverage.

Study carefully the map in this section (Fig. 148) which illustrates some of the principal features of air transport.

Space forbids a detailed consideration of the various air communications which link the different parts of the world together but special mention may be allowed of the services of the British Overseas Airways Corporation (B.O.A.C.). The chief long-distance services are: (*a*) the trans-atlantic route to Canada and the United States via Gander in Newfoundland; (*b*) the West African service, via Bathurst, to Lagos in Nigeria; (*c*) the service to South Africa, the so-called "Springbok route," via Cairo, Nairobi, and Johannesburg; (*d*) the service to Calcutta, the "Tiger route," via Vienna, Lydda (Lebanon), and Karachi; (*e*) the "Dragon route" following the previous route to Calcutta and thence to Tokyo via Rangoon, Bangkok, Hongkong, and Shanghai; and (*f*) the service to Australia and New Zealand, the "Kangaroo route," via Lydda, Karachi, Singapore, and Darwin. Many of these routes are worked in co-operation with other Commonwealth airlines.

## EXERCISES

1. What factors influence the location of international airports? Illustrate your answer with examples of at least three airports which you have studied.
2. Examine and account for the railway pattern of *either* South America *or* Australia.
3. Write an essay on the importance of *either* the River Rhine *or* the St Lawrence Seaway to the economy of the adjacent areas.
4. Give an account of the principal railways and inland waterways of Canada, pointing out the geographical difficulties which limit the usefulness of these forms of transport to the Dominion. (*Union of Educational Institutions.*)
5. Explain why the trans-atlantic shipping route is the world's most important ocean highway.
6. Discuss the problems of adapting and improving Britain's road and rail networks to meet the needs of modern traffic.
7. "The geographical and economic factors that encourage the growth of air transport in a given region are precisely those which restrict its surface competitors." Discuss the truth of this statement with reference to specific examples taken from Europe or one of the southern continents.
8. Discuss comparatively the advantages of road and railway transport.
9. Discuss the importance to the country concerned and indicate the traffic moved along *two* of the following routes: (*a*) the Canadian Pacific Railway; (*b*) the River Rhine; (*c*) the Trans-Siberian Railway; (*d*) the Yangtse-kiang. (*Northern Counties Technical Examinations Council.*)
10. What factors determine the pattern of air routes in Europe? How does this compare with the pattern of domestic air routes in the U.S.A.? What effect will this have on the cost of air transport in the two areas?

11. Discuss the characteristics and importance of ocean trade between the northern and southern hemispheres.

12. Compare the St Lawrence-Great Lakes and the Mississippi river system under the following heads: (*a*) position; (*b*) commercial importance; (*c*) traffic carried. (*Northern Counties Technical Examinations Council.*)

# *Chapter XXVI*

# SETTLEMENTS AND URBANISM

## *FACTORS INFLUENCING SETTLEMENT*

MAN is by instinct a gregarious creature, preferring to live in close contact with his fellow men rather than alone. Occasionally he has to live alone, or at least in a family unit, mainly because of economic conditions, *e.g.* the Scottish crofter, the Welsh hill sheep-farmer. But usually he congregates in communities of various sizes which run the whole range of settlement types from the hamlet to the vast, sprawling urban areas.

Settlement is not a haphazard business. One important fact to note is that settlements, whether large or small, have sites which may be said to be due to the working of fundamental geographical laws. Seldom, very seldom, is a settlement site due to an accident. In earlier days, as for example in medieval times, communities were very largely self-supporting and they chose localities which enabled them to be as self-sufficient as possible. Our ancestors, more alive than we are to the basic needs of life, had a greater appreciation of the countryside and used it to the best advantage.

In selecting a site for settlement, man, in earlier days, bore a number of factors in mind. The chief of these factors affecting his choice of location for settlement were:

*Water supplies.* First and foremost man had to have access to a supply of pure drinking water. Water is a primary human need and it may be taken for granted that man would be most likely to settle near a sure supply of water. This is most obviously illustrated in desert or semi-arid lands where people congregate around springs or wells. Settlements based upon such water sources are called wet-point settlements.

*Farming land.* After water, food is man's next basic need, hence sites offering fertile land for cultivation or good pasture for animals were looked for and a locality which provided a varied assortment of farming land, *e.g.* water meadows for cattle, arable land for the plough, and rough grazing for sheep, together with a coppice for timber and fuel, was much prized.

*Defence possibilities.* In former times security from attack by hostile neighbours, marauders, and pirates was often a very necessary condition with respect to siting. Hence man sought out defensible sites such as hilltops, river loops, peninsulas, rocky outcrops, islands, etc., all of which lent themselves to easy defence. Defensive sites of this kind are sometimes called fortress points.

*Dry land.* In areas where the land was liable to be marshy or suffered seasonal inundation through river floods, upstanding sites were chosen to avoid both damp and disaster. For instance, in the Low Countries the *terpen* or mounds raised above the flood lands formed the sites of the earliest settlements. And in the English Fenland Ely and other early settlements occupied elevated sites.

*Shelter.* Other things being equal, man would choose a sun-catching site and one sheltered from cold winds or clammy river mists. Very often, it is true, the other factors, which were more necessitous, outweighed this consideration. But in many upland areas, *e.g.* in the Yorkshire Dales and in the Alps, settlements tend to be located on the sunnier, southward-facing hill slopes or valley sides.

## *SITE AND POSITION*

In many examinations, questions are set involving a discussion of the site and situation of a settlement, usually a town. There is much confusion over these two terms and it will be useful to clarify their meaning. First, let us emphasise that they are not synonymous. Site means, and refers to, the specific location of a settlement; situation in the broader sense and especially in respect of accessibility may be said to define position. For example, a small island in the River Seine, which provided defence but which also assisted bridge-building in early times, pin-pointed the site of Paris (the Cité); but the geographical position of Paris is its situation in the centre of a fertile basin, the Paris Basin, where routes from all directions converged. Likewise, the site of New York is Manhattan Island at the mouth of the Hudson estuary whereas the position of New York implies a mid-Atlantic coastal situation in relation to the most important and easiest route across the barrier of the Appalachian Mountains.

Originally, site was more important than position. The factors influencing settlement, noted above, were in the early stages of man's development of decisive importance. The need for the basic necessities of life—water, food, shelter, and security—overshadowed all other considerations. Only at a later stage, when commercial and/or political interests and expansion began to become significant did general geographical position begin to assume any real importance. Geographical position implies a number of things: absolute location (as defined by latitude and longitude), situation in relation to other places and areas (whether land or water), and accessibility. The last two aspects of geographical position are

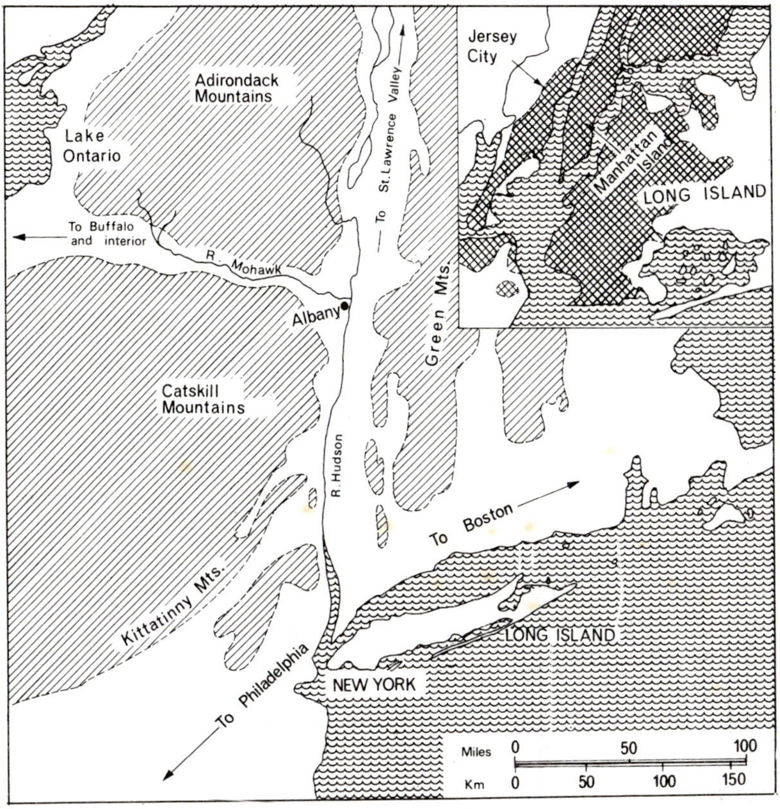


FIG. 149.—NEW YORK: ITS SITE AND SITUATION

New York affords an excellent example of the influence exerted by geographical conditions upon the siting and development of a town. Geographically, it owes its growth and importance to three main factors: (i) it lies at the end of the shortest all-weather trans-atlantic route from Western Europe; (ii) it is at the end of the lowland corridor, provided by the Mohawk and Hudson Rivers, giving the only really easy land route across the mountains to the continental interior; and (iii) the drowned estuary of the Hudson River provides an excellent roomy sheltered deep-water harbour with miles and miles of water frontage.

Although these three factors provide a very fortunate location for New York, the actual city site is far from ideal. The original site was Manhattan Island, which remains the heart of the city, but its limited area has forced building upwards instead of outwards and the southern end is now almost a solid mass of skyscraper blocks. The boroughs of New York are to a large extent separated by stretches of water. The construction of ordinary bridges would destroy New York's role as a port, hence the numerous high-level bridges, under-water tunnels, and ferries.

of major value in settlement location. For example, from the very earliest times a settlement, first Hissarlik, then Troy, and finally Constantinople, occupied a strategic position at the mouth of the Bosphorus–Dardenelles channel, a position commanding the great crossroads of the Old World (*i.e.* between Asia Minor and the Balkans and between the Black and Mediterranean Seas).

## *TYPES OF SETTLEMENT*

Settlements may be said to fall into two main groups: (*a*) rural settlements, and (*b*) urban settlements. In the early days of man's existence, when he was a collector and hunter and lived a mainly nomadic life, and among a few wandering groups at the present time, settlements of a permanent character did not, and do not, normally exist. Man may, in some cases, have occupied caves which formed permanent or semi-permanent habitations. In general, however, the need to search for food or to follow migrating animals prevented man from having a permanent dwelling. Not until man had discovered the arts of cultivation could he, or did he, have permanent habitations. Agriculture affected settlement in two ways: first, it relieved man of the necessity of constant roaming in search of food, and, second, it made it necessary for him to be near the scene of his activity. These two factors, therefore, predisposed towards permanent settlement and somewhere around 5000 B.C., or perhaps a little earlier, man began to settle down in a given place and to build himself a permanent habitation. Groups of dwellings came to form villages, for the village is the characteristic form of settlement in farming areas. Towns, or urban units, came later with the beginnings of trade and the specialisation of labour.

Permanent settlements are of varying kinds. Seven main types can be distinguished: (*a*) the isolated building; (*b*) the hamlet; (*c*) the village; (*d*) the small town; (*e*) the city; (*f*) ribbon settlement; and (*g*) the conurbation. The first three of these categories, together with the small town—usually a market town—are typically rural settlements. This rural settlement may be either dispersed, *i.e.* farmsteads scattered about the countryside, or nucleated, *i.e.* farms and dwellings concentrated together into a fairly compact group. The small town and the last three categories of settlement form, essentially, urban settlements.

*The isolated building*. The farm, the manor house, and the large country residence are the more common types of isolated buildings. Isolated farmsteads are found practically the world over; the crofts in the Highlands of Scotland, the livestock farms of the Pennines, the prairie wheat farms of Canada, the estancias of the Argentinian pampas are examples.

Such settlements may be rather remote and cut off from other isolated settlements, frequently they may be many miles apart, hence they must be self-sufficing to a very large extent.

*The hamlet*. The distinction between the hamlet and the village is not clear-cut. Typically, however, the hamlet is smaller and the settlements

[*Courtesy: National Film Board of Canada.*

FIG. 150.—VANCOUVER: AERIAL VIEW

This aerial view of the port of Vancouver illustrates well the grid-iron street plan which is characteristic of most North American cities. The skyscraper is another feature typical of North American cities, although it is no longer confined purely to North America. It is rapidly becoming a feature of most of the world's great cities.

composing it more scattered. The presence of a church, an inn, a post office, or school is no criterion of hamlet status: a hamlet may not, often does not, possess any of these things. One tiny, beautiful hamlet, Hubberholme in upper Wharfedale in Yorkshire, comprises three buildings only: a church, an inn, and a farm. In the British Isles hamlets tend to be associated with livestock-rearing.

*The village*. Villages are of variable size; they may have anything from a few dozen to several thousand people. Typically, the village has certain social features, *e.g.* church, inn, school, store with post office, village hall, etc., though each of these is by no means always to be found. The farming village is the oldest, as well as the most characteristic, village type. But there are, also, fishing villages and mining villages, the last being almost invariably a product of the Industrial Revolution.

*The small town.* Villages located in areas of good farming land and well placed for communication were likely to develop nodality; in other words, they became foci of routes. Because of their function as marketing centres, they grew in size, often developing into market towns. While some small towns may have grown up for alternative reasons, *e.g.* fishing,

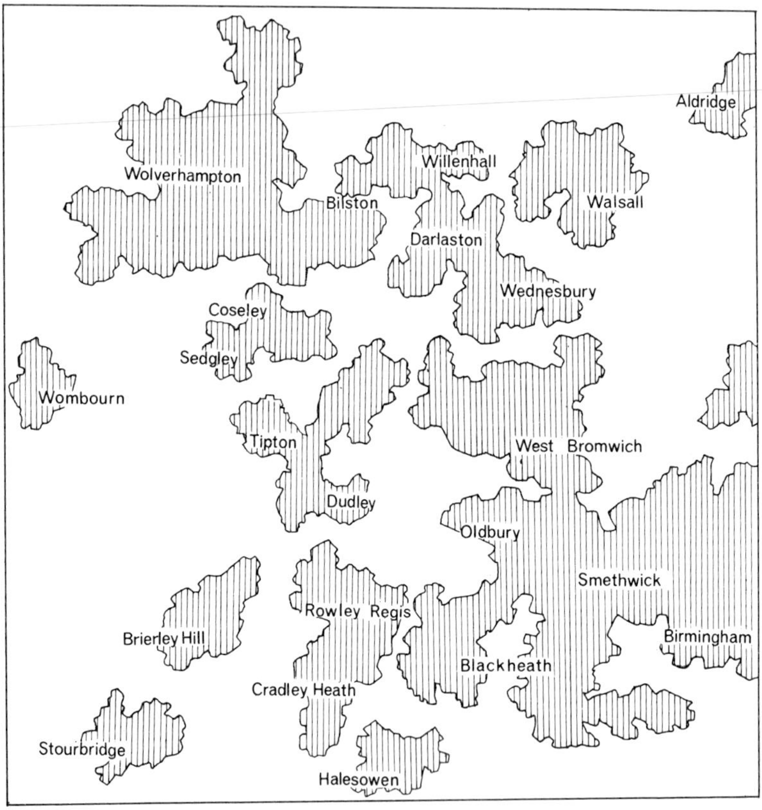


FIG. 151.—A CONURBATION

This map shows part of the "Black Country" in the English Midlands, one of Britain's greatest conurbations. A number of towns, once separated from their neighbours by stretches of countryside and farmland, now merge with one another producing, in effect, one huge continuous built-up area. The chief conurbations in Britain are: Greater London, West Midlands, West Riding, South-east Lancashire, Merseyside, Tyneside, and Clydeside.

quarrying, mining, manufacture, most British townships have their origins as local agricultural marketing centres. This, in fact, is true of towns the world over.

*The city.* The term "city" is loosely used nowadays to describe the large town but, in Britain, it refers in the strict sense to a town which has received its charter of incorporation. The fact that a town has a cathedral or university has nothing to do with its status as a city, although it may, of course, possess one or both. Neither has city status anything to do with size, although most large towns are, in fact, cities. The large town is very much a product of industrialism and modern commerce.

*Ribbon development.* A settlement feature of modern growth (largely since 1920) is the ribbon settlement, long, straggling tendrils of settlement which have grown up along the roads leading from towns. Such ribbons of houses, one deep on either side of the road, are a product of increased transport facilities, especially the development of the internal combustion engine which brought bus and motor car transport. Ribbon development is often the prelude to subsequent suburban housing estates.

*The conurbation.* The tentacles sent out by one town may eventually touch and interlock with those sent out by others. In due course, the spaces between the ribbons become largely filled in and the countryside becomes blotted out by built-up areas. Villages formerly on the outskirts of towns may become engulfed in the drab spread of bricks and mortar. Ultimately, the suburbs of one town run into those of another, there being no perceptible break or boundary between them. When this stage has been reached, as in the case of parts of the West Riding of Yorkshire or the Manchester–Salford–Stockport area, the continuously built-up urban area is described as a conurbation.

## *THE GROWTH OF TOWNS*

As we have indicated above, towns have grown up, in the main, from villages. Certain villages, advantageously situated in rich farming country and commanding more important routes than their neighbours, thus giving them greater nodality, would tend to grow into marketing centres and then into towns. Trade, in fact, throughout the ages and throughout the world has been *the* main factor behind the growth of towns. Increasing trade leads to more and improved communications which, in turn, give the trade centre greater nodality. Increased nodality brings a greater volume of trade. More trade attracts new means of transport. And so on. The result is the ever-increasing growth of the town. Norwich in East Anglia provides a good example of an original trade centre located in a good farming area and well placed for communications within the region; as a result it has grown, (and continues to grow) into an important marketing centre. Furthermore, through its function as a great regional market, it has developed varied manufactures based upon the commodities it markets and the needs of the surrounding farming communities.

Towns, then, grew out of villages, the first nucleated settlements, and they may be said to mark an important advance in social life. This concentration of people in towns, known as the "urban revolution," took

FIG. 152.
THE "NEW TOWNS" OF BRITAIN

The New Towns Act, 1946, designated various areas in Britain as "new towns." Twelve of these were to be in England and four in Scotland. The overcrowding in the London area, following upon the industrialisation of the London region and the drift of population to the south-east, led to the desirability of "decanting" some of this excess population into new towns—either completely new settlements such as Crawley and Harlow or expanded urban units such as Welwyn.

place, it is generally thought, around 5000 B.C., although recent excavations at Jericho in the Jordan valley have shown that there was a town on that site at least a thousand years prior to this date. The earliest towns appear to have grown up in the great river lowlands of the Near and Middle East. The idea of city life spread from Ancient Egypt and Mesopotamia through the Mediterranean and the establishment of towns was due mainly to the Phoenicians, Greeks, and Romans. Outside the Mediterranean, civic development was first undertaken by the Romans and the earliest English towns resulted from Roman rule. Places such as London, Lincoln, York, Chester, and Bath are of Roman origin.

Although many towns, especially in north-western Europe, grew up out of the Roman military camps, or *castra*—the names Win*chester*, Glou*cester*, Utto*xeter*, *Caistor*, are merely corruptions of *castra* and so betray their Roman origin—most early towns probably developed through trade and grew up as marketing centres. Later, towns came into being to undertake administrative, ecclesiastical, educational, or other social functions. Some, too, developed as centres of specialised craft industries, *e.g.* spinning and weaving towns, metal-working towns, saw-milling centres, and the like. And much later on in history the Industrial Revolution gave rise to the growth of many mining and manufacturing towns. During relatively recent times several new types of town have come into existence; for example, resort towns, dormitory towns, and the so-called "over-spill" towns which have been deliberately planned and created by the government (*see* Fig. 152).

Although many towns grew up originally on account of their performing a particular function, in course of time they frequently developed other activities, thus becoming multi-functional. London, for example, is at once an administrative, commercial, industrial, educational, cultural, and tourist centre as well as being a great port. Most sizable towns these days perform more than one function.

## *CLASSIFICATION OF TOWNS*

Because towns nowadays are usually multi-functional, attempts to classify towns by their function are not easy. Few towns fall into a clear-cut category and this is a point to be remembered. Even so, the geographer attempts to classify towns according to their function or the role they perform in the business and social life of the community.

In a broad way towns may be divided into two groups: ports and inland towns. Such a division recognises certain basic differences in their respective activities but, again, there is no clear-cut division; for example, how are we to class Manchester or Duisburg or St Louis or Hankow—as towns or ports?

### PORTS

There are many different kinds of port and these are distinguished according to the specialised work they do. The chief types are as follows:

*Passenger ports.* These, such as Southampton and Cherbourg, are concerned mainly with passenger traffic. They are located at the termini of important trans-ocean routes and have good rail links with the interior offering speedy carriage to the metropolitan centres. They are deep-water ports since they must be capable of accommodating large passenger liners.

*Commercial ports.* These deal principally with the import and export of goods, though they may have certain passenger services or other functions

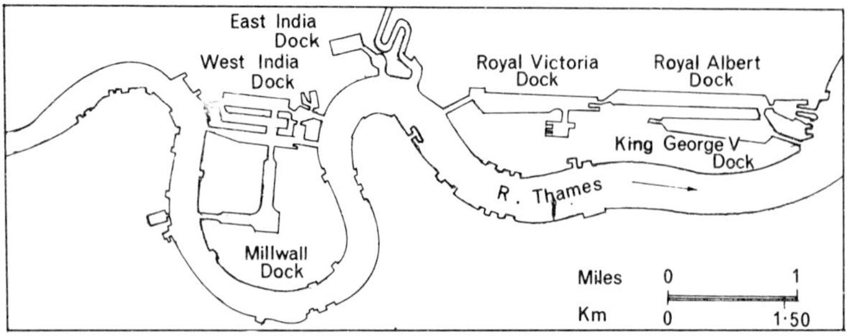


FIG. 153.—THE PORT OF LONDON

London vies with New York and Rotterdam for first place amongst the world's ports. The growth and importance of London as a port is due to (i) the sheltered navigable estuary of the Thames; (ii) the ease with which docks could be dug out of the soft rocks; (iii) the easy accessibility of London to other parts of the English plain; (iv) its location opposite the most important part of the Continent; (v) its position near to the important North Atlantic–North Sea trade routes; and (vi) the enormous demands of London itself for foodstuffs and raw materials.

The Port of London extends from Teddington, the tidal limit on the river, some 70 miles (112 km) downstream to the Thames estuary. Above London Bridge goods are transported mainly by barge and lighter. The St Katharine and London docks opened at the beginning of the nineteenth century and dealt with short-haul and continental trade. Here was the largest bulk wine berth in the country with a storage capacity for half a million gallons (22,725,000 litres). There are warehouses for tea, rubber and wool.

The Surrey Commercial docks on the south bank, opened in 1806, concerned themselves with timber imports especially from Scandinavia and the Baltic. 75% of the storage capacity is given over to softwoods. The docks are busiest during the May–December period when the timber exporting ports of the Baltic are ice-free. Motor-cars are also crated here.

The West and East India docks, opened in 1802 and 1806 respectively, and the Millwall dock, opened in 1868, are largely concerned with general trade, mainly with Scandinavia, the Mediterranean, and the Far East. Sugar, rum, bananas, and hardwoods from the West Indies are traditionally dealt with in the West India dock. Grain is important at the Millwall dock: there is storage capacity for 24,000 tons. The East India dock is more especially concerned with coastal traffic. Some of these older docks have now been closed down.

The Royal Docks—the Victoria, Albert, and King George V—form the most extensive area of enclosed docks in the world. There are berths for fifty ocean-going vessels. There are specialist plants for dealing with perishable commodities such as fruit and meat.

The Tilbury docks, which are a further 15 miles (24 km) downstream, were opened in 1886. They are mainly concerned with commodities from Australia, New Zealand, and the Far East. Passengers from these eastern countries also disembark here. Tilbury is also being developed as a container port.

(such as fishing) as in the case of Hull. They may be large ports, such as Glasgow or Hamburg, or small ports, such as King's Lynn or Preston. Small ports dealing with coastal carriage are sometimes classed as a separate type.

*Entrepot ports.* Entrepots are centres to which goods are brought for distribution to other countries rather than their own. They usually have ample storage accommodation for the temporary deposit of goods. Copenhagen, for instance, has long been a collector of Baltic produce, which is then shipped elsewhere. Rotterdam is another great European entrepot. Singapore is a notable Asiatic example.

*Outports.* These are comparatively recent developments; they are deep-water ports, built farther downstream, serving "parent" ports which have become progressively inaccessible due to either silting or the growth in the size of ships. They are very much a feature of north-western Europe. Well-known examples are Avonmouth, the outport of Bristol, St Nazaire for Nantes, Bremerhaven for Bremen, Leixoes for Oporto.

*Packet-stations.* Alternatively called ferry ports, these ports are almost exclusively concerned with the conveyance of passengers and mail over short-sea passages. Note the term "packet" has nothing to do with packages, but derives from boats named "packets" which in earlier times carried State letters and dispatches. Packet-stations occur in pairs, facing one another across narrow seas, *e.g.* Dover to Calais, Newhaven to Dieppe, Larne to Stranraer.

*Inland ports.* Though often located a considerable distance inland, through the agency of rivers or canals such ports are accessible by certain types of sea-going ships and are usually served by barges and other specialised river craft. Manchester, for example, was turned into a major port by the building of the Ship Canal. Notable river ports are Duisburg and Mannheim on the Rhine, Memphis on the Mississippi, and Hankow on the Yangtse.

*Fishing ports.* These form a self-explanatory group. They vary greatly in size, ranging from major fishing such as Grimsby and Bergen and Nagasaki in Japan to small harbours such as Brixham, Concarneau in Brittany, and Biloxi in the United States on the coast of the Gulf of Mexico.

*Naval ports.* Located primarily for strategic reasons, they serve as bases for ships of war, but may also sometimes undertake naval repair work. Spacious harbourage, shelter, and good anchorage are necessary requirements. Unlike most other ports, the hinterland is of little significance. Examples of naval ports are Portsmouth, Toulon in southern France, and Key West in Florida.

*Ports of call.* A number of ports have come into existence through their function as calling points on the main sea routes of the world; they are re-fuelling, watering, and victualling points. Plymouth, Funchal (Madeira), Kingston (Jamaica), Aden, Honolulu are well-known examples.

*Oil ports.* A new type of port which has come into existence during recent times is the port dealing with the processing and shipping of oil. The importance of petroleum in the modern world economy and the vast

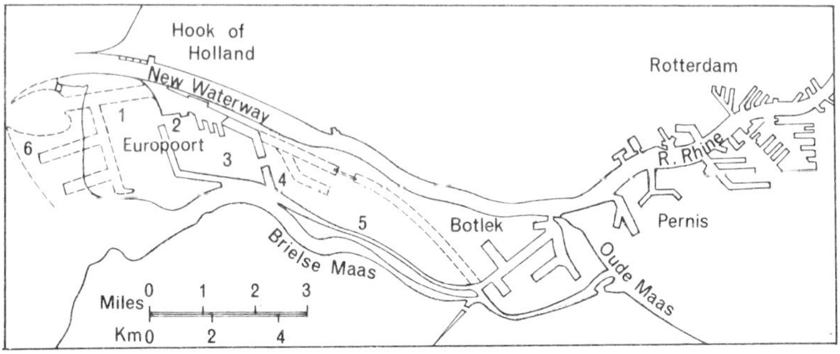


FIG. 154.—EUROPOORT

The new port of Europoort is really an extension of the great Dutch port of Rotterdam, the largest port in Europe and the second largest in the world. Rotterdam handles 20,000 vessels and over 70 million tons of freight annually. It is the natural outlet for the Rhine basin and it has, therefore, a vast and productive hinterland. Botlek and Pernis are largely oil ports. The numbers in the diagram explain the organisation of the new port: 1. Blast furnaces, iron and steel mills. 2. Storage depot for bulk goods. 3. Petroleum depot, distribution by pipeline. 4. Ship repair yard for large vessels. 5. Planned link with inland waterways. 6. Allocation and demarcation of this area not yet fixed.

quantities consumed have given rise to certain specialist ports concerned almost solely with oil; these may be termed tanker ports and refinery ports: Maracaibo in Venezuela, Skhira in Tunisia, and Tripoli in Lebanon are examples of the former, Wilhelmstadt on Curaçao, Abadan on the Persian Gulf, and Milford Haven in Wales of the latter.

## TOWNS

Most towns have varied activities and perform a variety of functions but, characteristically, either in their origin or their present day "work" one activity or function distinguishes them and it is upon this basis that the following classification is made.

*Market towns.* Many villages grew into towns because they were favourably placed with respect to communications and developed as centres of exchange. Norwich, Market Harborough, and Taunton are good examples of English market towns. Rennes in Brittany, Aleppo in Syria, and Kumasi in Ghana are other examples.

*Commercial towns.* Certain towns, probably trading centres in the first place, subsequently grew into cities more especially concerned with commerce, finance, and banking. Historically, Augsburg, Frankfurt, and Zürich were famous as financial centres. At present, towns like Amsterdam and Düsseldorf are essentially commercial centres though they have other functions.

*Mining towns.* Some towns have grown up where there is an abundance of a particular mineral commodity which is readily obtained. The mining town is more especially a growth of the past two hundred years, although some centres are much older, *e.g.* Salzburg ("Salt town") in Austria and Falun (copper) in Sweden. Present-day examples are Genk in Belgium (coal), Kiruna in Swedish Lappland (iron), Johannesburg (gold), Khouribga in Morocco (phosphate).

*Industrial towns.* Primarily concerned with the processing of raw materials and the manufacture of goods in mills and factories and workshops, a large proportion of industrial towns are of relatively recent growth and a product of the Industrial Revolution, *e.g.* Middlesbrough (iron and steel), Bradford (woollen textiles), Stoke (pottery), Eindhoven in Holland (electrical goods).

*Transport centres.* A number of towns connected with important routeways and closely linked with particular methods of transport function primarily as transport centres. Such towns as Crewe and Swindon in England and Hamm in West Germany are pre-eminently railway centres while many towns in the United States owe their origin to the railways. Sault Ste Marie, Welland, and Suez are canal towns. And of course there are many towns lying at the head of navigation on rivers; to some extent these may be said to fall into this group.

*Defensive towns.* Many old towns grew up where easily defended sites occurred in strategic positions. Many gap-towns fall into this category. The famous trio, Metz, Toul, and Verdun in eastern France were built as fortress cities. Edinburgh and Peking had their beginnings as fortresses commanding strategic routeways. Military towns such as Aldershot may be included in this category.

*Administrative towns.* Some towns, such as regional "capitals" and national capitals, are much concerned with government and organisation and these administrative activities distinguish them. Wakefield, Berne, the Hague, and the "new capitals" such as Canberra and Brasilia are good illustrations of this type.

*Ecclesiastical centres.* Many towns may be classed as religious centres, either because they are "holy" cities, such as Rome, Mecca, Jerusalem,

Benares, Lhasa, or because they owed their origin to the building of an early church and have continued to be primarily ecclesiastical centres, *e.g.* Canterbury, Southwell, St Asaph.

*Educational centres.* A few towns owe their origin and development very largely to the fact that they are seats of learning. In England, Oxford and Cambridge are obvious examples; in Scotland, St Andrews is an old university town, although it also has religious origins. Heidelberg in West

[*Courtesy: Swedish National Travel Association.*

FIG. 155.—VÄLLINGBY

The modern suburb of Vällingby, really a dormitory satellite of Stockholm, is reached by subway from the centre of Stockholm in about thirty minutes.

Germany, Leiden in Holland, Turku in Finland, and Patna in India are all old centres of learning.

*Resort towns.* These fall into two broad groups: seaside resorts and inland resorts. Both have also tended to become residential towns. The inland resorts are, again, of two main kinds: they are either spas, *i.e.* where natural medicated waters occur, *e.g.* Bath, Tunbridge Wells, Vichy, or scenic and sports centres, *e.g.* Keswick, Zermatt in Switzerland, Banff in Canada (*see* Chapter XXIII).

*Dormitory and Over-spill towns.* These form a new type of town. They are not towns of natural growth but have been developed deliberately to relieve the growing congestion in the major urban centres. Several towns of this type, *e.g.* Basildon, Crawley, Stevenage, have begun to ring London. Many workers travel daily to the city where they work. Satellite towns of this type are growing around all the big cities, *e.g.* Vällingby, a modern town housing 25,000 people, outside Stockholm (Fig. 155).

## *MILLION CITIES*

Historically there have been great cities with large populations, but the large city is very much a creation of the twentieth century. The number of cities having over one million inhabitants has steadily grown: in pre-war days there were about 25, just after the war about 36, in 1953 about 64, and in 1960, 83. It is possible that at the present time (1972) there are over 100. These "million cities", or "millionaire cities" as they are sometimes called, are an urban phenomena of every continent but they are least conspicuous in Africa and Australia.

The locations and the general distribution of these million cities are of considerable interest. If we study a map of the world showing their *distribution* we find: (*a*) relatively few lie within the Tropics; (*b*) relatively few are in the southern hemisphere; (*c*) a large proportion have coastal locations; and (*d*) the majority are found within the most densely populated areas of the earth (*see* Fig. 156).

If we make a careful study of their *locations* we find: (*a*) the overwhelming majority lie in lowland areas, although there are exceptions, *e.g.* Bogota, São Paulo, Mexico City, Madrid; (*b*) a very high proportion are sea-, lakeside-, or river-ports; and (*c*) many possess exceptional advantages of site, *e.g.* New York, Paris, Vienna, Istanbul, Cairo.

The earlier million cities were predominantly either old-established State capitals or advantageously situated ports; they had, as D. C. Money* has said, "some outstanding attributes, of location, or history" which enabled them to dominate other towns in the same general area. The more recent million cities, however, show a high proportion of basically industrial towns; this is true of many of the new Soviet, Chinese, and Indian million cities, *e.g.* Harbin, Taiyuan, Ahmadabad, Kanpur, Lahore.

The question which naturally arises is: why have so many cities in recent decades grown to be million cities? It is difficult to be precise or sure as to the reasons but, without doubt, some of the following conditions and factors have helped:

1. In many regions the growth in industrialisation has promoted the migration of the excess rural labour force into the towns.

2. In some countries, notably Japan, shortage of arable land has compelled people to seek alternative employment in the towns (note, this, a compelling factor, differs from (1) where industry, with perhaps higher wages, is an attracting factor).

* *Introduction to Human Geography*, University Tutorial Press, 1954, p. 156.

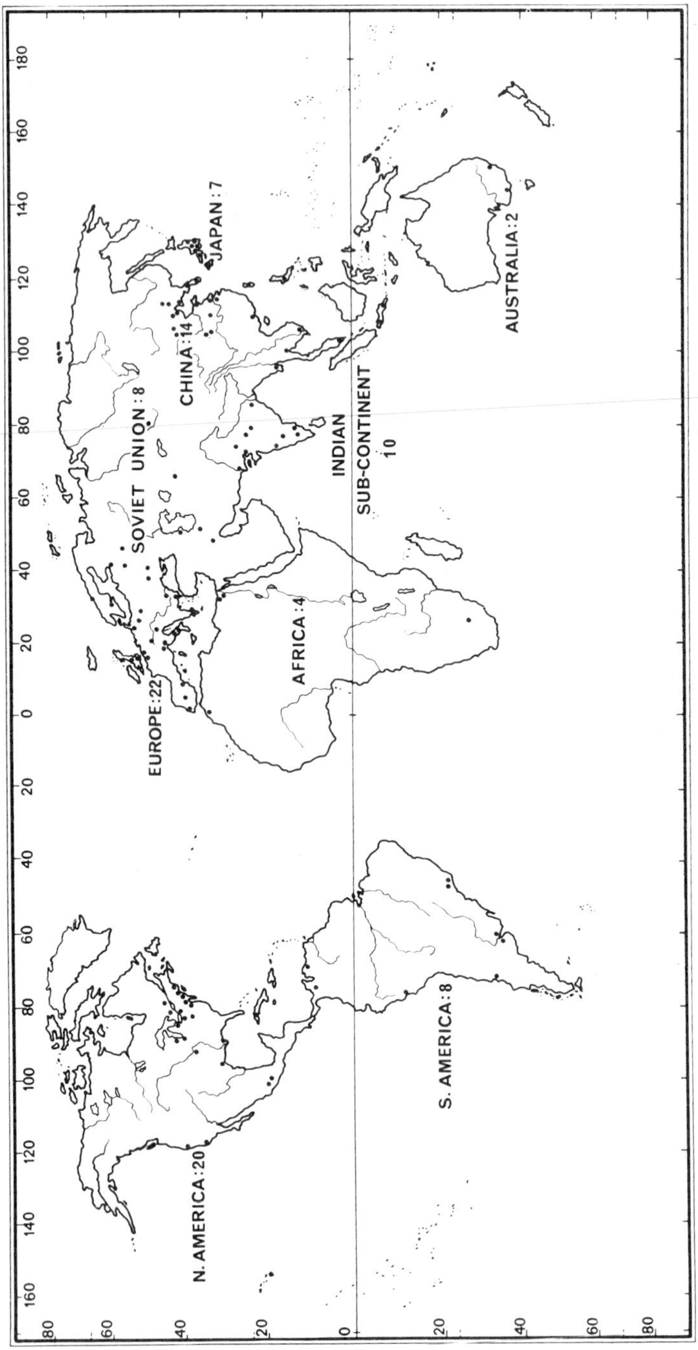


FIG. 156.—THE MILLION CITIES OF THE WORLD

3. In some areas the mechanisation of agriculture has led to rural unemployment and the people have thus been compelled to find work in the towns.

4. As people become more educated they desire "better jobs" and the opportunities for such are practically confined to the urban areas.

5. The social amenities and attractions of town life have led to a marked drift in many areas into the urban centres.

In conclusion, two points of considerable social importance may be noted. First is the exaggerated growth of many metropolitan centres. This is particularly the case in Latin America where many of the capital cities, *e.g.* Buenos Aires, Montevideo, Santiago, Caracas, contain an undue proportion of the population of their respective countries. Secondly, in the less well-developed countries there has been an influx of the native rural peoples into the towns, with attendant overcrowding and the growth of sordid slums. Around many of the great cities in South Americas, for instance, unorganised, unplanned shanty-towns (called *favelas* in Brazil) have sprung up. And the same is true of Tunis, Johannesburg, and Calcutta.

## EXERCISES

1. Choose any large city that you have studied, except for ocean ports. Analyse its site, position, and function. Illustrate with sketch-maps and diagrams. (*Institute of Bankers.*)

2. The five largest cities in Australia are seaports. Describe *two* of them and account for their location and functions. (*Institute of Bankers.*)

3. Choose one of the following ports, Glasgow, Liverpool, or Bristol; draw a sketch-map to show its position and give an account of its hinterland and the trade that passes through it. (*Chartered Institute of Secretaries.*)

4. Name four of the "new towns" created in Great Britain since the Second World War. Explain the reasons for the creation of the "new towns."

5. Suggest a classification of ports according to their function. Illustrate your answer with specific examples.

6. What is meant by the term "conurbation"? List the chief conurbations in Great Britain. Choose one and attempt to explain its growth.

7. Study Fig. 156 and attempt to explain the world distribution of million cities.

8. "Increasing urbanisation, or concentration in towns. has been one of the outstanding characteristics of the last two centuries." Consider the geographical importance of this fact.

9. Examine the factors that lead to the growth of very large ports. Select one example and show how these factors apply.

# *Appendix I*

## ADDITIONAL QUESTIONS

1. Describe, for any African state, the chief farming regions in that state and show how the differences in farming are related to differences of relief and climate. (*Chartered Institute of Secretaries.*)

2. Explain the measures proposed for the development of *either* the North-east Industrial Area *or* the Central Lowlands of Scotland. Illustrate your answer with a sketch-map. (*The Institute of Bankers.*)

3. In deciding whether or not a country is over-populated, what are the main points to be considered? Illustrate your answer by reference to particular areas.

4. Write a short geographical account of the various districts in Great Britain, outside of Yorkshire, where wool industries are carried on. (*Royal Society of Arts.*)

5. Africa's place in the commercial world is basically that of a producer of foodstuffs and raw materials for export. Examine the export trades of *two* countries in inter-tropical Africa and account for their principal features. (*The Institute of Bankers.*)

6. Discuss the statement, "The North Sea traffic route is the busiest in the world." (*The Institute of Transport.*)

7. Discuss the geographical background of the export trade of Australia and New Zealand. (*The Institute of Transport.*)

8. Examine the importance of irrigation in the agricultural development of the Indian sub-continent, and indicate some of the problems caused in this connection by political and social issues. (*Union of Educational Institutions.*)

9. Many of our present-day world problems are caused by the inequality in standards of living as between various countries. (*a*) What determines the standard of living in a country? (*b*) What efforts have been made since the last war to assist the under-developed countries of the world?

10. Discuss the importance as a means of communication of *three* of the following: the Trans-Siberian Railway, the River Congo, the Straits of Malacca, the Kiel Canal.

11. Name three states in Europe that make great use of hydro-electric power and describe the conditions in *one* of them that make cheap hydro-electric power possible. What industries are associated with hydro-electric power in the country you have chosen? (*Chartered Institute of Secretaries.*)

12. For any large country in the tropics you have studied explain: (*a*) Why it is backward, (*b*) What measures are being taken to remedy this. Illustrate with a sketch-map. (*Institute of Bankers.*)

13. What geographical difficulties confront those who propose to "reclaim" the Sahara Desert? (*Institute of Bankers.*)

14. What advantages does Canada possess over Australia in the future development of manufacturing industries on a large scale? Give details of existing manufactures in each area.

15. Give a brief systematic account of the economic geography of *one* of the following: Chile, Egypt, Belgium, Iraq.

16. Industrial development has been slow in the countries of Southern Europe. Suggest reasons for this situation in a geographical description of any *one* of these countries except Italy. (*The Institute of Bankers.*)

17. Give an account of *either* (*a*) hydro-electric power development in Scotland *or* (*b*) the reasons for the lack of industrial development in Eire (Southern Ireland). (*Northern Counties Technical Examinations Council.*)

18. Write brief notes on *three* of the following, giving their positions and importance, and nature of trade: Singapore, Karachi, Cape Town, Auckland, Galveston, Buffalo.

19. Write an essay on *one* of the following topics: (*a*) Californian industry; (*b*) Canadian forest resources; (*c*) Hydro-electric power in North America. (*The Institute of Bankers.*)

20. How has the Netherlands dealt with the problem of land reclamation? Illustrate your answer with a sketch-map.

21. What is meant by *natural resources*? Outline, and contrast, the natural resources of *either* Sweden and Italy *or* Chile and Egypt. (*Northern Counties Technical Examinations Council.*)

22. Write an essay on the industrial geography of *either* Japan *or* France.

23. Discuss the importance of the Great Lakes of North America from the commercial standpoint. (*The Institute of Transport.*)

24. Compare the trade of London with that of New York. (*Institute of Transport.*)

25. Compare the savannas of northern Australia and Venezuela from the standpoint of their productivity, and account for the differences.

26. Give an account of *either* cereal cultivation *or* dairying in England, indicating the chief areas of production and the local conditions favouring production.

27. Describe and account for the distribution of population in *either* Argentina *or* Japan.

# *Appendix II*

## EXAMINATION HINTS TO CANDIDATES

THE examination candidate, if he (or she) is to be successful, must (*a*) obey the rubric; (*b*) understand the questions asked; (*c*) arrange his material satisfactorily and avoid irrelevancy.

The rubric means the instructions at the beginning of the examination paper. These instructions should be read very carefully and fully understood. Frequently an examination paper is divided up into sections and the candidate is required to answer some questions from each section. If this is stated then the candidate should answer what is asked; no more, no less. It is pointless answering two questions from a section if only one is asked for: no matter how well answered the extra question is, it will gain the student nothing since it is superfluous; he has ignored the rubric. If the instructions say four questions must be attempted, this means four not five. Neither does it mean three; it may happen that a candidate can answer only three of the questions, but sometimes candidates think that if they answer three well instead of tackling four they will be able to make up the deficiency of the fourth by a good three. This seldom happens: if each question has a maximum of 25 marks, the candidate is scoring out of 75 not 100. Again, if the instructions say that sketch-maps and diagrams should be drawn to illustrate answers wherever possible, the student should observe this direction, first because the examiner is looking out for such illustrations and will give credit for them and, secondly, it may happen that a certain proportion of the marks for any given question is reserved for sketches and diagrams.

The student should read the question paper through very very carefully and spend a few minutes thinking about the questions before beginning to answer them. Do not rush into answering the first question you spot which you think you can answer well. It is a wise plan to read the questions through, tick off the ones you feel capable of answering and then re-read these and select the four or five that you decide you will answer. Before answering a particular question make sure you understand the examiner's demands in the question; it is a good idea to read the question through and underline the salient or key points requiring attention. Take, for example, the following question:

> "Describe very briefly the origin of petroleum. Name the principal producing areas in the world. Indicate the chief methods by which oil is transported and the chief movements of oil."

The crucial points here are: *briefly—origin of Petroleum—principal producing areas —methods of transport—movements of oil*. The examiner requires a brief account—a paragraph or so, not a page or more—of the origin of petroleum; he wants named the chief areas—not countries (though these may be named in addition to the areas)—of oil production; he asks for methods of transport (*i.e.* pipeline and tanker); and he wants to know the routes by which the oil is moved from the centres of production to the consuming centres.

If a question asks for the "industrial geography" of a country, this means industry not agriculture; if a question asks "*either/or*" this means one or the other, not both; if a question asks for "two of the following," it means two, not three; and so on. It is surprising how many candidates trip up on small points of this nature. Avoid superfluous and irrelevant "padding"; by padding, the student is not fooling the examiner, only himself. Obey the injunction: "Answer the question, the whole question, and nothing but the question."

Organise the material of your answer in an orderly, systematic, and logical way. Be precise; give figures, if possible; give examples. Avoid such meaningless phrases as "a fertile soil," "the right type of soil," "a good climate," "cheap labour," "in many other areas," which at this level just cannot be tolerated. In discussing rainfall and temperature figures required for the growth of a particular crop, give approximate inches of rainfall and degrees of temperature: not plentiful rainfall and high temperatures, for these can have different meanings under different circumstances. Likewise, in quoting cropping areas give precise locations: cacao in Brazil, jute in India, wheat in the United States, sugar-cane in Australia, coal in the Soviet Union, etc., are almost valueless. The student should pay particular attention to those questions which ask him to deal with special areas of production of a commodity: in such cases he or she must refer to what the conditions are and not to what the commodity requires.

A common type of question is that which asks for a description of the geographical conditions of production of a commodity. Answers to this kind of question should, generally speaking, deal with: (*a*) the geographical conditions of production (*i.e.* those of climate, soil, labour, transport facilities); (*b*) the methods of preparation or processing involved; (*c*) the world distribution of the commodity and the precisely located areas of production; and perhaps (*d*) the markets to which the commodity goes to be consumed. World distribution can be most conveniently shown on an outline map of the world, if this is provided. Regional distributions can be shown on simple sketch-maps.

Another common type of question involves the analysis of the factors which have assisted the growth of a particular industry, *e.g.* the Lancashire cotton industry, the German chemical industry, the United States motor car industry. In answering such questions base your account upon the following factors: (*a*) the supplies of raw materials; (*b*) the availability of power supplies; (*c*) labour supplies, quantity and quality; (*d*) markets; (*e*) transport facilities. But do not forget the possibility of the influence of historical, human, and political factors. Do not forget, either, to use the knowledge you have learned in your history, economics, or commerce classes; it is often applicable to geographical questions. Students are all too prone to keep their subject knowledge into watertight compartments; let one subject fertilise the other.

Sometimes a question is set which involves comparison and contrast, perhaps of countries, regions, crops, or towns. In such cases, take a point at a time and consider the similarity or dissimilarity. Two quite separate and distinct descriptions do not necessarily constitute a comparison or contrast. Moreover, if the question specifically asks for a comparison, the candidate may be penalised if he does not fulfil the instruction.

"Write a geographical essay upon such and such a country" or "Give a systematic account of the economic geography of such and such a country" are fairly common types of question. Systematic simply means orderly and logical and in answering questions of this kind the student should consider (*a*) the natural conditions, *i.e.* geographical position, relief and drainage, structure and minerals, climate, vegetation, and soils, which have influenced man and his activities; (*b*) the people, their numbers, character, culture, and stage of development, and their economic activities, especially in relation to the use which they have made of the available natural resources; and (*c*) the communications, towns, ports, foreign trade, and trading relationships.

Illustrate your answers with maps and diagrams. But remember, these should add to your written work and they should be used in lieu of a page of written matter (in other words maps should save time—precious time). There is not much point in drawing a map which merely repeats in pictorial form what you have said in the written word. Keep maps clear, simple, and as accurate as possible; there is no need to embellish them with blue-shaded seas, etc. And do not forget that if an outline map of the world is provided this can be used to help you get the shapes and sizes of countries reasonably correct.

Finally, and this should not need emphasising, write legibly and neatly and use good English, paying some respect to grammar, punctuation, and spelling. Avoid using slang, abbreviations, and ungeographical expressions. Exemplify and amplify statements. The quality, not the quantity, of your answer matters to the examiner. Allot your time carefully so that you do not over-write on any particular question.

The good student and the one who scores heavily in examinations is the one who can give that little bit extra which places him above the general run of candidates. Examiners purr with delight when they come across some (relevant) fact, illustration, idea, or approach which is different and unexpected. The best way to widen, deepen, and particularise one's geographical knowledge is to read books. You may not have much time to do this, but at least you can read your daily paper intelligently, and from this you can gain much.

# INDEX


A

ABACA, 195, 203, 343
Aborigines, 5
Accessibility, 12, 57, 385, 440
Afforestation, 60, 233
Agriculture, 28–30, 32, 41, 47–9, 53, 55, 61–4, 67, 70–2, 74–7, 79–82, 89–91, 95–101, 107–8, 115–17, 118–30, 140–1, 169, 175, 206, 365, 442
Aircraft industry, 334–6
Air transport, 368, 402, 404, *431–7*
Alcohol, 44, 268
Algeria, 107, 148–9, 203, 253, 291
Alloy metals, 314
Allspice, 167
Alluvial deposits, 243, 261
Aluminium, 40–1, *262–3*, 297
Amazonia, 14, 17, 40, 67–8, 148, 211, 214
Ambergris, 4, 192
Ammonia, 52, 355
Amsterdam, 4, 342, 429, 451
*Andenes*, 29
Angola, 157, 203, 259
Aniseed, 168
Anthracite, 274
Antwerp, 429
Apples, 148, 151
Arabia, 286, 288
Argentina, 6, 31–2, 55, 107, 134, 138, 150, 160, 164, 173–4, 177–8, 198, 210, 219–22, 225, 228, 269, 280, 290, 310, 331, 367, 371, 379, 389
Artesian wells, 106
Artificial fibres, 204–5, 343, 346, 349–50, 352, 355, 358
Asbestos, 41, 242, 247–8
Asphalt, 287
Assam, 71, 154
Aswan High Dam, 104, 294, 330
Australia, 6, 8, 17, 28, 35–6, 51, 74, 106, 134, 149, 158, 164, 174, 176–7, 180, 218, 225, 233, 263, 265, 279, 297, 299, 310, 330, 367
Austria, 136, 237, 249
Automation, 311
Avocado pears, 4

B

Babassu oil, *208*
Balance of trade, 374–5
Balata, 4, 238
*Bajra*, 139
Bananas, 5, 67, 70, 98, 119, 147–8
Barley, 96, 135–6
Bauxite, 60, 243, *262–3*, 297
Beans, 67, 70, 77
*Bêche de mer*, 31
Bedouin, 27, 109
Beef, 110, *172–4*
Beet sugar, 164–6
Belgium, 36, 60, 118, 193, 200–1, 254, 276, 315, 336, 383, 392
Benelux, 376
Bessemer process, 314–15
Birth control, 21
Birth rate, 18–19
Bison, 28, 45
Bituminous coal, 274–5
Blackfellows, 6, 26, 53
Bolivia, 51, 244, 261, 367
*Börde*, 14
Boro, 29–30
Bradford, 5, 95
Brazil, 6, 12, 25–6, 70–1, 77, 93, 99, 107, 110, 116, 138, 142–3, 149, 158, 164, 174, 196, 199, 202–3, 208, 211, 225, 233, 269–70, 331, 365, 370, 379, 412
Brazil nuts, 67
Britain, 2, 4, 11, 33, 36, 47, 51, 81, 176, 178, 185, 188, 191, 206, 233, 247, 276, 291, 296, 298, 302–3, 316, 357, 360, 369, 371–3, 385, 407, 427
British agriculture, 126–7
British Columbia, 81, 151, 181, 260
British Travel Authority, 392
Broken Hill, 260
Brown coal: *see* lignite

Buenos Aires, 12
Buffalo, 78, 104, 170, 225
Bulgaria, 36, 224
Burma, 235, 369
Bushmen, 9, 26
Butter, 110, 175, 177

C

Cableways, 402, *414–15*
Cacao, 67, 101, 152, *158–60*
*caliche*, 249
California, 6, 12, 16, 75, 108, 119, 149, 152, 165, 180, 250, 265, 335
Camels, 27, 74, 112
Camphor, 77, 238
Canada, 6, 12, 22, 61–2, 81–2, 85, 89, 118, 133, 136, 145, 158, 164, 177, 179, 210, 215–16, 237–8, 245–6, 248, 253, 258, 265, 270, 278, 283, 289, 299, 367, 370, 372, 399, 408, 431, 433, 452
Cane sugar, 167
Cardamon, 167
Caribou, 28, 85, 170
Carnauba wax, 41
Carraway seed, 168
Cashmere, 202
Cassava, 30, 145
Cassia, 167
Cattle, 27–8, 31, 38, 59, 69–74, 78, 99, 129, 136, 169, *171–7*
Cattle tick, 69, 174
Caviar, 4
Cellulose, 204–5
Cement, 247, 414
Ceylon, 12, 19, 99, 153, 157, 167, 192, 203, 248
Channel tunnel, 4
Cheese, 109–10, 175–7
Chile, 14, 75, 81, 84, 107, 116, 134, 150–1, 160, 178, 186, 232, 248–9, 256, 258, 280, 322, 331, 378–9
China, 2, 4, 12, 15, 19, 31, 42, 51, 72, 77, 104, 107, 111, 116, 134, 138–9, 142–3, 146, 149, 154–5, 167, 174, 179, 186–7, 197, 207, 210, 216, 221, 223, 229, 247, 270, 275–6, 293, 310, 315, 328, 353, 364, 369, 403, 409, 412, 421
China Clay, 243, 246–7
Chromium, 41, 255, 314
Cinchona, 239
Cinnamon, 167
Climate, 12, 14, 61, 63, 65–85, 88, 132, 137, 141, 156, 162, 196, 199, 211, 336, 385, 434
Clover, 4, 167
Coal, 4, 6, 16, 34, 51–2, 80, 264, 268, 271, *273–80*, 351–2, 354–9, 368, 371, 376, 417, 427
Cobalt, 46, 255, 314
Cochineal, 4
Cocoa: *see* cacao
Coconuts, 145, 148, 203, 207–8
Coffee, 4, 32, 67, 70, 93, 96, 99, 152, *155–8*, 368, 370
Coir, 203, 207–8
Collectivisation, 27, 98, 115
Colombia, 36, 157–8, 233, 280, 379
Comecon Bloc, 287, 377
Commercial agriculture, 32–3
Commercial fishing, 30–1
Commercial grazing, 31–2
Communications, 2, 5, 7, 19, 57, 90, 394–9
Conservation, 47–9, 92, 193–4
Containerisation, 405
Co-operative farming, 98, 115–16, 175
Copper, 14, 35, 242, 245, 252, *255–8*, 366, 379, 430
Copra, 41, 206–8
Coriander, 4, 168
Cork oak, 232, 239
Costa Brava, 389
Cottage industry, 35
Cotton, 5–7, 30, 32, 41, 70, 72, 74, 96, 100, 107, 195–9, 205, 217, 243–5, 347–54, 422, 427
Cotton manufacturing, *347–54*
Cotton-seed oil, 210
Cuba, 36, 51, 148, 161, 164, 215, 331, 368
Cumin seed, 168
Czechoslovakia, 136, 145, 165, 201, 299

D

Dairying, 35, 81–2, 126, *176–7*
Dairy produce, 31, 44, 110, 170, 174–7, 367, 427
Damodar Valley, 60, 248, 329

Datepalm, 74, 148
Denmark, 35–6, 60, 93, 100, 116, 118, 125, 135, 172, 175, 179, 190, 193, 378, 392
Density of population, 10–11
Diamonds, 4, 74, 243
Diet, 53, 92–3
Diseases, 57
Domestication of animals, 27
Domestic industries, 35
Donetz Basin, 278, 326
Double-cropping, 121
Douglas fir, 231
Drainage, 56, 91
*Durra*, 139
Dutch Guiana, 263
Dyestuffs, 193, 234

E

East Germany, 205, 250, 275, 299
East Indies, 67
Economic nationalism, 99
Ecuador, 5, 26, 147, 160, 379
Eggs, 6, 43–4, 124, 175, 179
Egypt, 12, 14, 36, 74, 89, 104, 142, 162, 196–7, 199, 302, 330, 343, 363, 369, 446
Eiderdown, 6
Eire, 116, 172, 176, 179, 193
Electro-chemical industry, 297, 355
Elephants, 170, 228, 235
El Salvador, 22
Emigration, 20
Engineering, *332–42*
England, 93, 102, 397, 416
Epiphytes, 67
Eskimo, 5, 14, 26, 109, 181, 425
Esparto grass, 41
*Estancias*, 116
Europoort, 450
Extensive farming, 118, 123, 133

F

F.A.O., 90
Falkland Islands, 85
*Favellas*, 455
*Fazendas*, 115, 155–6
Fertilisers, 28, 31–2, 44, 118, 124, 175, 192, 228, 247, 249–50, 297, 355
Finland, 83, 232, 237–8, 296, 452
Fish, 38, 44, 53, 92–3, 109, 145, 181
Fish farming, 4, 92, 187
Fishing, 7, 24, 29–31, 81, 92, 95, *181–90*
Fish oil extraction, 31
Flax, 199, 343, 346
Flood control, 60, 91, 296
Flooding, 69, 140
Fodder crops, 96, 107, 136
Food collecting, 24, 26, 29
Food industries, 306
Forest products, *230–9*
Forests, 67
Forestry, 7, 23–4
Formosa: *see* Taiwan
Fragmentation of holdings, 98
France, 6, 11, 15, 36, 114, 135, 141, 145, 150, 165, 168, 174, 186, 189, 193, 199, 200–2, 205, 214, 222, 247, 249, 253, 258, 291, 315, 323–4, 335–6, 341–2, 347, 358, 373, 390, 399, 416, 421, 449, 451
Franco-Belgian coalfield, 278, 324
*Frigorificos*, 172–3
Fruit, 6, 75, 81–2, 107, 122, *148–52*, 427, 435
Furs, 38, 43–4, *226–7*, 364

G

Galeria forest, 69
Geothermal energy, 4, 44, 268
Germany, 6, 15, 111, 179, 214, 247
Gezira, 104, 198
Ghana, 120, 159
Ginger, 167
Glasgow, 61
Glass, 355, 358
Goats, 27–8, 74, 114, 129, 169, 225
Gold, 14, 34, 39, 46, 242–3, 263–5, 368, 375
Grapes, 96, 148
Graphite, 247–8
Grand Banks, 82
Great Lakes, 30
Greece, 106, 132, 149, 192, 209, 216, 389
Groundnuts, 62–3, 70, 72, 146, *205–6*
Guano, 44, 247, 249
Gums, 38, 42, 77, 195, 238
*Gur*, 164
Guyana, 70, 130, 262–3

H

*Haciendas*, 116
Hair, 43, 170
Haiti, 32
Hamburg, 341, 365, 449
Hankow, 353, 447
Hanseatic League, 365
Hardwoods, 5, 42
Hawaiian Islands, 4, 148
Hemp, 44, 195, *202*, 349–50
Henequen, 195, *203*
Hides, 31, 38, 110, 170, 173, *225–6*
Holland, 35, 55, 60, 93, 99, 118, 125, 160, 172, 176–7, 179, 191, 193, 200–1, 323, 325, 336, 342, 359, 372
Honey, 26, 44, 179–80
Horn, 38, 44, 227
Horses, 27, 168
Hungary, 179, 202
Hunting, 7, 24, 26, 29, 53
Hwang-ho scheme, 107, 293
Hydraulic mining, 4, 262
Hydro-electric power, 39–40, 60, 107–8, 237, 249, 273, *272–9*, 328, 351, 354–5, 358

I

Iceland, 84, 96, 181, 186, 193, 386
India, 12, 22, 36, 71, 99, 134, 142–3, 154–5, 164, 167, 172, 174, 177, 197, 202, 207, 210, 216, 224, 228, 248, 279, 296, 302, 329–30, 344, 353–4, 370
Indo-China, 211
Indonesia, 154–5, 167, 208, 284
Industrialisation, 20, 269, 311–12, 366, 370, 451, 453
Insect pests, 69, 174
Intensive agriculture, 120, 122
Inter-cropping, 121
Invisible trade, 374
Iron ore, 7, 16, 38, 44, 244, *253–5*, 257
Iron and steel industry, *313–31*
Irrigation, 20, 56, 70, 75–6, 80, 98, *101–8*, 121, 123, 140–2, 145, 196–7, 293, 296, 352
Israel, 15, 177, 250
Italy, 11, 22, 116, 131–2, 143, 177, 199, 202, 205, 208, 224, 244, 270, 291, 323, 325, 339, 349–50, 373, 376, 390, 392, 408, 416
Ivory, 44, 227–8
Ivory Coast, 157, 160
Ivory nuts, 239

J

Jamaica, 167, 215, 263, 365, 387
Japan, 12, 22, 31, 35–6, 76–7, 81, 119, 122, 133, 142, 177, 186, 191–2, 199, 205, 209, 222, 224, 238, 247–8, 270, 279, 295, 315, 328, 339–40, 343–4, 352–3, 370, 373, 426
Java, 10–11, 15, 67, 146, 153, 157, 168, 412
Jews, 9, 15, 111, 179
Jivaro Indians, 26–7
Johannesburg, 262, 451
*Jowar*, 139
Jute, 44, 195, *201–2*

K

Kalmuks, 27–8, 109
*Kaoliang*, 138
Kaolin, 243
Kapok, 41, 44, 67, *203*, 238
Karaganda, 80
Kariba Dam, 92, 245
Katanga, 245, 256, 259
Kenya, 154, 157, 203, 239, 400
Kirghiz, 27–8, 78, 109, 129
*Kok sagyz*, 4, 40
Kola nuts, 67
Korea, 12, 111, 186, 209, 224
Kuwait, 286
Kuznetsk Basin, 327

L

Labrador, 231, 257, 322–3
Lake District, 247, 387
Language, 9–10, 398–400
Languages of commerce, *398–400*
Lapps, 28, 104, 129
Latent resources, 39–40
*Latifundia*, 116
Latin America, 10, 55, 93, 116, 167
Lead, 44, 243, 258–61
Leather, 6, 225, 305, 354
Lebanon, 104
Liberia, 253
Lignite, 274–5

Limes, 149
*Lingua franca*, 399–400
Linen, 6, 205, 343
Linen industry, 346–8, 350
Linseed, 199, 206
Linseed oil, 209–10
Liquorice, 4
Livestock-rearing, 32, 109, 111
Living standards, 58–9
Location of industry, *303–10*
Locust bean, 4
Locusts, 69, 98, 129
Lodes, 242–3, 261
London, 309, 334, 358, 387, 436
Lorraine, 254
Lumbering, 34, 81, 83, 95
Luxembourg, 254

M

Mace, 167
Madagascar, 70, 142, 168, 233, 248
Mahogany, 34, 230–1, 236
Maize, 30, 70, 77, 93, 96, 99, 131, *135–8*, 172
Malaria, 19
Malawi, 154, 216
Malaya, 51, 148, 208, 211–14, 238, 243, 261
Manchester, 238, 344, 445
Manganese, 255, 314
Manioc, 20, 70, 119, *144–5*
Manufacturing, 35, *301–3*
Manures, 44
Marble, 246
Market-gardening, 32, 75, 123
Marseilles, 336, 358
Marquis wheat, 61–2
Meat, 6, 31, 43, 93, 170
Mechanisation of agriculture, 123, 132
Medicinal plants, 34, 67
Menhaden, 4
*Metayage*, 98, 115
Mexico, 16, 36, 138, 148, 164, 189, 197, 199, 248, 265, 289, 310, 331, 353, 379
Mica, 247–8
Middle East, 9, 104, 113, 168, 225, 286, 415, 427
Middle West, 136
Milk, 43, 53, 93, 109, 124, 170
Millet, 30, 70, 72, 119, *138–9*
*Milpa* agriculture, 4
Mineral wealth, 14, 34, 46, 55, *241–63*
Mixed farming, 124
Molybdenum, 255, 314
Monoculture, 32, 121
Morocco, 6, 107, 148, 250, 253, 390
Mosquitoes, 19, 85
Mozambique, 70
Mustard, 168
Mutton, 110, *177–8*
Myxomatosis, 4, 43

N

Natal, 77, 154, 162, 239
Natural gas, 41, 44, 55, 248, 268, 273, 289–91, 355, 359, 416
Natural resources, 20, 38–9
Neolithic Revolution, 18, 27, 131, 169
Newfoundland, 82, 84, 135, 183, 185, 189, 231, 437
New Guinea, 5, 15, 26
New South Wales, 12, 76–7, 163, 279
New Zealand, 12, 22, 81, 107, 110, 116, 125, 174, 176–7, 190, 219–20, 225, 232, 280, 297, 425, 437
New York, 177, 335, 356, 437, 440–1
Nicaragua, 22
Nickel, 46, 255, 314
Nigeria, 12, 30, 101, 160, 207, 261, 280
Nitrate, 14, 46, 74, 242, 249, 297
Nitrogen, 39, 41, 47
Nomadism, 27, 80, 95, 112–13, 123, 128–9
Norway, 36, 51, 83, 95, 181, 186–7, 233, 237, 249, 296
Nuclear energy, *298–300*, 368
Nuclear fuels, 268, 299
Nutmeg, 167
Nuts, 34, 38, 42, 148

O

Oats, 131, *135–6*
Oil palm, 67, 195, 207
Oil pipelines, 51, 290–1, 403
Oil refining, *284–7*
Oil seeds, 72
Olive oil, 208–9, 358
Olives, 96
Ontario, 82

Over-grazing, 42, 110
Over-population, 21–2

P

Pakistan, 12, 105, 131, 134, 146, 154, 164, 197, 199, 202, 296, 353, 368
Palm oil: *see* oil palm
Pampas, 31, 173
Panama, 142, 430
Panama Canal, 430
Paper, 42, 147, 236–8, 297, 354–5
Paraguay, 15, 22, 239, 423
Paraná pine, 231
Paris, 368, 387, 440
Pearls, 44, 192
Peas, 146
Peat, 268
Peppers, 30, 70, 166–7
Permafrost, 84
Peru, 14, 74, 89, 107, 186, 189, 248, 265, 369, 379
Petro-chemical industries, 355–6, 359
Petroleum, 5, 15, 34, 40, 44, 54, 74, 78, 268, 273, *280–9*, 354, 415, 450
Philippines, 148, 164, 203, 208
Phormium, 203
Phosphate, 247, 249–50, 355
Phylloxera, 88
Physiological drought, 85
Pie-graph, 142
Pigs, 59, 72, 99, 121, 136, 170, 178–9, 225
Pineapples, 70, 148
Pipelines, 286–9, 403, *415–16*
Plantation agriculture, 32, 67, 120, 152, 206
Plastics, 356, 360
Poland, 15, 136, 145, 165, 193, 200, 202, 367
Polygamy, 15, 19
Population, 10, 23, *87–90*
Population growth, 15, 19
Pork, 110–11, 179
Portugal, 132, 150, 189, 193, 209, 239, 378
Potash, 247, 250, 356, 359
Potatoes, *145*, 179
Potteries, 247
Poultry, 44, 121
Primary production, 6, 35
*Primeurs*, 125
Pulses, *146*
Pygmies, 67

Q

*Quebracho*, 44, 225, 239
Queensland, 4, 70–1, 107, 163, 174, 279
Quinine, 4

R

Rabbits, 43
Races of man, 8–10
Raffia, 44, 203, 239
Railways, 4, 375, 384, 402, 404, *409–14*
Ramie, 203
Rapeseed, 210
Rayon, 42, 205, 343, 347–9, 352–3
Reindeer, 28, 84–5, 129
Relief, 12, 14, 29, 55, 57, 59, 65, 70, 77, 79, 88, 90, 97–8
Religion, 9, 59
Resin, 77, 195
Rhine, 417, 419, 421
Rhodesia, 216, 280
Rhubarb, 5, 95
Rice, 70, 72, 77, 92, 121, 131, 135, *139–43*
Ring-barking, 120
Roads, *406–9*
Ropeways, 402, 414–15
Rotation of crops, 28, 30, 32, 45, 91, 118, 124
Rotterdam, 287, 290, 342, 359, 450
Rubber, 32, 34, 38–9, 51, 67, 120, 194, *210–14*, 247, 358, 371, 427
Ruhr, 278, 324–5, 412, 419
Rumania, 6, 138, 291
Rye, 131, *135–6*

S

Sago, 4
Sahara Desert, 4, 10, 14, 72–3, 283, 288, 291
Sal, 72
*Saladeros*, 173
Salt, 35, 44, 46, 242, 247, 249, 354–9, 363
Samoyeds, 28
Sarawak, 167, 211–12
Sarsaparilla, 4
Scotland, 11, 135, 187, 385, 442

Sea cucumber, 4
Sealing, 31
Seaweed, 44, 183, 192
Secondary production, 7, 35
*Selva*, 14, 67, 69
Sennar Dam, 104
Shea nuts, 30
Sheep, 27–8, 31, 38, 72, 80, 112, 114, 123, 129, 169, *217–23*, 225, 343
Shifting agriculture, 14, 28, 67, 70, 120
Sigatoka disease, 98
Silk, 6, 44, 205, 217, *221–5*, 343, 347–50, 352–3, 368
Silver, 6, 44, 243, 265, 365, 379
Singapore, 3, 449
Sisal, 70, 195, *203*, 343
Slash and burn agriculture, 28, 120, 230
Snowy River scheme, 107
Soap, 5, 195, 209
Softwoods, 230–1, 233
Soil, 14, 38, 41, 47, 95, 98, 121, 133, 153, 156, 196–7, 212–13
Soil erosion, 41–2, 48, 60, 91
Solar energy, 44, 268
South Africa, 78, 138, 149–50, 219–20, 225, 265, 299, 310
Soviet Union, 15, 35, 89, 100, 113, 116, 132, 136, 138, 152, 154, 174, 186, 193, 197, 200, 205, 224–5, 237, 248–9, 259, 261, 265, 270, 275, 278–9, 288–9, 291, 323, 326–8, 338, 342, 352
Soya beans, 77, 146, 206, *209*
Spain, 36, 116, 132, 144, 150, 158, 162, 186, 189, 203, 208, 239, 250, 321, 325, 339, 365, 389–90
Spices, *166–8*, 368
Sponges, 31, 44, 192
St Lawrence Seaway, 237
Strip-cropping, 45
Suez Canal, 427–9
Sugar, 6, 44, 51, 107, 366, 368
Sugar-cane, 67, 70, 72, 74, 77, 120, *160–4*
Sui, 291
Sulphur, 35, 44, 46, 242, 247–9, 355, 358, 367
Sulphuric acid, 248, 355
Sunflowers, 206, 210
Sweden, 51, 83, 193, *233–4*, 237–8, 243, 249, 254, 270, 296, 316, 318, 325, 337–8, 342
Sweet potatoes, 119, *145*
Switzerland, 51, 199, 225, 249, 296, 338, 378, 385, 388–9, 392

T

Tagua nuts, 4, 239
*Taiga*, 82–3
Taiwan, 12, 154–5, 202, 238
Tannin, 223, 239
Tanzania, 67, 157, 400
Tapioca, 4, 145
Tasmania, 81, 297
Tea, 44, 72, 77, 120, *151–2*, 367, 370, 427
Teak, 34, 72, 230–1
Terracing, 29, 97–8, 140–1
*Terra roxa*, 156
Tertiary activities, 7
Textiles, 7
Thailand, 72, 142, 202, 213–14, 261, 368, 390, 392
Tibet, 5
Tidal energy, 268
Tierra del Fuego, 85, 181
Timber, 6, 76, *229–39*, 364
Tin, 34, 46, 51, 243, 252, 260–1, 367
Titanium, 40
Tobacco, 4, 30, 41, 44, 70, 77, 195–6, *214–16*, 310, 366, 427
Toquilla palm, 239
Tourist trade, 51, *382–93*
Trade, 6, 59, *363–80*
Trade blocs, *376–9*
Trading Estates, 309
Transhumance, 75, 113–15, 123
Transport, 2, 5, 45, 55–6, 307–8, 394, *401–37*, 445
Trans-Siberian Railway, 308, 414
Triangular trade, 366
Tsetse fly, 69, 98, 111
Tundra, 84–5
Tung oil, 206, 210
Tungsten, 255, 314
Tungus, 28
Tunisia, 107, 203, 209, 250
Turkestan, 15, 79–80, 169, 197, 274

Turkey, 106, 135–6, 149, 209, 216, 219, 270
Turmeric, 167
T.V.A., 60, 291, 296

U

Under-population, 22
United Kingdom, 222, 226, 238, 258, 270, 274, 289, 316, 332, 339, 345–6, 357, 360
United States, 2, 16, 35, 42, 53, 56, 60, 77, 81, 89, 93, 102, 108, 111, 118, 134, 138, 145, 165, 174, 185–6, 196–7, 205, 207, 210, 216, 218, 225, 245, 247, 248, 250, 256, 265, 270, 276–7, 288–9, 295, 299, 303, 322–3, 334–5, 341, 350–2, 355–6, 367, 369, 372–3, 377, 385, 387, 398, 409, 412, 414–6, 426, 431
Uranium, 40, 44, 299
Uranium City, 15, 299
Uruguay, 160, 172–3, 178, 210, 225, 379, 387

V

Vanadium, 4, 46, 255, 314
Vanilla, 4, 166–7
Vegetation, 12, 14, 41–2, 47, 67, 69–70, 72, 74–8, 81, 83, 85, 112–14
Vegetable oils, 41, 146, 195, *205–10*, 358
Venezuela, 21, 174, 253, 288–9, 323, 331, 450
Victoria Nyanza, 30
Vietnam, 143, 213–14
Vines, 96, 123
Vodka, 4, 136
Volga, 420–1
Volta scheme, 295
Vulcanisation, 211, 248

W

Water, 14, 46–7, 55–6, 242, 292, 308, 344–5, 355, 383, 387, 403
Water-power, *292–97*
Water transport, 402, *416–31*
Waxes, 42, 195
Wealth, 37–9, 59
West Africa, 207, 322
West Germany, 11, 36, 135–6, 145, 151, 165, 174, 193, 199, 202, 205, 214, 222, 237–8, 258, 270, 274, 289, 299, 323–5, 336–7, 341, 347–8, 356–7, 370, 451
West Indies, 149, 162, 167, 233, 344
Whales, 45, 183, 190–1
Whaling, 31, *190–2*
Wheat, 32, 53, 61–2, 72, 93, 96, 100, *131–4*, 367
Whisky, 134, 136, 367
Wild life, 45, 69, 92
Wine-making, 76, 150
Witwatersrand, 259, 264, 299
Wood-pulp, 42, 83, 205, *236–8*, 291, 367
Wool, 6, 31, 38, 110, 170, 178, *217–21*, 366–7, 371, 427
Woollen manufacturing, 307, 345–8, 350–3

Y

Yaghans, 26
Yams, 30, 67, 70, 119, *144*
Yangtse, 183, 421
*Yerba maté*, 4, 77, 152, 239
Yucatan, 203
Yugoslavia, 138, 152, 385

Z

Zaire Republic, 258
Zambia, 256, 259
Zamenhof, Ludwig, 400
Zanzibar, 167, 400
Zebu cattle, 172
Zinc, 243, 258–61